PRAISE FOR

'THE AUTONOMOUS FINANCE REVOLUTION'

In my exploration of emerging technology and innovation, I've always been captivated by the power of forward-thinking to enact real change with a nod to the historical events that shaped how we got to today. Anuj Saxena's 'The Autonomous Finance Revolution' does exactly this and is an important guide for the future of finance, merging ethical considerations, data security, and groundbreaking technologies such as AI and blockchain into a transformative narrative that can serve as a roadmap for leaders. Anuj crafts a vision that is both a forecast and a call to action for financial pioneers, providing the tools needed to navigate and lead in the evolving financial landscape. This work illuminates one through the complexities of AI, machine learning, big data, and cybersecurity, and guiding us toward a future ripe with opportunity. Reflecting on the barriers and opportunities for innovation, 'The Autonomous Finance Revolution' emerges as an important work. It offers solutions and imagines a world where finance is both autonomous and deeply human-centric, resonating with my commitment to impactful technological advancement. Saxena invites us to think boldly and lead with vision. Join me in embracing 'The Autonomous Finance Revolution.' equipped with Saxena's insights and foresight on the important role of emerging technologies.

Elizabeth McCalley, Emerging Tech Leader, TED Speaker, CEO of StartStak

In 'The Autonomous Finance Revolution - Shaping the Future of Finance,' Anuj Saxena masterfully navigates the complex landscape of Autonomous Finance, presenting it as the beacon of the future of finance. This book not only demystifies the intricate workings of digital transformation within finance but also serves as a crucial roadmap for leaders aiming to harness the power of innovation for sustainable growth. Anuj's insights offer a compelling vision of a future where finance is more accessible, efficient, and secure. A must-read for anyone committed to shaping the next era of Finance in the age of Artificial Intelligence and Machine Learning." .

Tony Saldanha, President, Transformant

Anuj Saxena's book skillfully draws parallels between the groundbreaking realms of Autonomous Driving and Autonomous Finance, illustrating the innovative pathways both sectors navigate amidst evolving regulatory and ethical landscapes. By delving into how finance is poised to leverage AI and ML for unparalleled data-driven insights,

Anuj charts a visionary course for technology-driven finance transformation. This narrative, enriched by an in-depth look at finance transformation initiatives, positions itself as a pivotal read for leaders at the forefront of technology and finance. It distills complex concepts into actionable insights, providing a comprehensive roadmap for navigating the digital finance landscape with foresight and integrity.

Johnny Vincent, ex-CFO, HighRadius Corporation

A transformative guide for the digital era, this book offers a visionary outlook on the future of finance. It is an indispensable resource for anyone looking to navigate the complexities in the rapidly changing Finance landscapes. Anuj's ability to demystify the future of finance through practical insights and actionable strategies makes this a must-read for industry leaders.

Amir Goldman, Growth Investor, Susquehanna Growth Equity

'The Autonomous Finance Revolution' by Anuj Saxena is an impressively comprehensive and authoritative exploration of how Autonomous Finance is reshaping the future of Finance. Anuj's work is not just insightful; it is a pivotal read for anyone looking to grasp the full potential of technology in transforming the traditional Finance domain in the B2B space. His mastery in weaving together complex concepts with real- world applications make this book an invaluable resource for industry leaders and innovators alike.

Ben Ryan Weinberg, Growth Investor, Susquehanna Growth Equity

In his book, Anuj has skillfully navigated through several complex topics such as the future of finance, the future of work, data security and privacy in the age of big data and analytics, and the complex web of ever-evolving regulatory and ethical compliance. He offers a rich perspective on the synergies between these concurrently evolving areas, making this book an indispensable read for pioneers at the nexus of these fields. Anuj has distilled complex concepts into actionable insights, charting a course for navigating the Autonomous Finance landscape with foresight and integrity.

Kay Rogers, VP of Finance, Ferrero

Anuj Saxena masterfully blends AI's transformative power in Finance with a keen eye on ethics, compliance, and data privacy and security, presenting a compelling vision for customer-centric solutions. His narrative skillfully tackles bias, advocating for technologies that ensure fairness, transparency, and trust. This profound analysis marks a pivotal step towards ethically responsible, technologically advanced financial services, offering a blueprint for innovation that honors both efficiency and ethical integrity in the rapidly evolving Autonomous Financial landscape.

Sajeesh Kumar, Global Service Line Leader for Customer-to-Cash, Genpact

Anuj Saxena offers a visionary look into finance's future, blending ethical and regulatory considerations, data security, and groundbreaking technologies into a narrative that promises to reshape economies and financial ecosystems. This eloquent synthesis provides an essential roadmap for financial pioneers, equipping them with the insights needed to navigate and lead through the evolving global financial landscape.

Olivier Novasque, Founder and CEO, Sidetrade

Anuj's insightful exploration into the convergence of rapidly emerging technologies such as AI, ML, RPA, Big Data, and Analytics, and their impact on the Finance domain is both timely and transformational. His narrative on Autonomous Finance as a catalyst for industry-wide innovation speaks directly to technology enthusiasts and practitioners alike, offering a roadmap to a future that is both visionary and actionable. This excellent read not only demystifies complex technologies but also empowers finance professionals with practical strategies for leveraging these advancements to drive real-world innovation and efficiency.

Morris Kupfer, VP of Finance, Imbuit

THE AUTONOMOUS FINANCE REVOLUTION

Shaping the Future of Finance

Anuj Saxena

1st edition 2024

ISBN 979-8-88940-812-3 (Paperback)
ISBN 979-8-88940-815-4 (Hardcover)
ISBN 979-8-88940-814-7 (Kindle eBook)

To the silent muse
whose spirit danced through these pages,
infusing them with profound wisdom
and effervescent energy.

You are both the whisper and the echo,
the question and the answer,
guiding this exploration
into the fascinating intricacies
of Autonomous Finance.

You know who you are...
With you know what...
& you know why.

This one's for you!

Table of Contents

PREFACE xxiii

INTRODUCTION xxv

SECTION I: TIME TRAVEL IN FINANCE | 1

CHAPTER 1: FINANCE IN THE PAST 3

1.0 **Key Insights** 4

1.1 **Introduction** 4

1.2 **Introducing the Age of Disruptions** 5

1.3 **The Domino Effects of Disruptions** 7

From Theory to Reality: Case Studies of Disruptions 8

Caught in the Crosshairs: The Finance Team's Dilemma 10

1.4 **Resilience in the Face of Disruptions** 11

Understanding Business Resilience 12

Financial Resilience Defined 13

The Elements and Role of Financial Resilience in Navigating Disruptions 14

Harmonizing Fiscal Fortitude: Strategies for Financial Resilience 15

Barriers to Financial Resilience and Overcoming Strategies 17

Survive-Revive-Thrive: An Approach to Bolster Financial Resilience 18

1.5 **Conclusion** 21

1.6 **Key Recommendations** 22

1.7 **References** 22

CHAPTER 2: FINANCE IN THE PRESENT 25

2.0 **Key Insights** 26

2.1 **Introduction** 26

2.2 **The Current State of Finance** 27

2.3 **Unleashing the Power of Agility in Finance** 29

2.4 **Core Elements of Agile Finance** 30

Real-Time Data and Analytics 30

Technology Integration 30

	Cross-functional Collaboration	31
	Continuous Learning and Improvement	32
	Navigating the Regulatory Landscape	32
	Measuring Agility: KPIs and Metrics	33
2.5	**Implementing Agile Finance: Practical Steps**	34
	Assessing Organizational Readiness	35
	Building an Agile Team: Skills, Roles, and Mindsets	35
	Implementing Agile Methodologies and Tools	36
	Driving Cultural Change for Agility	37
2.6	**Future Trends & Opportunities in Agile Finance**	38
	Advanced Technology Integration	39
	Emphasis on Data Literacy	39
	Growing Focus on ESG Factors	39
	Expansion of Cross-Functional Roles	39
	Compliance with Evolving Regulations	40
	Continuous Learning and Development	40
	Embracing Growth and Innovation	40
2.7	**Conclusion**	41
2.8	**Key Recommendations**	41
2.9	**References**	42

CHAPTER 3: FINANCE IN THE FUTURE		45
3.0	**Key Insights**	46
3.1	**Introduction**	46
3.2	**Adaptability: The Quintessential Virtue**	47
3.3	**Reimagining the Future of Finance**	50
3.4	**Transforming Finance for the Future**	52
	The New People Agenda: Skills, Capabilities, and Mindsets	52
	Rethinking Processes: From Traditional to Agile Finance	53
	Technology: The Enabler of the Future of Finance	54
3.5	**Gaps in Financial Management Solutions Today**	55
3.6	**Characterizing Future-Ready Finance**	56
	Building Resilience: Financial Management's Shield against Volatility	57
	Agile Finance: The New Normal in a Fast-Paced World	58
	Adaptability: The Key to Thriving in Uncertainty	59
	Self-Execution, Self-Learning, and Self-Regulation: The Triad of Autonomy	59
	Modularity and Redundancy: Enhancing Flexibility and Robustness	60
	Prudence and Innovation: Balancing Risk and Reward	60
	Diversity: A Multifaceted Lens for Finance	61
	Systemic Integration: Connecting the Dots in an Interconnected World	62
3.7	**Conclusion**	64
3.8	**Key Recommendations**	64
3.9	**References**	65

SECTION II: THE FUTURE IS AUTONOMOUS | 67

CHAPTER 4: AUTONOMOUS FINANCE ... 69

4.0 **Key Insights** ... 70
4.1 **Introduction** ... 70
4.2 **The Concept and Origin of Autonomous Finance** ... 71
4.3 **The Need for Autonomous Finance** ... 72
4.4 **The Fundamentals of Autonomous Finance** ... 73
Principles of Autonomous Finance ... 73
Building Blocks of Autonomous Finance ... 75
4.5 **Understanding the Shift from Automation to Autonomous Finance** ... 79
4.6 **Autonomous Finance: A Comparative Advantage** ... 80
4.7 **Overcoming the Barriers to Autonomous Finance** ... 81
4.8 **Envisioning the Future of Autonomous Finance** ... 84
4.9 **Conclusion** ... 85
4.10 **Key Recommendations** ... 86
4.11 **References** ... 87

CHAPTER 5: SHOWCASING SUCCESS ... 89

5.0 **Key Insights** ... 90
5.1 **Introduction** ... 90
5.2 **Top Use Cases in the Private Sector** ... 91
Financial Accounting ... 92
Management Accounting ... 94
Financial Planning & Analysis ... 96
Treasury Management ... 97
Financial Risk Management ... 99
Corporate Finance ... 100
Financial Systems and Technology ... 102
Compliance and Regulatory Reporting ... 103
5.3 **Top Use Cases in the Public Sector** ... 105
5.4 **The Future of Autonomous Finance Use Cases** ... 106
5.5 **Conclusion** ... 107
5.6 **Key Recommendations** ... 108
5.7 **References** ... 108
5.8 **Appendix A: Autonomous Finance Use Cases** ... 110

CHAPTER 6: INSIDE THE ENGINE ... 139

6.0 **Key Insights** ... 140
6.1 **Introduction** ... 140
6.2 **Core Technologies in Autonomous Finance** ... 142
The Quintessential Role of AI & ML in Autonomous Finance ... 142
RPA: The Evolutionary Leap in Autonomous Finance ... 143

	Blockchain's Transformative Impact on Autonomous Finance	144
	Cybersecurity's Role in Financial Autonomy	146
	Big Data and Advanced Analytics: The Catalysts of Autonomous Finance	146
	The Architectural Backbone: Cloud Computing in Autonomous Finance	147
	Streamlining Operations: The Role of SaaS in Autonomous Finance	148
	Bridging the Digital Divide: APIs in Autonomous Finance	149
	Concluding Reflections on Technologies Powering Autonomous Finance	150
6.3	**Emerging Technological Disruptors**	151
	Generative AI: The Vanguard of Financial Transformation	152
	DeFi: The Financial Democratization Movement	153
	Cryptocurrencies and Digital Assets: Unseen Tech in Autonomous Finance	154
	Internet of Things (IoT): The Nexus of Interconnected Financial Ecosystems	155
	Biometrics: The New Face of Financial Security	155
	Regulatory Technology: The Sine Qua Non of Financial Compliance	156
	Quantum Computing: The Paradigm Shift in Computational Power	157
	Augmented Reality & Virtual Reality: The Immersive Financial Experience	158
	Concluding Reflections on Emerging Technologies	159
6.4	**The Panorama of Interconnected Ecosystems**	160
	Open Banking	161
	Open Finance and Autonomous Finance	161
	Open Data and Autonomous Finance	162
	Embedded Finance and Autonomous Finance	163
	DeFi and Autonomous Finance	164
	Autonomic Finance and Autonomous Finance	165
	Concluding Reflections on the Symbiosis of Interconnected Ecosystems	166
6.5	**Key Criteria for Technology Selection & Integration**	167
6.6	**Ethical and Responsible Use of Technology**	168
6.7	**Conclusion**	169
6.8	**Key Recommendations**	170
6.9	**References**	170

SECTION III: BLUEPRINT FOR SUCCESS | 173

CHAPTER 7: THE AGE OF AUTONOMOUS FINANCE		175
7.0	**Key Insights**	176
7.1	**Introduction**	176
7.2	**The Roadmap to Autonomous Finance Transformation**	177
7.3	**Vision, Alignment, & Commitment: Setting the Course**	178
	Creating a Shared Vision	179
	Building Alignment	179
	Gaining Organizational Commitment	180
7.4	**Conduct an Assessment using the FOCUS Framework**	181
	Overview of the FOCUS Framework	181
	Conduct Maturity Assessments	184

7.5 **Strategic Prioritization & Planning** ... 186
Strategic Prioritization: Data-Driven Comparative Analysis of Initiatives ... 186
Planning: Orchestrating the Transformation Journey ... 187
7.6 **Pilot Projects, POCs, and Experimentation** ... 187
7.7 **Implementation, Rollouts, and Scaling** ... 188
Implementation Strategy ... 189
Rollouts and Scaling ... 189
Challenges and Best Practices ... 189
7.8 **Support, Maintenance, and Continuous Improvement** ... 190
Ongoing Support and Maintenance ... 190
Continuous Improvement and Innovation ... 191
7.9 **Pragmatic Steps and Best Practices** ... 192
7.10 **Conclusion** ... 193
7.11 **Key Recommendations** ... 194
7.12 **References** ... 194
7.13 **Appendix B: Dimensions of the FOCUS Framework** ... 195
Dimension #1: Strategic Alignment ... 195
Dimension #2: Data Intelligence ... 197
Dimension #3: Technology Enablement ... 199
Dimension #4: Human Capital Development ... 201
Dimension #5: Operational Architecture ... 202
Dimension #6: Scalable Adoption ... 204

CHAPTER 8: SYNERGISTIC STRATEGIES ... 207
8.0 **Key Insights** ... 208
8.1 **Introduction** ... 208
8.2 **Dimensions of Strategic Partnerships** ... 209
8.3 **The Value Proposition of Strategic Partnerships** ... 213
8.4 **Ecosystem Dynamics: Partner Selection & Integration** ... 214
8.5 **Maintaining & Optimizing Partner Relationships** ... 215
8.6 **Risk Management in Strategic Partnerships** ... 215
8.7 **Building & Managing Strategic Partnerships** ... 216
8.8 **Conclusion** ... 217
8.9 **Key Recommendations** ... 217
8.10 **References** ... 218

CHAPTER 9: COHESIVE CONNECTIVITY ... 219
9.0 **Key Insights** ... 220
9.1 **Introduction** ... 220
9.2 **Interplay in Autonomy: Integration & Interoperability** ... 221
9.3 **Mastering the Art of Integration** ... 222
9.4 **Tech Giants on the Horizon** ... 224
9.5 **Security, Compliance, & Standardization** ... 225
9.6 **Decentralization & Scalability** ... 226

	Decentralization: A Paradigm Shift in Finance	227
	Scalability: Fueling Growth in Autonomous Finance	227
9.7	**Crafting Superior User Experiences**	228
9.8	**The Future of Integration and Interoperability**	229
9.9	**Conclusion**	230
9.10	**Key Recommendations**	230
9.11	**References**	231

CHAPTER 10: THE INFLUENCE OF INSIGHT 233

10.0	**Key Insights**	234
10.1	**Introduction**	234
10.2	**Data Management**	235
10.3	**Data Quality and Integrity**	236
	The Data Quality Dilemma	238
10.4	**Data Governance**	239
10.5	**Unification: Integration and Interoperability**	241
10.6	**The Beacon of AI and ML**	241
10.7	**The Periscope of Data Visualization**	242
10.8	**Data Security and Privacy**	243
	Data Protection	244
	Compliance with Data Privacy Regulations	244
	Privacy by Design	245
	Data Breach Response	246
	Data Minimization	247
	Data Disposal	247
	Audits and Reviews	248
	Transparency	248
	Third-Party Risk Management	249
	Employee Training	249
10.9	**Predictive Analytics and Forecasting**	250
10.10	**Surfing the Data Wave: The Significance of Big Data**	251
10.11	**Conclusion**	253
10.12	**Key Recommendations**	253
10.13	**References**	254

CHAPTER 11: BEYOND TRANSACTIONS 255

11.0	**Key Insights**	256
11.1	**Introduction**	256
11.2	**Journey Beyond Transactions: The Paradigm Shift**	257
11.3	**Fashioning Finance: The Power of Personalization**	258
11.4	**Anytime, Anywhere: The Power of Accessibility**	259
11.5	**A Step Ahead: Proactive Financial Management**	261
11.6	**Seamless Integration with Lifestyle Apps**	262
11.7	**Instant Assistance: AI Chatbots**	263

11.8 Security, Privacy, & Trust in Autonomous Finance 264
11.9 The Future of Customer Experience 265
11.10 Conclusion 266
11.11 Key Recommendations 267
11.12 References 268

CHAPTER 12: PLAYING BY THE RULES 269
12.0 Key Insights 270
12.1 Introduction 270
12.2 Navigating the Landscape: The Essential Lenses 271
The Privacy Paradox: Balancing Data Protection and Utility 271
The Financial Maze: Navigating Regulatory Complexities 272
The AI Ethical Dilemma: Striving for Fairness amid Bias 273
Cybersecurity: Safeguarding the Digital Treasure Trove 274
Interoperability: Bridging the Autonomous Islands 274
Compliance Conundrum: Traversing the Regulatory Jungle 275
12.3 The Multidimensional Regulatory Landscape 276
Geography: Mapping the Global Financial Terrain 277
Industry: The Vertical Spectrum 277
Functional Areas within Finance: The Niche Navigators 279
Governing Ecosystems: The Overarching Framework 280
Summary of the Regulatory Terrain 281
12.4 The Ethical Compass in Autonomous Finance 282
12.5 Cultural Values and Autonomous Finance 283
12.6 Roadmap to Compliance in Autonomous Finance 284
12.7 Future of Regulatory and Ethical Compliance 286
Anticipating Regulatory Developments 286
Globalization vs. Localization: The Symphony of Regulation 287
Ethical Considerations: The Soul of Financial Evolution 287
Technology's Transformative Role: The Alchemy of Compliance 288
Navigating the Unexpected: The Art of Resilience 288
Functional Area-Specific Future Insights: The Tapestry of Complexity 289
12.8 Conclusion 289
12.9 Key Recommendations 290
12.10 References 291

CHAPTER 13: UNVEILING THE MYSTIQUE 293
13.0 Key Insights 294
13.1 Introduction 294
13.2 Decoding AI & ML: From Complex to Clear 295
13.3 Explainability: The Key to Trust in AI and ML 296
13.4 Battling Bias: Ensuring Fairness in AI and ML 297
13.5 Steering AI & ML Towards Good Governance 298
13.6 Compliance with AI and ML Rules 298

13.7 The Guardians of AI Ethics ... 300
13.8 Lessons from the Field: Ethics in AI and ML ... 300
13.9 Conclusion ... 302
13.10 Key Recommendations ... 303
13.11 References ... 304

CHAPTER 14: THE DIGITAL FORTRESS ... 305
14.0 Key Insights ... 306
14.1 Introduction ... 306
14.2 Cybersecurity Threats in Autonomous Finance ... 307
14.3 The Autonomous Finance Threat Landscape ... 308
14.4 The Top 10 Cybersecurity Threats ... 312
14.5 Proactive Strategies for Cybersecurity ... 314
Regulatory and Industry Standards Requirements ... 315
Data Protection and Privacy Threats ... 316
Risks to Infrastructure & Secure Transactions ... 317
Risks with Leveraging Advanced Technologies ... 317
14.6 Implementing a Cybersecurity Framework ... 318
Key Elements of a Robust Cybersecurity Framework ... 318
Implementing the Cybersecurity Framework ... 320
14.7 Fostering a Cybersecurity-Conscious Culture ... 322
Cybersecurity Awareness and Training for Autonomous Finance ... 322
Importance of Cybersecurity Education and Training ... 323
Fostering a Culture of Security Awareness ... 323
Insurance Considerations in Cybersecurity ... 323
14.8 Collaborative Measures in Cybersecurity ... 324
Collaboration and Partnerships in Cybersecurity ... 324
Collaborative Approach to Enhancing Cybersecurity Resilience ... 325
Partnerships and Information Sharing for Effective Cybersecurity ... 325
14.9 Conclusion ... 326
14.10 Key Recommendations ... 326
14.11 References ... 327

CHAPTER 15: THE SYMBIOTIC DANCE ... 331
15.0 Key Insights ... 332
15.1 Introduction ... 332
15.2 Autonomous Finance Reshaping the Future of Work ... 333
15.3 The Future of Work Steering Autonomous Finance ... 334
15.4 The Human-Machine Dance in Autonomous Finance ... 335
The Symphony of Combined Strengths ... 336
A Pas de Deux of Shared Control ... 336
The Beauty of Feedback Loops ... 336
The Art of Adaptive Automation ... 337
The Elegance of Explainable Systems ... 337

15.5 **Navigating the Cross-Roads: Challenges & Strategies** ... 337
15.6 **A Case Study of Accounts Receivables** ... 339
Change in Roles and Responsibilities in A/R ... 340
The Skill Evolution in A/R ... 341
A Shifting Organizational Structure in A/R ... 342
The Regulatory Ballet in A/R ... 343
The Impact of Using AI in A/R ... 344
Cybersecurity Considerations in A/R ... 345
Data Security and Privacy in A/R ... 345
Governance of AI Models in A/R ... 346
Change Management in A/R ... 347
15.7 **Conclusion** ... 347
15.8 **Key Recommendations** ... 348
15.9 **References** ... 349

SECTION IV: THE HORIZONS AHEAD | 351

CHAPTER 16: UNFOLDING THE FUTURE ... 353
16.0 **Key Insights** ... 354
16.1 **Introduction** ... 355
16.2 **The Philosophical Underpinnings** ... 356
16.3 **Strategic Foresight: Preparing for an Uncertain Future** ... 357
16.4 **Navigating the Challenges** ... 360
Customer Education and Transparency in Autonomous Finance ... 361
Platforms and Ecosystem Considerations in Autonomous Finance ... 362
Technological Resilience and Adaptability in Autonomous Finance ... 363
Data Security & Privacy Considerations in Autonomous Finance ... 363
Regulatory and Policy Considerations in Autonomous Finance ... 364
Ethical Considerations in Autonomous Finance ... 365
Financial Inclusion Considerations in Autonomous Finance ... 366
Concluding Reflections on Navigating the Challenges ... 367
16.5 **The Autonomous Finance Revolution** ... 367
16.6 **Key Recommendations** ... 369
16.7 **References** ... 369

LIST OF ABBREVIATIONS ... 371
INDEX ... 375

List of Tables

TABLE 1.1: Examples Of Past Disruptions 8

TABLE 4.1: Potential Solutions to Barriers of Autonomous Finance 82

TABLE 5.1: Autonomous Financial Accounting Use Cases 110

TABLE 5.2: Autonomous Management Accounting Use Cases 113

TABLE 5.3: Autonomous Financial Planning & Analysis Use Cases 116

TABLE 5.4: Autonomous Treasury Management Use Cases 121

TABLE 5.5: Autonomous FRM Use Cases 125

TABLE 5.6: Autonomous Corporate Finance Use Cases 128

TABLE 5.7: Autonomous Financial Systems & Technology Use Cases 131

TABLE 5.8: Autonomous Compliance and Regulatory Reporting Use Cases 135

TABLE 7.1: Detailed Assessment of Strategic Alignment Across Maturity Levels 196

TABLE 7.2: Detailed Assessment of Data Intelligence Across Maturity Levels 197

TABLE 7.3: Detailed Assessment of Technology Enablement Across Maturity Levels 199

TABLE 7.4: Detailed Assessment of Human Capital Development Across Maturity Levels 201

TABLE 7.5: Detailed Assessment of Operational Architecture Across Maturity Levels 203

TABLE 7.6: Detailed Assessment of Scalable Adoption Across Maturity Levels 205

TABLE 9.1: Integration vs. Interoperability 222

TABLE 12.1: Regulatory Landscape by Geography 277

TABLE 12.2: Regulatory Landscape by Industry Vertical 278

TABLE 12.3: Regulatory Landscape by Functional Areas within Finance 279

TABLE 12.4: Guiding Principles for Ethical Compass in Autonomous Finance 282

TABLE 14.1: Top 10 Cybersecurity Threats in Autonomous Finance 313

TABLE 15.1: Challenges in the Human-Machine Dance & Strategies to Address Them 338

TABLE 16.1: Preparation Strategies for Potential Future Scenarios 358

List of Figures

FIGURE A-1:	Parallels between Autonomous Driving and Autonomous Finance	xxiv
FIGURE 1.1:	Types of Disruptions	6
FIGURE 1.2:	Domino Effects of Disruptions	7
FIGURE 1.3:	Top 5 Negative And Top 5 Positive Impacts Of Disruptions	10
FIGURE 1.4:	The Pillars of Business Resilience	12
FIGURE 1.5:	The Key Elements of Financial Resilience	15
FIGURE 1.6:	Strategies to Fortify Financial Resilience	16
FIGURE 1.7:	Barriers to Achieving Financial Resilience	18
FIGURE 1.8:	The Survive-Revive-Thrive Framework for Bolstering Financial Resilience	19
FIGURE 2.1:	Key Challenges in Finance Today	28
FIGURE 2.2:	Transition from Traditional Finance to Agile Finance	30
FIGURE 2.3:	Technology Integration in Traditional Finance vs. Agile Finance	31
FIGURE 2.4:	Implementation Methodology for an Agile Finance Implementation	35
FIGURE 2.5:	Future Trends and Opportunities in Agile Finance	38
FIGURE 3.1:	Short-Term Hurdles Faced by Finance Teams	48
FIGURE 3.2:	Medium-Term Hurdles Faced by Finance Teams	49
FIGURE 3.3:	Long-Term Hurdles Faced by Finance Teams	50
FIGURE 3.4:	The Future of Finance - PACE Framework	50
FIGURE 3.5:	Makeover Elements of New Finance	52
FIGURE 3.6:	Shortcomings of Today's Financial Management Systems	55
FIGURE 3.7:	Future Ready Finance	56
FIGURE 3.8:	Impact of Resilience on the Finance Domain	57
FIGURE 3.9:	Impact of Agility on the Finance Domain	58
FIGURE 3.10:	Impact of Adaptability on the Future of Finance	59
FIGURE 4.1:	Principles of Autonomous Finance	73
FIGURE 4.2:	Building Blocks of Autonomous Finance	76
FIGURE 4.3:	The Shift from Automation to Autonomous	79
FIGURE 4.4:	Advantages of Autonomous Finance	80
FIGURE 4.5:	Barriers to Autonomous Finance	82
FIGURE 4.6:	The Future of Autonomous Finance	85
FIGURE 5.1:	Functional Areas and the Processes/Functions Within Finance	92
FIGURE 5.2:	The Bedrock of Financial Accounting	93
FIGURE 5.3:	Navigating Managerial Accounting	94
FIGURE 5.4:	The Financial Planning & Analysis Odyssey	96
FIGURE 5.5:	Treasury Management Tapestry	98
FIGURE 5.6:	The Fortress of FRM	99
FIGURE 5.7:	The Centers of Excellence in Corporate Finance	100

FIGURE 5.8:	Financial Systems and Technology Landscape	102
FIGURE 5.9:	The Key Cornerstones of Compliance and Regulatory Reporting	104
FIGURE 5.10:	The Future of Autonomous Finance Use Cases	106
FIGURE 6.1:	Emerging Technologies in Autonomous Finance	151
FIGURE 6.2:	The Autonomous Finance Ecosystem	160
FIGURE 6.3:	Technology Selection and Integration	168
FIGURE 7.1:	Roadmap for Autonomous Finance Transformation	178
FIGURE 7.2:	Maturity Levels in the FOCUS Framework	182
FIGURE 7.3:	Dimensions of FOCUS Framework	183
FIGURE 7.4:	FOCUS Framework For Journey to Excellence in Autonomous Finance	184
FIGURE 7.5:	Synergy between Autonomous Finance and Other Business Domains	185
FIGURE 7.6:	Dimensions of Ongoing Support & Maintenance	190
FIGURE 7.7:	The Continuous Improvement and Innovation Cycle	191
FIGURE 8.1:	Partnerships in Autonomous Finance	210
FIGURE 8.2:	Value Proposition of Strategic Partnerships in Autonomous Finance	214
FIGURE 8.3:	Primary Criteria for Partner Selection	214
FIGURE 8.4:	Roadmap for Maintaining and Optimizing Partnership Relationships	216
FIGURE 9.1:	Integration & Interoperability in Autonomous Finance	221
FIGURE 9.2:	The Art of Integration in Autonomous Finance	223
FIGURE 9.3:	Technology Giants on the Horizon in Autonomous Finance	224
FIGURE 9.4:	The Future of Integration & Interoperability	229
FIGURE 10.1:	Data Management in Autonomous Finance	235
FIGURE 10.2:	Data Quality and Integrity in Autonomous Finance	237
FIGURE 10.3:	The Dilemma: To Cleanse or Not To Cleanse?	239
FIGURE 10.4:	Data Governance in Autonomous Finance	240
FIGURE 10.5:	Data Visualization in Autonomous Finance	242
FIGURE 10.6:	Data Security & Privacy in Autonomous Finance	243
FIGURE 10.7:	Predictive Analytics and Forecasting in Autonomous Finance	250
FIGURE 10.8:	Navigating the Big Data Challenges in Autonomous Finance	252
FIGURE 11.1:	The Paradigm Shift	258
FIGURE 11.2:	Tailored Approach to Customer Engagement and Decision-making	259
FIGURE 11.3:	"Finance Anytime Anywhere" via Autonomous Finance	260
FIGURE 11.4:	"Finance at your Fingertips" via Autonomous Finance	261
FIGURE 12.1:	A Multidimensional Mosaic of Regulations in Autonomous Finance	276
FIGURE 12.2:	A Roadmap to Compliance in Autonomous Finance	285
FIGURE 12.3:	The Future of Regulatory and Ethical Compliance in Autonomous Finance	286
FIGURE 13.1:	Explainability in Autonomous Finance	296
FIGURE 13.2:	Battling Bias in Autonomous Finance	297
FIGURE 13.3:	Battling Bias in Autonomous Finance	298
FIGURE 13.4:	Case Studies of Bias in AI & ML in Autonomous Finance	302
FIGURE 14.1:	The Threat Landscape in Autonomous Finance	308
FIGURE 14.2:	Sources of Threats in Autonomous Finance Landscape	310
FIGURE 14.3:	Areas of Vulnerabilities in Autonomous Finance Landscape	311
FIGURE 14.4:	Proactive Strategies for Cybersecurity in Autonomous Finance Landscape	315
FIGURE 14.5:	Cybersecurity Framework in Autonomous Finance Landscape	319

FIGURE 14.6: Implementing a Cybersecurity Framework in Autonomous Finance 321
FIGURE 14.7: Fostering a Cybersecurity-Conscious Culture in Autonomous Finance 322
FIGURE 14.8: Collaborative Measures in Autonomous Finance Cybersecurity 325
FIGURE 15.1: The Impact of Autonomous Finance on the Future of Work 333
FIGURE 15.2: The Impact of the Future of Work on Autonomous Finance 335
FIGURE 15.3: The Human-Machine Dance in Autonomous Finance 336
FIGURE 15.4: Guiding Principles for the Human-Machine Dance in Finance 337
FIGURE 15.5: Accounts Receivable in Spotlight 340
FIGURE 16.1: Impact of Adaptability on the Future of Finance 356
FIGURE 16.2: Preparing for an Uncertain Future using Strategic Foresight 358
FIGURE 16.3: Navigating the Challenges 360

Preface

Selecting a quote to inaugurate this scholarly investigation into the rapidly advancing fields of Finance and Technology required thoughtful consideration. One might ponder, *"What reference could aptly capture the essence of one of the most vital, convoluted, yet fascinating developments in finance?"* The pursuit of the ideal quote, a form of productive procrastination, became a distraction from the enormity of the task ahead.

However, after an extensive examination of cultural references from film, literature, music, history, and television, a thought settled on the concluding line from a preeminent work of American fiction, 'The Great Gatsby' by F. Scott Fitzgerald. Specifically, this evocative line was: *"So we beat on, boats against the current, borne back ceaselessly into the past"* (Fitzgerald, 1925, p. 110).

This sentence reflects a return to a theme permeating the book: the significance of history in shaping dreams of the future and the struggle of human beings to achieve goals by transcending and re-creating the past. However, why is this significant to Autonomous Finance? Fitzgerald composed his masterpiece during the Roaring Twenties, an era marked by rapid industrialization, burgeoning consumerism, and societal transformation.

Innovations such as the automobile and the telephone forever altered the rhythm of life during that period.

Today, the world witnesses analogous seismic shifts as it navigates a period of rapid changes fueled by technology. Traditional financial systems, previously centralized and resistant to change, are experiencing relentless disruption and restructuring. The era of **Autonomous Finance** has arrived, marking a radical transformation in how money is managed, invested, and interacted with in both personal and professional spheres.

This book offers a guided tour through this enthralling revolution. It strives to unravel the potential and implications of Autonomous Finance in an interconnected global landscape. The journey includes an exploration of the evolution of financial technology (FinTech) and the transformative power of AI, ML, Robotic Process Automation (RPA), Big Data, Advanced Analytics, SaaS, Cloud Computing, and Application Programming Interfaces (APIs), delving into the disruptive forces that are remolding the business landscape.

Each chapter presents a detailed exploration of a specific facet of Autonomous Finance, imbued with insights, real-world examples, and practical applications. The reader will traverse the inner workings of Autonomous Finance, the core technologies empowering it, use cases, challenges posed by big data, cybersecurity, data security and privacy, regulatory and ethical considerations, and more.

Through rigorous analysis, insightful discussions, and a touch of storytelling, this book aspires to serve as a comprehensive guide to Autonomous Finance. It targets Finance and Technology professionals seeking to stay abreast of a rapidly changing business landscape and business leaders interested in the tectonic shifts foretelling the future of Finance.

The excitement of embarking on this journey is palpable. The time has come to turn the page and take the first steps into the world of Autonomous Finance.

References

Fitzgerald, F. S. (1925). About the American Dream. In *The Great Gatsby* (p. 110).

Introduction

What is Autonomous Finance and Why is it needed?

In a world defined by swift technological advancements, where AI, ML, RPA, Big Data, and Analytics pervade nearly every sphere of existence, the finance domain finds itself at the cusp of a transformative revolution.

Visualize a future wherein personal and corporate finances are flawlessly managed, optimized, and orchestrated without continual human oversight. This future represents the burgeoning world of **Autonomous Finance**: a realm where machines assume control, thus enabling individuals and organizations to make wiser financial decisions, amplify growth, and attain financial equilibrium.

Embarking on this fascinating voyage into Autonomous Finance requires drawing an analogy that might initially seem peculiar yet uncovers the essence of this extraordinary metamorphosis.

Autonomous Finance is the orchestrated integration of technologies such as AI, ML, RPA, Big Data, Advanced Analytics, SaaS, Cloud Computing, and APIs into financial ecosystems, creating a symbiotic nexus where human expertise and machine intelligence converge. This fusion cultivates data-driven resilience, agility, and adaptability, transcending traditional paradigms to offer a self-evolving, ethically compliant framework that champions real-time insights, strategic foresight, and sustainable profitability.

The Road to Autonomy: A Parallel Universe of Finance and Driving

In the throbbing nucleus of contemporary innovation, the vibrant beat of autonomy guides two converging universes: the electrifying sphere of Autonomous Driving and the transformative world of Autonomous Finance.

Picture the enchantment of being in an autonomous vehicle, an engineering wonder that does more than merely follow the road; it comprehends, adapts, learns, and reacts, morphing the previously inflexible act of driving into an intelligent, fluid voyage.

This poetic image now finds a correspondence in the field of Finance. Autonomous Finance transcends mere number crunching or transaction execution. It alters the very essence of financial decision-making into a self-executing, harmonious orchestration.

From levels of autonomy to insights, strategic navigation, adaptational response to change, compliance safety, and human interaction, Autonomous Finance emerges as a profound alteration. It is a vision where Finance rises above the ordinary to become a lively, intelligent force propelling toward a future replete with potential, creativity, and boundless opportunities. Figure A-1 below illustrates the parallels between Autonomous Finance and Autonomous Driving's evolution, emphasizing shared traits such as varied automation levels and adaptational prowess. Both areas stress combining technology and human touch in utilizing algorithms, sensors, safety measures, compliance, and personalized experiences.

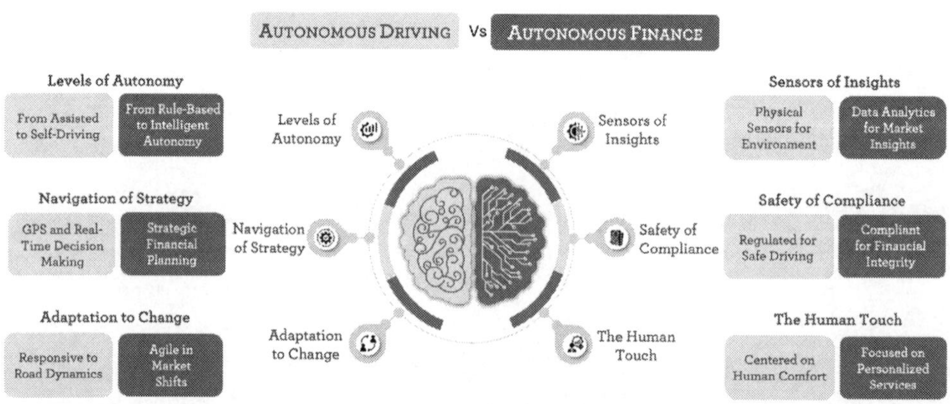

FIGURE A-1: Parallels between Autonomous Driving and Autonomous Finance

While reflecting on this vivid analogy, consider the following pertinent statistics that emphasize the relevance and significance of Autonomous Finance:

- Salesforce's "Trends in Financial Services" Report asserts that 76% of Finance leaders regard Autonomous Finance as vital for their organization's success in the upcoming five years (Agrios, 2023).
- Deloitte's survey reveals that 73% of millennials favor self-service FinTech tools over conventional services (*The Deloitte Global 2022 Gen Z and Millennial Survey*, 2023).
- According to a recent Vantage Market Research report, the global Autonomous Finance market is expected to grow from $8.13 billion in 2022 to $27.03 billion in 2030 (Vantage Market Research, https://www.vantagemarketresearch.com/, 2022).

Autonomous Finance symbolizes the fusion of AI, ML, RPA, Big Data, Advanced Analytics, SaaS, Cloud Computing, APIs, and automation within the financial ecosystem. This alignment leverages cutting-edge technologies to mechanize financial processes, obliterate manual errors, and render proactive, data-driven determinations. The objective is both simple and profound: to empower entities and commercial endeavors with personalized, real-time financial insights and counsel, facilitating superior choices, cash flow optimization, and the attainment of financial goals.

What distinguishes this transformative structure is its unparalleled ability for self-learning and self-correction, transcending the mere employment of advanced technologies. Autonomous Finance goes beyond traditional rule-based operations to proffer a dynamic and forward-thinking financial strategy. It shapes rules based on emerging trends, adapts to evolving parameters, and proffers predictive insights.

With Autonomous Finance, it becomes conceivable to automatically categorize, analyze, and reconcile financial transactions, absolving one from the mundane task of manual bookkeeping. It opens a world where algorithms scrutinize market trends, investment prospects, and risk elements, furnish tailored recommendations to augment wealth, and shield assets. Autonomous Finance evolves from a mere concept to a tangible reality, revolutionizing how finance is perceived and engaged.

The imperative for Autonomous Finance emanates from the drawbacks and constraints of conventional financial management practices. Human inaccuracies, delays in insights, and the intricacies of a perpetually changing financial terrain often obstruct sound financial decision-making. Autonomous Finance addresses these issues by leveraging technology to augment accuracy, efficiency, and foresight.

Adopting Autonomous Finance demands more than mere technological amalgamation; it calls for a comprehensive transformation involving a shift in mentality. It requires allegiance to a robust technological trajectory, a culture nurturing data and innovation, and recognition of advanced analytics. This mental pivot propels the transition from a stage of automation to a new era of autonomy across diverse operational domains of Finance, encompassing front-, middle-, and back-office functions.

What emerges from this radical alteration is a holistic view of financial operations, ushering in an epoch of augmented consistency, amplified data security, intelligent analytics, and optimized resource utilization. It cultivates an environment that advocates strategic tasks and enriches customer experiences. It aligns business growth with technological advancement, laying the groundwork for scalability, enhanced business Key Performance Indicators (KPIs), elevated operational efficiency, and cost reductions.

Moreover, the escalating intricacy of financial regulations and compliance obligations significantly encumbers both large and small enterprises. Autonomous Finance offers streamlined compliance, diminishing the risk of penalties and empowering organizations to channel more resources toward innovation and expansion.

In the ensuing chapters, this exploration will delve into the multifaceted dimensions of Autonomous Finance, scrutinizing its applications in personal finance, business operations, and governmental institutions. Together, the expedition will uncover the merits and potential challenges, study instances of successful implementations, and garner insights from experts sculpting the future of Finance.

Embark on this transformative voyage into Autonomous Finance, where algorithms metamorphose into allies, data evolves into guidance, and financial growth and sustained profitability become tangible.

References

Agrios, E. (2023). What do customers want most from their financial institutions? we asked them. *Salesforce*. https://www.salesforce.com/blog/financial-services-trends/

The Deloitte Global 2022 Gen Z and Millennial Survey. (2023, May 17). Deloitte. https://www.deloitte.com/global/en/issues/work/genzmillennialsurvey-2022.html

Vantage Market Research,https://www.vantagemarketresearch.com/. (n.d.-b). *Autonomous Finance market size USD 27.03 billion by 2030*. Vantage Market Research. https://www.vantagemarketresearch.com/industry-report/autonomous-finance-market-1756

SECTION I

Time Travel in Finance

Resilience, Agility, and Adaptability

Section 1 of the volume lays the foundation for a stimulating and enlightening journey through the evolution of finance. In this text segment, the expedition will transcend time, scrutinizing the pivotal roles that resilience, agility, and adaptability have played in sculpting today's financial universe.

Chapter 1, *"Finance in the Past: Lessons for Building Financial Resilience,"* delves into the intricate concept of financial resilience, guiding the reader through remarkable financial perturbations that have marked history. An examination ensues of how past crises have interrogated and molded finance, bequeathing invaluable insights into the robustness and adaptability that characterize the field.

Chapter 2, *"Finance in the Present: Embracing Agility in Contemporary Finance,"* aims to captivate and direct the reader toward the current epoch of transformation. Within these pages, an exploration unfolds of the agile metamorphoses transpiring in the financial landscape of our time. A discourse develops, elucidating how innovation and adaptability steer the course toward a more proficient and customer-attuned financial system.

Chapter 3, *"Finance in the Future: A Vision of an Adaptable Financial Framework,"* addresses the imperative need for adaptive, visionary financial frameworks capable of adept responses to the ceaseless flux of the business milieu. These reflections aid in setting the scene for subsequent dialogues regarding the emergence and potential ramifications of Autonomous Finance.

Upon the conclusion of Section 1, readers will have cultivated a robust comprehension of how finance has deftly maneuvered through its historical landscape, how it is in a state of dynamic evolution at this juncture, and what prerequisites it may necessitate to flourish in the future. This understanding paves the way, preparing them for an in-depth dissection of Autonomous Finance that will unfold in the successive portions of this literary work.

CHAPTER 1

Finance in the Past

Lessons for Building Financial Resilience

"In the middle of difficulty
lies opportunity."

- ALBERT EINSTEIN (ISAACSON, 2007)

1.0 Key Insights

- **Seizing Opportunity Amid Chaos:** Financial resilience is not merely a defense mechanism but the visionary foresight to recognize and seize opportunity even in the most tumultuous times, akin to how Edison capitalized on the Panic of 1893.
- **Navigating the Age of Unprecedented Disruptions:** In a world where change is the only constant, financial resilience is the compass, enabling organizations to adapt, innovate, and thrive amid relentless technological, social, and economic shifts.
- **The Dual-Purpose Shield and Catalyst:** More than a protective barrier, financial resilience acts as a dynamic catalyst that propels growth and innovation. It fortifies an organization while also inspiring a creative renaissance in business strategies.
- **Overcoming the Barriers with Strategy and Insight:** Facing barriers like constrained access to finance and market volatility requires not just resources but innovative thinking and strategic planning. Financial resilience is a symphony of analytical rigor, creative problem-solving, and risk management.
- **Embracing Technological Disruption with Vision:** The future of finance is intertwined with technological advancements. Financial resilience requires a visionary embrace of technologies like AI, ML, RPA, Big Data, Advanced Analytics, SaaS, Cloud Computing, and APIs, transforming them from challenges into tools for efficiency and innovation.

1.1 Introduction

The United States economy faced a severe economic depression in 1893. This downturn resulted in the collapse of the stock market, widespread unemployment, and the demise of businesses. However, amidst this turbulent storm, an ambitious young inventor named Thomas Edison perceived an invaluable opportunity. Edison invested decades in developing electric light bulbs and other electrical apparatuses, driven by the conviction that electricity harbored the power to transform societal norms and daily existence. However, he also understood that this nascent technology required significant investment to move from theory to practical application.

Ironically, the Panic of 1893 offered Edison a chance to establish General Electric (GE), his burgeoning enterprise centered around this groundbreaking technology. While the economic horizon seemed merciless, investors were possessed with a vision and a willingness to embrace risk. According to investing wisdom, the most auspicious buying opportunities arise when the scent of blood is on the streets. With tenacity and a resolute vision, Edison marshaled the required capital to establish GE, promptly elevating it to a zenith among the world's eminent electrical equipment manufacturers. Success was sculpted by the prevailing economic conditions, necessitating affordable and reliable electricity and Edison's indomitable will.

The tale of GE serves as a timeless testament that vast opportunities may indeed be found within the turbulent folds of the most chaotic epochs. In contrast to Edison's era, today's economic canvas, though not as dire, is continuously shaped and reshaped by relentless upheavals. From unparalleled global phenomena that bring the world to a halt to rapid technological advancements and the precarious balance of many economies on the cusp of recession, a plethora of complexities abound. Modern organizations must endure these disruptions and adapt, learn, and emerge with enhanced vigor, recognizing and grasping the opportunities concealed within these tumultuous waves. A key element in unlocking this potential lies in fortifying the organization's financial resilience.

This chapter offers an excavation into the essence of financial resilience, unraveling its foundational aspects, its cardinal role in steering through uncertainties, and offering practical methodologies for its cultivation. A sturdy mechanism for financial resilience serves as both a shield in turbulent periods and a catalyst for growth and innovation. Nevertheless, the journey toward financial resilience is strewn with substantial impediments. Failings in financial planning, limited avenues for financing, and market volatility stand as formidable barricades. The ensuing sections of this chapter shall dissect these intricacies, meticulously scrutinize each barrier, and delineate efficacious strategies for their circumvention.

By absorbing the insights presented herein, finance teams may elevate the financial resilience of their organizations, enhancing their capacity to thrive during disruption. Thus, this chapter constitutes the initial step on an expedition toward comprehending the vigor of financial resilience. Its ambition is to illuminate organizations on how financial resilience may empower survival and prosperity in a disruptive and exacting business environment.

1.2 Introducing the Age of Disruptions

Change remains the perennial constant. These disruptions, which permeate every facet of human existence, may manifest as social, economic, political, technological, legal, ethical, or environmental shifts. By their very nature, disruptions signify profound transformations that alter traditional paradigms.

Within finance, these disruptions command a reassessment of strategies, a challenge to entrenched norms, and, at times, a reconstruction of the foundational pillars of operations. Comprehending their myriad forms is the initial stride toward navigating this era of disruptions. Figure 1.1 elucidates the diverse manifestations of disruptions that have influenced the finance domain over time. A subsequent exploration will shed light on these forms and their consequential impact on the finance function.

a) **Social Disruptions:** These are rooted in cultural, demographic, or societal changes that influence how people interact, work, or behave (Eccles, 2014). For instance, the ongoing global push for diversity, equity, and inclusion has caused substantial changes in organizational structures and practices. Companies are now encouraged and expected to cultivate diverse workforces and foster inclusive cultures. Such societal shifts can affect finance teams, from talent acquisition to performance management metrics and how corporate social responsibility initiatives are funded and reported.

b) **Economic Disruptions**: These are drastic shifts in economic circumstances or structures, often occurring at a macro level. A poignant example is the global financial crisis of 2008, which caused a significant shake-up in the finance domain. The crisis, the worst economic disaster since the Great Depression of 1929, impacted businesses worldwide. It changed how finance teams viewed risk and emphasized the need for better liquidity management and contingency planning.

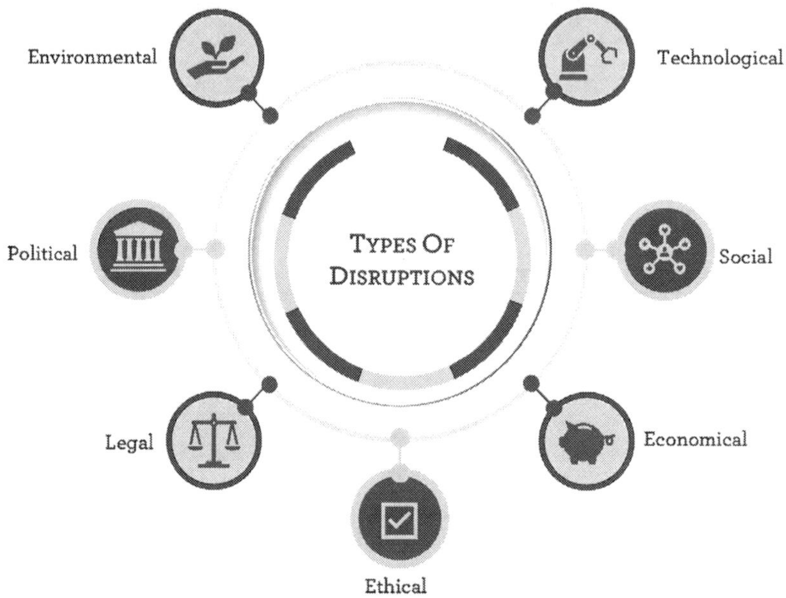

FIGURE 1.1: Types of Disruptions

c) **Political Disruptions:** Examples of such disruptions are changes in government, political climate, policy, or international relations. A recent example would be the geopolitical tensions between the U.S. and China, which have led to trade disputes and tariffs impacting global supply chains. Finance teams have grappled with the increased costs, altered supply and demand dynamics, and potential risks to overseas investments that these political disruptions can cause.

d) **Technological Disruptions:** The disruptions brought about by technological innovations are often associated with digital transformation initiatives. The rapid evolution of technologies such as AI, cloud computing, and blockchain has led to considerable disruption in how finance functions operate (Arner et al., 2015). These technologies have automated tedious manual tasks, enhanced data analysis capabilities, and streamlined financial processes, thus redefining the very nature of finance work.

e) **Legal Disruptions:** Changes in law or regulations that require organizations to alter their operations or strategies are called Legal Disruptions. A significant, recent example is the General Data Protection Regulation (GDPR) introduced in the European Union (Voigt & Von Dem Bussche, 2017). This regulation had widespread implications for organizations worldwide, forcing them to change how they manage and protect data, significantly impacting financial reporting and controls.

f) **Ethical Disruptions:** Ethical disruptions stem from societal shifts in the understanding and expectation of corporate ethics. The increased focus on corporate social responsibility and sustainability is an example. This shift is not merely a matter of public image but also affects the bottom line. Finance teams must now factor in ethical considerations in investment decisions, as evidenced by the increased inflow into ESG (Environmental, Social, Governance) funds (Eccles, 2014).

g) **Environmental Disruptions:** This final category of disruptions includes climate change and other environmental phenomena that have tangible impacts on business operations. Environmental disruptions can lead to physical risks to assets, operational risks, and regulatory risks as companies are pressured to become more environmentally friendly. They underscore the need for finance teams to incorporate environmental risk into their financial planning and risk management strategies.

While each type of disruption is unique, they are all interconnected. They have the power to influence each other and often *compound* the effects of the others, resulting in complex and multifaceted impacts.

1.3 The Domino Effects of Disruptions

FIGURE 1.2: Domino Effects of Disruptions

In today's ever-changing landscape, organizations face a series of consequences resulting from disruptions. These consequences can range from financial difficulties, such as decreased revenue and limited capital, to increased risks, like cybersecurity threats, operational hazards, and physical damage to assets.

Organizations grapple with a spectrum of consequences spawned by disruptions in an era characterized by ceaseless transformation in an era characterized by ceaseless transformation. These range from financial and operational ramifications to strategic alterations and challenges to cultural ethos.

a) **Financial Impact:** Disruptions exert significant financial strain by their inherent nature. The repercussions can be severe and prolonged, from immediate financial stress, revenue contractions, liquidity crunch, and capital depletion.

b) **Operational Impact:** Interruptions to regular operations may entail severe ramifications. Such interruptions are particularly poignant in the aftermath of the COVID-19 pandemic, wherein operations worldwide were halted, forcing organizations to reconsider operational norms and innovate novel business models (Nicola et al., 2020).

c) **Strategic Impact:** The strategic course of organizations can be substantially realigned in response to disruptions. Strategic planning, consequently, must remain pliant to adapt to changing circumstances swiftly.

d) **Cultural Impact:** Organizational culture is not immune to disruptions. From modifying employee behavior, revising organizational values, and redefining performance metrics, cultural shifts require careful navigation.

e) **Reputational Impact:** An organization's reputation can be imperiled during disruptions, potentially affecting customer trust, investor confidence, and overall market perception. Navigating these waters is critical for long-term success.

From Theory to Reality: Case Studies of Disruptions

To build a more holistic understanding of the transformative power of disruptions across various domains, Table 1.1 below presents several real-world case studies. These examples were carefully selected to represent various scenarios across different sectors and regions, demonstrating the diversity and impact of disruptions on the financial landscape.

TABLE 1.1: Examples Of Past Disruptions

NO.	TYPE OF DISRUPTION	HOW DISRUPTIONS HAVE SHAPED THE FINANCIAL LANDSCAPE OVER TIME
1	Social	**World Wars and Economic Crises:** World wars and economic crises, such as the Great Depression in the 1930s and the 2008 global financial Crisis, caused widespread social and economic disruptions. These events highlighted the importance of financial regulation, risk management, and the need for stability in the financial system. **Social Impact Investing and Philanthropy:** The growing trend of social impact investing and philanthropy has reshaped the finance function. As investors increasingly consider ESG factors, Finance teams have adapted their strategies to incorporate them into their financial decision-making processes (Ziolo et al., 2019). They have played a crucial role in analyzing the financial implications of ESG investments and ensuring transparent reporting on social impact and sustainability metrics (Muigua, 2022).
2	Economic	**Financial Crisis of 2008:** Triggered by the collapse of the subprime mortgage market in the U.S. the global financial crisis exposed the financial system's vulnerabilities, leading to a drastic decline in economic growth. Finance teams reacted by reinforcing their risk management practices and prioritizing financial resilience (*World Economic Outlook, April 2009: Crisis and Recovery*, 2016). **COVID-19 Pandemic:** The pandemic resulted in global economic shocks, with disrupted supply chains, reduced consumer spending, and financial market volatility. Finance teams proactively managed liquidity, secured funding, and devised strategies to mitigate economic fallout (*World Economic Outlook, April 2021: Managing Divergent Recoveries*, 2021). Simultaneously, organizations had to speed up their digital transformation initiatives to manage the sudden digitization of daily life, including remote work. The finance function had to remodel its service delivery options to accommodate more self-service alternatives and digital support systems. (Battisti et al., 2022).

NO.	TYPE OF DISRUPTION	HOW DISRUPTIONS HAVE SHAPED THE FINANCIAL LANDSCAPE OVER TIME
3	**Political**	**Trade Tariffs and Geopolitical Tensions:** Recent trade disputes between the U.S. and China resulted in trade tariffs that disrupted businesses operating in those sectors. Finance teams successfully navigated these disruptions by assessing financial implications, mitigating risks, and ensuring the financial resilience of their organizations (Zeng et al., 2022). **Dodd-Frank Act:** The implementation of the Dodd-Frank Act introduced stricter compliance requirements and reporting standards, requiring finance teams to ensure compliance, adapt systems and processes, and provide accurate reporting to regulatory authorities (Zeng et al., 2022).
4	**Technological**	**RPA:** RPA has introduced an opportunity for finance teams to streamline operations, reduce errors, and focus more on strategic activities (Rutaganda et al., 2017). **AI and ML:** The proliferation of AI and ML technologies has transformed the finance function, enabling finance professionals automate processes, enhance data analysis capabilities, and facilitate predictive analytics (Kommunuri, 2022). AI's impact is a story that continues to unfold prodigiously, with the full ramifications still yet to emerge.
5	**Regulatory**	**GDPR:** The GDPR introduced stringent data protection and privacy regulations, prompting finance teams to enhance data security and respect user consent, thereby adhering to the highest standards of privacy rights (Hoofnagle et al., 2019). **Anti-Money Laundering (AML) and Know Your Customer (KYC) Regulations:** AML and KYC regulations mandate finance teams to conduct thorough due diligence on customers and monitor transactions to prevent illicit activities. Finance professionals ensure adherence to these regulations through robust internal controls and compliance procedures (Van Vliet, 2023).
6	**Ethical**	**Unethical Behavior in Credit Rating Agencies and Investment Advisory Services:** Unethical behavior has led to significant financial losses and a lack of trust among investors. Finance teams have responded by prioritizing ethical conduct and implementing robust governance structures and internal controls (Sharifani & Amini, 2023). **Payday and Predatory Lending Practices:** Predatory lending practices targeting vulnerable individuals have raised ethical challenges. Finance teams have navigated these challenges by ensuring fair lending processes, responsible loan products, and compliance with consumer protection regulations (McKee et al., 2011).
7	**Environmental**	**Transition to a Low-Carbon Economy and Renewable Energy Investments:** As the world shifts toward a low-carbon economy, finance teams are crucial in securing funding for renewable energy projects and incorporating environmental considerations into investment decisions (Martí-Ballester, 2019). **ESG Investing and Disclosure Requirements:** ESG investing has necessitated finance teams to integrate ESG factors into financial analysis, risk assessment, and reporting practices. This transition has enhanced the company's reputation and contributed to long-term financial sustainability (Riahi et al., 2007).

By understanding these real-world examples, finance professionals can glean insights into the challenges and opportunities associated with various disruptions. This knowledge will enable them to successfully devise effective strategies and navigate the changing landscape.

Caught in the Crosshairs: The Finance Team's Dilemma

Finance teams are encountering many obstacles and uncertainties due to the disruptions in their surroundings. These disruptions, which can stem from factors such as the economy, technology, regulation, or the environment, have a far-reaching impact on organizations, especially the finance function. Figure 1.3 depicts the top 5 negative and top 5 positive impacts of disruptions on finance teams. This section aims to illuminate the impact of these disruptions on finance teams and how they can, and are, navigating them.

a) **Financial Instability:** Disruptions often lead to financial instability due to sudden shifts in market dynamics, consumer behavior, and supply chains. The 2008 financial crisis, for instance, disrupted the stability of many organizations globally (*World Economic Outlook, April 2009: Crisis and Recovery*, 2016). Finance teams must carefully monitor the organization's financial health, finding ways to maintain stability amidst these storms.

b) **Forecasting Challenges:** Disruptions introduce an air of unpredictability that makes accurate forecasting challenging. Disruptions to traditional patterns and trends make conventional forecasting models unreliable (Steindel et al., 2005). Finance professionals must adapt to this uncertainty, incorporating scenario planning and sensitivity analyses.

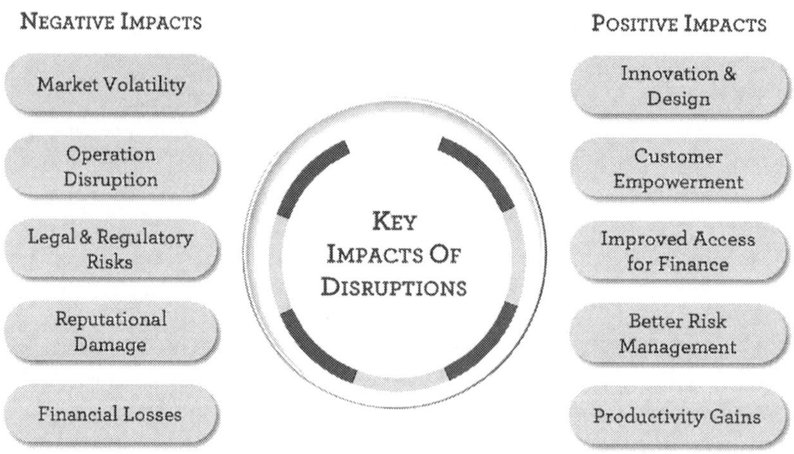

FIGURE 1.3: Top 5 Negative And Top 5 Positive Impacts Of Disruptions

c) **Limited Access to Credit:** During disruptions, accessing credit can become more complex, as lenders may tighten their lending criteria due to increased risk. For instance, during the COVID-19 pandemic, many small businesses struggled to secure necessary funds (Thetechyhub, 2022). Finance teams must explore alternative funding sources and establish strong relationships with financial partners to secure necessary credit.

d) **Decreased Revenue and Cash Flow:** Disruptions can decrease revenue streams due to shifts in the market, changes in consumer preferences, and logistical challenges. Finance teams must adapt quickly, reevaluating budgets, optimizing working capital, and ensuring sufficient liquidity. The experience of retail organizations during COVID-19 exemplifies this, where many had to adjust their operations and financial strategies rapidly (Nandi et al., 2021).

e) **Increased Costs:** Disruptions can also increase costs. Whether sourcing materials from alternative suppliers or implementing contingency plans, the associated expenses can be significant. Finance teams play a vital role in analyzing and managing these costs, as seen in the aftermath of Brexit, where companies had to navigate additional supply chain costs (Natarajarathinam et al., 2009).

f) **Compliance Complexity:** Regulatory changes often accompany disruptions, adding to the compliance landscape's complexity. The introduction of GDPR, for instance, imposed new compliance demands on organizations, with finance teams playing a crucial role in navigating these changes.

g) **Challenges in Financial Reporting:** Disruptions can significantly complicate the financial reporting process due to uncertainties, changes in business operations, and additional disclosure needs. For example, the transition to remote work during the COVID-19 pandemic posed significant challenges for financial reporting due to the distributed nature of teams and data security concerns (Ahrens & Ferry, 2021).

In the face of these challenges, an organization's financial resilience emerges as a pivotal asset. It constitutes a multifaceted capability that embodies financial stability, adaptability, and innovation, ensuring the organization remains robust and agile to navigate challenges and leverage opportunities.

The subsequent sections unravel the essence of financial resilience, its integral role in navigating disruptions, and practical strategies for its fortification. The chapter culminates in delineating a phased approach to guide finance teams through financial stress and disruptions.

1.4 Resilience in the Face of Disruptions

The beginning of the 21st century marked a shift into an era characterized by frequent disruptions, aptly named the 'Age of Disruptions' (Frank & Goyal, 2009). Rapid technological advancements, volatile economies, climate change, and sociopolitical shifts have radically altered the business landscape, presenting significant challenges and unprecedented opportunities for organizations worldwide (Sheffi, 2007). In this turbulent era, business resilience has emerged as a vital organizational competency for survival, sustainability, and success.

Business resilience refers to an organization's capacity to anticipate, prepare for, respond, and adapt to incremental change and sudden disruptions while maintaining continuous business operations, protecting people and assets, and ensuring long-term success (Douglas, 2021).

One critical facet of this resilience is financial resilience, the focus of our exploration in this chapter. We will explore the essence of financial resilience, its implications in the Age of Disruptions, and the strategic path to nurturing it. Our exploration will lay the foundation for subsequent discussions on Autonomous Finance, mapping the transformative trajectory of finance in response to the challenges of our time.

Understanding Business Resilience

The concept of resilience originated in ecology, with Holling (1973) defining it as *"the capacity of a system to absorb disturbances, self-organize, and persistently endure change."* In business, we see resilience as an organization's ability to persist, adapt, and grow in the face of disruptions (*2019 Global Risk Management Survey - Results | AON*, 2019). It stretches beyond mere survival, encompassing the ability to turn crisis into growth opportunities, ensuring long-term sustainability.

Resilience in business is an amalgamation of various aspects, encapsulated within five key pillars (as shown in Figure 1.4):

a) **Strategic Resilience:** Strategic resilience concerns an organization's capacity to recalibrate its strategy in response to evolving environments. According to Sutcliffe and Vogus (2003), this includes discerning strategic risks and opportunities, fostering strategic agility, and modifying the business model to adapt to emerging circumstances. Cultivating strategic resilience demands a forward-looking perspective, flexibility, and creativity.

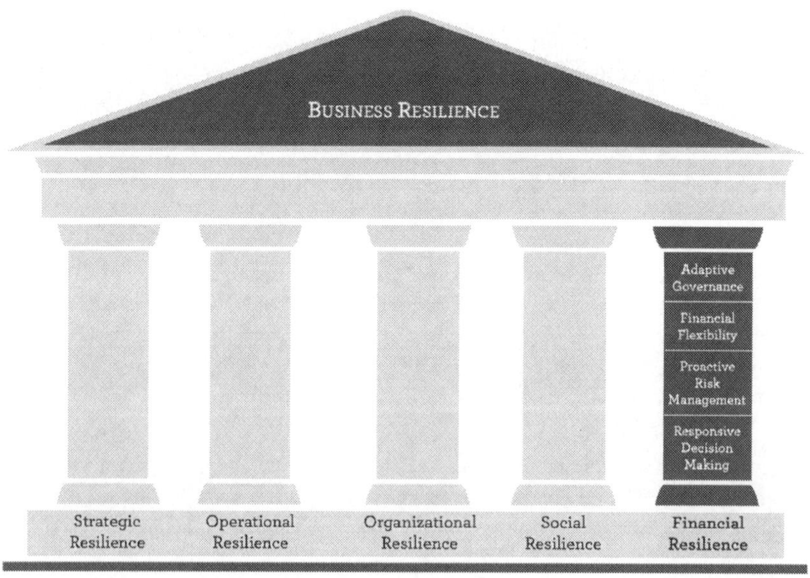

FIGURE 1.4: The Pillars of Business Resilience

b) **Operational Resilience:** Operational resilience reflects an organization's ability to sustain operations amidst internal or external shocks. Essuman et al. (2020) assert that it covers domains like supply chain management, business continuity planning, crisis management, and operational risk management. Ensuring operational resilience entails building robust, flexible operational structures and processes that respond effectively to disruptions.

c) **Organizational Resilience:** Organizational resilience is about the cultural, structural, and human factors that enable organizations to respond and adapt to disruptions effectively. It requires fostering a resilient culture, building adaptable structures, and developing human capital (Burnard & Bhamra, 2011). The strategies to strengthen organizational resilience include cultivating a mindset of resilience among employees, fostering leadership that promotes resilience, and building structures that enable rather than hinder resilience.

d) **Social Resilience:** This refers to a company's ability to build and maintain strong relationships with the communities in which it operates and to consider its operations' social and environmental impacts. It involves supporting community development projects, collaborating with local organizations, and being responsive to the needs and concerns of stakeholders.

e) **Financial Resilience:** The last, but by no means least, is financial resilience, which refers to an organization's ability to absorb and recover from financial shocks, ensuring financial stability and sustainability amidst disruptions.

With a basic understanding of the pillars of business resilience, we can now dive deeper into our primary area of interest—financial resilience.

Financial Resilience Defined

Financial Resilience refers to an organization's integrated capability to anticipate, withstand, recover from, and adapt to financial volatility. It encompasses robust financial management, efficient risk mitigation, and strategic agility, both as a protective mechanism and a catalyst for sustainable growth.

Financial Resilience is an organization's capacity to withstand and adapt to financial volatility, ensuring financial continuity, stability, and flexibility in facing adversities. It entails financial preparedness, robust financial management, efficient risk management, and the ability to capitalize on opportunities for financial growth and sustainability (Aon, 2019).

The ever-evolving Age of Disruptions continually underscores the importance of financial resilience. With fluctuating economies, unpredictable global events, and rapid technological changes, organizations must be prepared to weather financial shocks and seize emerging opportunities (Sheffi, 2007).

Financial resilience serves as both a protective shield and a springboard for businesses, enabling them to endure disruptions while leveraging them for growth and innovation. By providing stability amidst volatility, financial resilience becomes the foundation upon which organizations can build their strategies for growth, even in the face of uncertainties.

The Elements and Role of Financial Resilience in Navigating Disruptions

The role of financial resilience in an organization, especially in navigating the treacherous waters of disruptions, can be likened to the keel of a grand ship. As the vessel of commerce embarks on uncharted territories filled with unseen obstacles, the keel provides stability, ensures direction, and enables the ship to withstand unpredictable storms. In this metaphorical understanding, financial resilience serves as the organizational keel, imparting steadfastness, direction, and vigor, even in the face of the most chaotic economic and financial tempests. To dissect the role of financial resilience, it is quintessential first to understand its very structure. Financial resilience is not merely a monolithic attribute; it encapsulates three key components, as shown in Figure 1.5 below:

a) **Financial Stability:** This component encompasses the core functions that enable an organization to absorb shocks without significant alteration to regular operations.

   Disruptions, economic downturns, political upheavals, technological advancements, or unforeseen global events inevitably exert pressure on financial resources. For instance, the onset of the 2008 global financial crisis sent shockwaves through corporate balance sheets, draining liquidity and eroding capital bases (Reinhart & Rogoff, 2011). Financial stability is the anchor that holds the organization firm. It signifies having robust financial controls, adequate liquidity reserves, diversified revenue streams, and a prudent risk management framework. Such stability ensures that the organization can continue its essential operations, meet its obligations, and survive the immediate impact of the disruption.

b) **Financial Adaptability:** This component represents an organization's capacity to modify financial strategies and operations to respond to changing circumstances.

   The prevailing conditions may necessitate strategic realignment in the aftermath of a disruption. The COVID-19 pandemic is a salient illustration of altering consumer behavior, supply chain dynamics, and operational norms (Nicola et al., 2020). Financial adaptability is akin to the ship's rudder, allowing the organization to adjust its course according to the prevailing winds and currents. It involves reassessing financial strategies, reallocating resources, adjusting budgeting, and modifying financial goals per the altered environment. This adaptability ensures that the organization can survive the disruption and emerge aligned with the new reality.

c) **Financial Innovation:** This component measures an organization's ability to leverage financial resources to pursue new opportunities and drive growth.

   Disruptions, though often viewed through a lens of trepidation, can also be fertile ground for opportunities. For example, new technologies such as AI and blockchain have reshaped industries, creating new markets and innovative business models (Potts & Rennie, 2019). Financial innovation is the sail that captures the wind, propelling the organization forward. It involves leveraging financial resources creatively to invest in new opportunities, fund innovation, and drive growth. The organization can turn a disruption into a catalyst for advancement, seizing opportunities others may overlook.

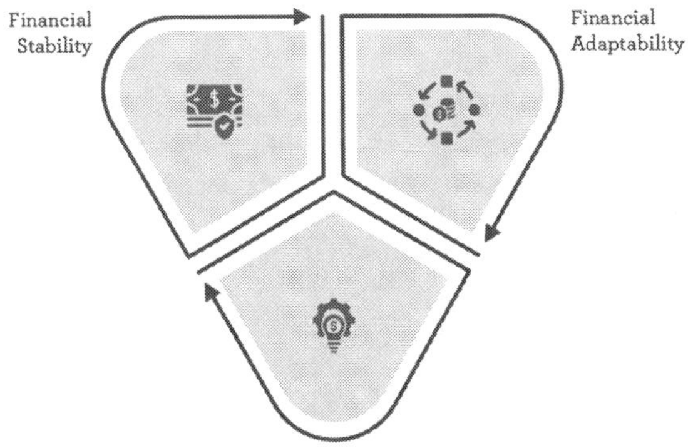

FIGURE 1.5: The Key Elements of Financial Resilience

In summary, financial resilience plays an instrumental role in navigating disruptions by anchoring the organization through financial stability, ensuring that it can withstand the immediate shock, steering through change through financial adaptability, realigning with the new reality, and seizing opportunities through financial innovation, turning disruption into a springboard for growth. These attributes are neither isolated nor static; they interact in a harmonious symphony guided by insightful leadership, prudent governance, and a well-crafted financial strategy. Through the cultivation of financial resilience, organizations can transcend mere survival and actively thrive, transforming disruptions from existential threats into opportunities for rejuvenation and growth.

Harmonizing Fiscal Fortitude: Strategies for Financial Resilience

The fortification of financial resilience within an organization is an art and science that transcends mere fiscal prudence. It demands an intricate choreography of strategies that tap into not just the financial fabric of the business but also its strategic, operational, and cultural dimensions. Figure 1.6 visually represents the eight most leveraged and practically potent strategies in fortifying financial resilience.

In an era where disruptions have become the rule rather than the exception, crafting a financially resilient organization is paramount. Below, the most leveraged and practically potent strategies are delineated.

a) **Risk Management & Hedging:** A robust risk management framework represents the first line of defense. Organizations can protect themselves from undue exposure by identifying, assessing, and mitigating financial risks. Strategies such as diversification of investments, hedging against currency or commodity price fluctuations, and implementing strict credit risk controls can ensure stability (Jorion & Risk, 2010).

b) **Liquidity & Capital Management:** Maintaining sufficient liquidity and strong capital reserves is akin to having well-calibrated ballast in a ship. It ensures the organization can navigate short-term financial obligations and invest in long-term growth. Tools such as cash flow forecasting, prudent working capital management, and maintaining committed credit lines provide flexibility and agility in capital deployment (Damodaran, 2016).

c) **Strategic Cost Management:** Cost management does not imply mere cost-cutting. It necessitates a strategic approach to evaluate the cost structure, identify areas of inefficiency, and redesign processes without compromising the quality of products or services. Implementing zero-based budgeting or utilizing activity-based costing techniques offers a systematic and transparent way to align cost structures with strategic objectives (Cokins, 2001).

d) **Diversification of Revenue Streams:** Relying on a single source of revenue can render an organization vulnerable to disruptions. Therefore, diversifying revenue streams across different markets, products, and channels enhances adaptability. Exploring alternative revenue models and tapping into new customer segments creates a well-balanced and resilient revenue portfolio (Markowitz, 1952).

e) **Technology Utilization & Automation:** In a world where data drives decision-making, leveraging technology and automation enhances efficiency and fosters adaptability. Incorporating Financial Planning and Analysis (FP&A) tools, utilizing predictive analytics, and employing AI-driven algorithms for portfolio management empower organizations to make proactive and informed decisions (Davenport & Harris, 2017).

f) **Building a Resilient Organizational Culture:** A resilient culture encourages an adaptive and innovative mindset. Through strong leadership, clear communication, and consistent training, organizations can foster a culture where financial resilience is fiscal responsibility and an intrinsic organizational value.

g) **Regulatory Compliance & Ethical Conduct:** Aligning with regulatory requirements and maintaining ethical conduct strengthens stakeholders' trust. Regular audits, transparent reporting, and adherence to ethical finance practices ensure that the organization is compliant and stands as a beacon of integrity (Paine, 1994).

FIGURE 1.6: Strategies to Fortify Financial Resilience

h) **Scenario Planning & Stress Testing:** Regular scenario planning and financial stress testing ensure preparedness. By simulating different economic, market, or industry conditions, organizations can gauge their financial resilience and proactively address any identified vulnerabilities (Apgar, 2006).

The fortification of financial resilience capability is not a mere financial exercise but a symphonic arrangement that harmonizes strategy, operations, technology, culture, and governance. By embracing these multifaceted strategies, an organization transforms into an architectural marvel, standing unshaken amidst disruptions and ready to leap forward when opportunity knocks.

Barriers to Financial Resilience and Overcoming Strategies

Financial resilience, while indispensable, can often prove challenging. The hurdles stem from many factors: internal and external, controllable, and uncontrollable. Here, we will explore the key barriers and how to surmount them. Figure 1.7 highlights the key barriers organizations face in achieving financial resilience.

a) **Inadequate Financial Planning and Management:** Many organizations struggle with effective financial planning and management. This deficiency may arise from insufficient expertise, inaccurate forecasting, poor budgeting, or inadequate financial controls (Aljohani et al., 2017). Overcoming this barrier requires investing in financial skills development, adopting robust financial planning tools, and ensuring a disciplined approach to budgeting and financial control.

b) **Access to Financing:** Access to affordable financing is a significant challenge for many organizations. This issue is often magnified during financial crises or economic downturns ("Financing SMEs and Entrepreneurs 2020," 2020). Companies can address this barrier by building strong relationships with various financiers, maintaining a good credit record, exploring different funding sources, and leveraging government assistance programs.

c) **Market Volatility:** Organizations operating in volatile markets face significant financial resilience challenges. Such volatility can arise from fluctuations in commodity prices, exchange rates, or interest rates. Managing this requires effective financial risk management (FRM) strategies, including hedging, diversification, and insurance.

d) **Technological Disruption:** Rapid technological advancements can disrupt traditional business models, leading to financial instability. Organizations not keeping pace with these changes can face obsolescence (Sørensen & Stuart, 2000). To overcome this, companies must embrace digital transformation, invest in research and development, and foster a culture of innovation.

e) **Regulatory Compliance:** Changes in regulatory environments can have financial implications. Compliance failures can result in penalties, reputational damage, and even business closure (*Navigating Regulatory Change for Emerging Technologies*, n.d.). Companies must have effective compliance programs, invest in compliance training, and closely monitor regulatory changes to mitigate these risks.

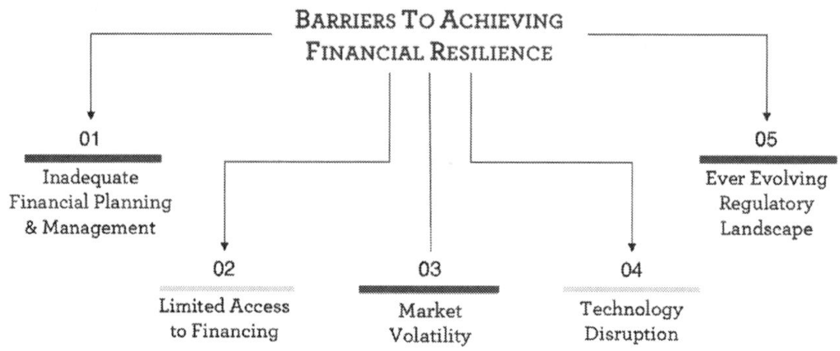

FIGURE 1.7: Barriers to Achieving Financial Resilience

Survive-Revive-Thrive: An Approach to Bolster Financial Resilience

Given the challenges and barriers associated with building financial resilience, organizations can adopt a survive, revive, and thrive framework as shown in Figure 1.8 below) to bolster their financial resilience muscle.

In the **Survive** phase, the main priority is to ensure immediate survival by securing liquidity, cutting costs, managing risks, and sustaining operations. This phase often necessitates tough decisions to guarantee that the organization has sufficient cash and liquidity to fulfill its immediate financial responsibilities and remain operational (Sheffi, 2007). Below are five essential tactics organizations can employ to weather a financial shock:

a) **Building Cash Reserves:** This can be done by setting aside a portion of profits to be used as a buffer during difficult times. For example, during the COVID-19 pandemic, companies with strong cash reserves could weather the economic downturn and continue to pay their employees and suppliers better.

b) **Maintaining a Strong Balance Sheet:** A strong balance sheet, which includes a healthy ratio of assets to liabilities, can help a business weather a financial shock by providing a cushion of cash and other assets. This cushion can include significant cash, investments, or property. Companies with a strong balance sheet could access financing during the pandemic when access to credit became more difficult.

c) **Managing Debt:** Keeping debt levels low gives businesses more flexibility to navigate a financial crisis. For example, during the pandemic, companies with low levels of debt were better able to weather the economic downturn and avoid defaulting on loans.

d) **Reviewing and Updating Risk Management Plans:** This can help a business quickly and effectively respond to a financial shock by identifying critical functions and developing a plan to maintain operations. It typically includes identifying key suppliers, customers, and employees and developing a plan to keep them informed and reassured during a crisis.

e) **Cost-Cutting Measures:** This can include reducing expenses, shedding non-core assets, and cutting unnecessary costs to preserve cash. For example, companies implementing cost-cutting measures during the pandemic, such as reducing staff and overhead costs, could better survive the economic downturn.

Walmart is an example of an organization that has implemented these strategies regularly. Walmart is well known for its substantial cash reserves and efficient supply chain operations, which have helped it weather economic downturns and natural disasters. The company also has a strong balance sheet, with a healthy ratio of assets to liabilities. In addition, Walmart has implemented cost-cutting measures, such as automation and technology, to improve its operational efficiency and reduce expenses. These strategies have helped Walmart survive financial shocks and thrive in the face of them.

During the **Revive** phase, organizations gradually move toward recovery. The emphasis here is on improving financial performance, optimizing operations, rebuilding supply chains, and re-engaging with customers. This phase may involve adopting new business models, investing in technology, and rethinking business strategies (Sheffi, 2007). Some key strategies organizations can use to revive after a financial shock include:

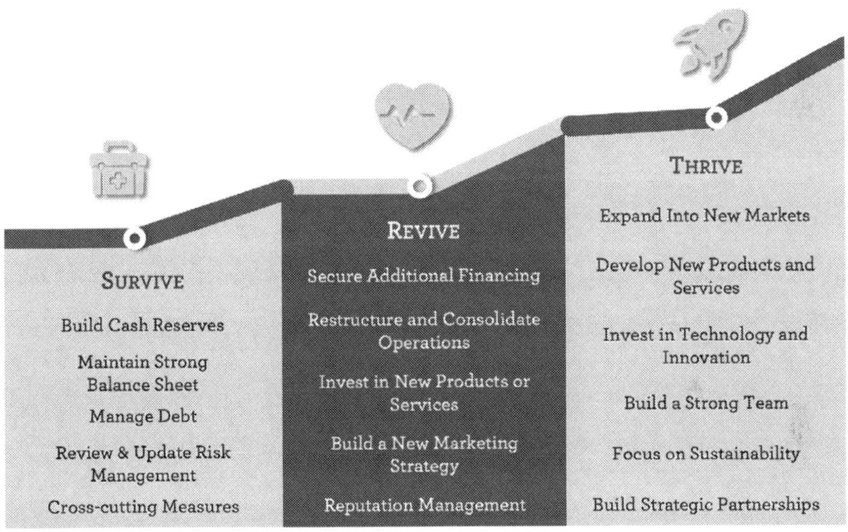

FIGURE 1.8: The Survive-Revive-Thrive Framework for Bolstering Financial Resilience

a) **Securing Additional Financing:** Obtaining loans, issuing new stock, or finding new investors. For example, during the COVID-19 pandemic, many companies turned to government-backed loans, such as the Paycheck Protection Program, to help them revive their operations.

b) **Restructuring and Consolidating Operations:** This can involve consolidating locations, streamlining operations, or divesting non-core assets. For example, many companies restructured their operations during the pandemic to focus on their core businesses and reduce costs.

c) **Investing in new products or services:** This can help revive a business by creating new revenue streams and diversifying the organization's customer base. For example, many companies turned to e-commerce and online sales during the pandemic to revive their operations.

d) **Building a new marketing strategy** can include developing new campaigns, launching new products, or targeting new customers. For example, many companies shifted their marketing strategies to focus on online and social media channels during the pandemic.

e) **Reputation management:** This can include rebuilding the organization's reputation with customers, investors, and other stakeholders by being transparent and honest about the company's challenges and the steps taken to overcome them.

Uber serves as a prime example of a company successfully implementing the strategies mentioned above to recover from a financial. The company experienced significant hardships during the pandemic, with travel restrictions and social distancing measures causing declining service demand. To combat this, Uber took proactive measures such as securing additional funding, restructuring their operations, and venturing into new markets like food and online grocery delivery. Moreover, they prioritized rebuilding trust with their customers and stakeholders by openly addressing the challenges they faced and outlining their steps to overcome them. Thanks to these efforts, Uber managed to bounce back and continue its growth trajectory.

Finally, in the **Thrive** phase, organizations move beyond recovery toward growth. Here, the focus is on innovation, expansion, and seizing new market opportunities. Organizations must invest in research and development, employee skills development, and sustainable business practices (Sheffi, 2007).

As seen in the Figure 1.8, some key strategies organizations can use to thrive after a financial shock include expanding into new markets, entering new geographic markets, or targeting new customer segments. For example, during the COVID-19 pandemic, many companies expanded their online sales and delivery capabilities to reach new customers.

a) **Developing new products and services** can help businesses capitalize on new market opportunities and diversify their revenue streams. For example, during the pandemic, many companies developed new products and services to meet the changing needs of their customers, such as virtual consultations, online classes, or online service delivery.

b) **Investing in technology and innovation:** This can include investing in new technology, such as automation, data analytics, and AI, to improve operational efficiency and gain a competitive edge.

c) **Building a solid team:** This can include hiring new employees, developing existing employees, and creating a positive work environment that fosters innovation and collaboration.

d) **Focusing on sustainability:** This can include implementing sustainable business practices, such as reducing carbon emissions and using renewable energy sources, which can help to attract and retain customers, as well as reduce costs.

e) **Building strategic partnerships:** This can include partnering with other companies, organizations, or individuals to access new markets, resources, or expertise.

For example, in the late 2000s and early 2010s, Microsoft was in a difficult situation with diminishing returns from its flagship products like Windows and Office Suite, and its attempts to compete in the mobile operating system market failed to make a significant impact.

However, under the leadership of Satya Nadella, who became CEO in 2014, Microsoft embarked on a transformation journey, focusing on areas where they could provide unique value and committed to becoming a leader in cloud services. They invested heavily in building and improving Azure, their cloud platform, and started transitioning their Office Suite into a cloud-based subscription service, Office 365. This strategy proved to be highly successful. When writing this book, Microsoft's commercial cloud business generated over $52.9 billion in quarterly revenue, and the company's market capitalization has grown to over $2.5 trillion (Microsoft News Center, 2023). The financial resilience and adaptability shown by Microsoft in the face of disruption have made it one of the most valuable companies in the world.

Adopting this phased approach can help organizations navigate through periods of financial stress and disruption and emerge stronger on the other side.

1.5 Conclusion

In the traverse through Chapter 1, readers have journeyed through the complex maze of historical financial disruptions, uncovering financial resilience's seminal role in steering the helm of financial destiny. In this symbolic landscape, the resilience of finance teams stands parallel to the mythical Phoenix, rising from the ashes of turmoil, rejuvenated, regenerated, and prepared to confront the uncertainties ahead.

The heart of this chapter beats with an unwavering pulse: the vital necessity for financial resilience. Illuminated through a tapestry of historical examples, it uncloaks an enduring truth: *"Disruptions, in their multifaceted manifestations, serve as forges for change and metamorphosis."* These disruptions invite us to peer into our deepest vulnerabilities, recalibrate our approaches, and ascend with reinforced vigor.

In this age, characterized by perpetual change and escalating disruptions, the cultivation of Financial Resilience transmutes from a mere mechanism for survival to an architectural blueprint for flourishing in the face of adversity. The vicissitudes of history are not merely obstacles but profound lessons in fortitude—a celebration of the human spirit's capacity to endure, adapt, and transcend.

Invoking the wisdom of the philosopher Friedrich Nietzsche (2000), *"What does not kill me, makes me stronger!"* emerges as more than a philosophical musing; it transforms into an actionable anthem for Financial Resilience. A symphony of past tribulations composes our collective wisdom, equipping us with the strength and discernment required to traverse the turbulent waves of the future.

As Chapter 1 draws its exploration to a dignified close, it serves as a prologue to the probing examination of contemporary trials and triumphs that lie ahead. The subsequent chapter extends an invitation to explore the intricate dance of Agility, an essential partner in navigating the fluid and ever-changing tapestry of today's financial world.

1.6 Key Recommendations

a) **Embrace a Holistic Perspective:** Recognize financial resilience as interconnected with operational, technological, human, and reputational resilience. Foster an organizational culture that values and invests in resilience across all aspects.

b) **Invest in Continuous Evolution:** In an age marked by relentless disruptions, organizations must actively adapt, innovate, and grow. A commitment to ongoing financial evolution is vital; stagnation is not an option.

c) **Leverage Technology with Insight:** The digital era is here, and technology is not merely a tool but a strategic partner. Invest in and harness technological innovations such as AI, ML, RPA, Big Data, Advanced Analytics, SaaS, Cloud Computing, and APIs to enhance financial operations, analytics, and decision-making.

d) **Implement Robust Strategies to Overcome Barriers:** Identify and systematically address barriers to financial resilience, such as inadequate financial management, restricted access to financing, and market volatility. Tailored and robust strategies must reflect the organization's unique context.

e) **Develop a Phased Framework for Navigation:** Create and follow a phased framework that encompasses surviving immediate financial stress, reviving through strategic response, and thriving by seizing new opportunities. This structure should guide organizations through the complexities of financial stress and disruption.

f) **Cultivate a Mindset that Sees Opportunities in Challenges:** Embrace a visionary perspective that looks beyond the apparent chaos of disruptions to find hidden opportunities. Encourage a culture of innovation and adaptability, turning potential threats into new avenues for growth.

1.7 References

2019 Global Risk Management Survey - Results | AON. (n.d.). https://www.aon.com/2019-top-global-risks-management-economics-geopolitics-brand-damage-insights/index.html.

Ahrens, T., & Ferry, L. (2021). Accounting and accountability practices in times of crisis: a Foucauldian perspective on the UK government's response to COVID-19 for England. *Accounting, Auditing & Accountability*, *34*(6), 1332–1344. https://doi.org/10.1108/aaaj-07-2020-4659.

Aljohani, A., Ahiaga-Dagbui, D., & Moore, D. (2017). Construction projects cost overrun: What does the literature tell us?. *International Journal of Innovation, Management and Technology, 8*(2), 137.

Apgar, D. (2006). *Risk intelligence: Learning to manage what we don't know*. https://openlibrary.org/books/OL8854632M/Risk_Intelligence.

Arner, D. W., Barberis, J., & Buckley, R. P. (2015). The evolution of Fintech: a new Post-Crisis paradigm? *Social Science Research Network*. https://doi.org/10.2139/ssrn.2676553

Battisti, E., Alfiero, S., & Leonidou, E. (2022). Remote working and digital transformation during the COVID-19 pandemic: Economic–financial impacts and psychological drivers for employees. *Journal of Business Research, 150*, 38–50. https://doi.org/10.1016/j.jbusres.2022.06.010.

Burnard, K., & Bhamra, R. (2011). Organisational resilience: development of a conceptual framework for organisational responses. *International Journal of Production Research, 49*(18), 5581–5599. https://doi.org/10.1080/00207543.2011.563827.

Cokins, G. (2001). Activity-Based Cost Management: An Executive's Guide. In *John Wiley and Sons eBooks*. http://ci.nii.ac.jp/ncid/BA55898387.
Damodaran, A. (2016). *Damodaran on valuation: Security Analysis for Investment and Corporate Finance*. John Wiley & Sons.
Davenport, T., & Harris, J. (2017). *Competing on Analytics: Updated, with a New Introduction: The New Science of Winning*. Harvard Business Press.
Douglas, S. (2021). Building organizational resilience through human capital management strategy. *Development and Learning in Organizations: An International Journal, 35*(5), 19-21.
Eccles, R. G. (2014, August 1). *Reputation and its risks*. Harvard Business Review. https://hbr.org/2007/02/reputation-and-its-risks.
Essuman, D., Boso, N., & Annan, J. (2020). Operational resilience, disruption, and efficiency: Conceptual and empirical analyses. *International Journal of Production Economics, 229*, 107762. https://doi.org/10.1016/j.ijpe.2020.107762.
Financing SMEs and Entrepreneurs 2020. (2020). In *Financing SMEs and Entrepreneurs*. Routledge. https://doi.org/10.1787/061fe03d-en.
Frank, M. Z., & Goyal, V. K. (2009). Capital Structure Decisions: Which Factors Are Reliably Important? *Financial Management, 38*(1), 1–37. http://www.jstor.org/stable/20486683.
Holling, C. S. (1973). Resilience and Stability of Ecological Systems. *Annual Review of Ecology and Systematics, 4*, 1–23. http://www.jstor.org/stable/2096802.
Hoofnagle, C. J., Van Der Sloot, B., & Borgesius, F. Z. (2019). The European Union general data protection regulation: what it is and what it means. *Information & Communications Technology Law, 28*(1), 65-98.
Isaacson, W. (2007). *Einstein: His Life and Universe*. Simon & Schuster.
Jorion, P., & Risk, G. A. O. (2010). *Financial Risk Manager Handbook: FRM Part I /*. John Wiley & Sons.
Kommunuri, J. (2022). Artificial intelligence and the changing landscape of accounting: a viewpoint. *Pacific Accounting Review, 34*(4), 585–594. https://doi.org/10.1108/par-06-2021-0107.
Markowitz, H. M. (1952). Portfolio selection. *Journal of Finance, 7*(1), 77. https://doi.org/10.2307/2975974.
Martí-Ballester, C. (2019). Do European renewable energy mutual funds foster the transition to a low-carbon economy? *Renewable Energy, 143*, 1299–1309. https://doi.org/10.1016/j.renene.2019.05.095.
McKee, K., Lahaye, E., & Koning, A. (2011). Responsible finance: putting principles to work. Focus Note, 73, 1-16.
Muigua, P. D. K. (2022, December 27). *Embracing Environmental, Social and Governance (ESG) principles for sustainable development in Kenya*. https://ssrn.com/abstract=4312699.
Nandi, S., Sarkis, J., Hervani, A. A., & Helms, M. M. (2021). Redesigning supply chains using blockchain-enabled circular economy and COVID-19 experiences. *Sustainable Production and Consumption, 27*, 10-22.
Natarajarathinam, M., Capar, I. D., & Narayanan, A. (2009). Managing supply chains in times of crisis: a review of literature and insights. *International Journal of Physical Distribution & Logistics Management, 39*(7), 535–573. https://doi.org/10.1108/09600030910996251.
Navigating regulatory change for emerging technologies. (n.d.). Deloitte United States. https://www2.deloitte.com/us/en/pages/advisory/articles/regulatory-change.html.
Nicola, M., Alsafi, Z., Sohrabi, C., Kerwan, A., Al-Jabir, A., Iosifidis, C., Agha, M., & Agha, R. (2020). The socio-economic implications of the coronavirus pandemic (COVID-19): A review. *International Journal of Surgery, 78*, 185–193. https://doi.org/10.1016/j.ijsu.2020.04.018.
Nietzsche, F. W., & Grünwald, M. (2000). *What doesn't kill me makes me stronger: Powerful Medicine for Overcoming Setbacks*.
Paine, L. S. (2014, August 1). *Managing for organizational integrity*. Harvard Business Review. https://hbr.org/1994/03/managing-for-organizational-integrity.
Potts, J., & Rennie, E. (2019). Web3 and the Creative Industries: How Blockchains are reshaping Business Models. *Social Science Research Network*. https://doi.org/10.2139/ssrn.3372108
Reinhart, C. M., & Rogoff, K. S. (2011). *This time is different: Eight Centuries of Financial Folly*. Princeton University Press.
Riahi, K., Grübler, A., & Nakicenovic, N. (2007). Scenarios of long-term socio-economic and environmental development under climate stabilization. *Technological forecasting and social change, 74*(7), 887-935.
Rutaganda, L., Bergstrom, R., Jayashekhar, A., Jayasinghe, D., & Ahmed, J. (2017). Avoiding pitfalls and unlocking real business value with RPA. *Journal of Financial Transformation, 46*(11), 104-115.
Sharifani, Koosha and Amini, Mahyar, Machine Learning and Deep Learning: A Review of Methods and Applications (2023). *World Information Technology and Engineering Journal, 10*(07), pp. 3897-3904, 2023, Available at SSRN: https://ssrn.com/abstract=4458723.
Sheffi, Y. (2007). *The Resilient Enterprise: Overcoming Vulnerability for Competitive Advantage*. MIT Press.
Sørensen, J. B., & Stuart, T. E. (2000). Aging, obsolescence, and organizational innovation. *Administrative Science Quarterly, 45*(1), 81-112.
Steindel, Charles, Cecchetti, Stephen G., & Chu, Rita, (2005). *The unreliability of inflation indicators*.

Sutcliffe, K. M., & Vogus, T. J. (2003). Organizing for resilience. In K. Cameron, J. E. Dutton, & R. E. Quinn (Eds.), *Positive Organizational Scholarship* (pp. 94-110). Berrett-Koehler.

Thetechyhub. (2022, January 18). *Struggle of Malaysian SMEs during the COVID-19 pandemic - KSI Strategic Institute for Asia Pacific.* KSI Strategic Institute for Asia Pacific. https://kasi.asia/publications/struggle-of-malaysian-smes-during-the-covid-19-pandemic/.

Van Vliet, B. (2023). Cryptocurrency Anti-Money Laundering (AML) and Know-Your-Customer (KYC) ManagementSystemStandard—Requirements.*SocialScienceResearchNetwork*.https://doi.org/10.2139/ssrn.4403529.

Voigt, P., & Von Dem Bussche, A. (2017). *The EU General Data Protection Regulation (GDPR): A Practical guide*. https://dl.acm.org/citation.cfm?id=3152676.

World Economic Outlook, April 2009: Crisis and Recovery. (2016, December 31). IMF. https://www.imf.org/en/Publications/WEO/Issues/2016/12/31/World-Economic-Outlook-April-2009-Crisis-and-Recovery-22575.

World Economic Outlook, April 2021: Managing divergent recoveries. (2021, March 23). IMF. https://www.imf.org/en/Publications/WEO/Issues/2021/03/23/world-economic-outlook-april-2021

Zeng, K., Wells, R., Wilkins, A., & Gu, J. (2022). Bilateral tensions, the trade war, and US–China trade relations. *Business and Politics, 24*(4), 399–429. https://doi.org/10.1017/bap.2022.8

Ziolo, M., Filipiak, B. Z., Bąk, I., & Cheba, K. (2019). How to design more sustainable financial systems: The roles of environmental, social, and governance factors in the decision-making process. *Sustainability, 11*(20), 5604.

CHAPTER 2

Finance in the Present

Embracing Agility in Contemporary Finance

"The sailor knows that the sea is dangerous and the storm terrible, but they have never found these dangers sufficient reason for remaining ashore."

- VINCENT VAN GOGH (KAFKA, 1916/1995, P. 119)

2.0 Key Insights

- **Agility as a Universal Business Mantra:** Once confined to software development, agility has transcended into a universal business principle. Finance represents a revolutionary shift from rigidity to flexibility, empowering organizations to adapt swiftly to change.
- **Resonance of Agile Finance Across Dimensions:** Agile Finance resonates across managing uncertainty, enhancing process efficiency, fostering accurate forecasting, and igniting cultural transformation. Its implementation is both an art and a science, requiring a blend of philosophical alignment and pragmatic execution.
- **The Essence and Core Elements of Agile Finance:** Agile Finance emphasizes real-time data analytics, technology integration, cross-functional collaboration, continuous learning, and proactive regulatory navigation. These elements mark a transformative shift in the financial operations model.
- **Practical Steps for Implementing Agile Finance:** Implementation requires assessing organizational readiness, building agile teams, adopting agile methodologies, and driving cultural change. Each step is vital in navigating the complex journey to financial agility.
- **Future Trends and Opportunities:** Agile Finance will continue to evolve with advanced technology integration, emphasis on data literacy, ESG factors, expansion of cross-functional roles, continuous learning, and a commitment to growth and innovation.
- **Key Recommendations for Success:** Success in Agile Finance calls for a unified vision, fostering empowered teams, embracing agile methodologies and technologies, championing continuous learning and cross-functional collaboration, aligning with ESG factors, navigating compliance with agility, and embracing growth and innovation.

2.1 Introduction

In the tapestry of contemporary finance, the threads of resilience and agility intertwine, crafting a fabric that is as complex as it is compelling. Chapter 1 laid the foundation, exploring the art of financial resilience, the anchor that stabilizes organizations amidst the storms of disruption. Chapter 2 heralds a new horizon, shifting the gaze from grounding stability to the dynamic dance of agility.

Vincent Van Gogh's insightful words, *"The sailor knows that the sea is dangerous and the storm terrible, but they have never found these dangers sufficient reason for remaining ashore"* (Kafka, 1916/1995, p. 119) resonate with profound wisdom. They capture the essence of the financial landscape—a vast sea, unpredictable and challenging, yet filled with opportunities for those who dare to embrace agility.

"Finance in the Present: Embracing Agility in Contemporary Finance" is more than a theme; it is a manifesto for the modern financial leader. It is about recognizing that the traditional finance paradigms are evolving, morphing into a new reality where agility is not a mere attribute but a necessity. In this dynamic world, finance teams become navigators, guided by vision, innovation, and an unyielding pursuit of excellence.

This chapter unravels the multifaceted nature of financial agility, examining how technological advancements, regulatory shifts, global dynamics, and ethical considerations have converged to redefine the financial landscape. It is an exploration that goes beyond numbers and spreadsheets, delving into the soul of finance, where strategy, creativity, and foresight merge into a harmonious symphony.

The reader will journey through the transformative role of technology, from AI to blockchain, unraveling their potential to reshape industries and create new paradigms. The importance of ethical governance, aligned with societal values and regulatory compliance, will be dissected, illuminating the path to responsible financial stewardship. Emerging trends such as sustainability, social impact investing, and the growing importance of data analytics will be explored, each a testament to the evolving role of finance in shaping not just the bottom line but the very fabric of society.

Chapter 2 reflects the present state of finance and is a beacon illuminating the path forward. It is an invitation to embark on a voyage where agility transforms from a mere skill into a philosophy, a way of thinking that transcends challenges and turns uncertainty into a canvas of possibilities. As the chapter unfolds, the insights gleaned will serve as guiding stars, leading the way to a future where finance is not just a function but a visionary force driving innovation, growth, and sustainable success.

This chapter is not merely a continuation of the exploration; it is a deepening, an elevation, a call to action. It is about embracing the present with wisdom, courage, and grace, forging a future where finance is agile, ethical, and profoundly impactful.

2.2 The Current State of Finance

It was economist John Maynard Keynes (1937) who said, *"The difficulty lies not so much in developing new ideas as in escaping from old ones."* This nugget of wisdom resonates with the essence of agility explored earlier in this chapter and captures modern finance teams' challenges.

In a rapidly changing corporate world, finance teams often face challenges from outdated practices, traditional mindsets, and an ever-evolving external environment. The legacy of stability and predictability has given way to a landscape characterized by volatility, uncertainty, complexity, and ambiguity, raising the expectations bar for Finance teams (*New Technology, New Rules: Reimagining the Modern Finance Workforce*, 2018).

The growing demand for real-time financial data, the increasing complexity of financial regulations, the constant pressure to optimize costs, and the relentless push for operational efficiency have put unprecedented pressure on finance teams (*CFOs Prioritize Digital Tech Investments in Finance Function*, 2022).

Function-specific challenges further deepen this complexity. In FP&A, teams struggle with data fragmentation, outdated forecasting methods, and limited analytical capabilities, hindering their ability to deliver accurate and timely financial insights (Prijic, 2022). Treasury Management and Risk Management face hurdles, including liquidity management, currency risk, compliance, and fraud detection (*2022 Global Corporate Treasury Survey*, n.d.).

Adding to the complexity are the rapid technological advancements. While technologies like AI, ML, RPA, Big Data, Advanced Analytics, SaaS, Cloud Computing, and APIs promise to transform finance functions, their successful implementation is a complex dance (see Figure 2.1). It requires a blend of technical skills, leadership support, and cultural readiness, elements that many organizations still strive to weave into their fabric (*Gartner Top Priorities for HR Leaders in 2021*, n.d.).

The global COVID-19 pandemic has further intensified these challenges, forcing Finance teams to operate constantly, managing remote work, fluctuating market conditions, and the ever-looming threat of an economic downturn (*The territorial impact of COVID-19: Managing the crisis across levels of government*, 2020).

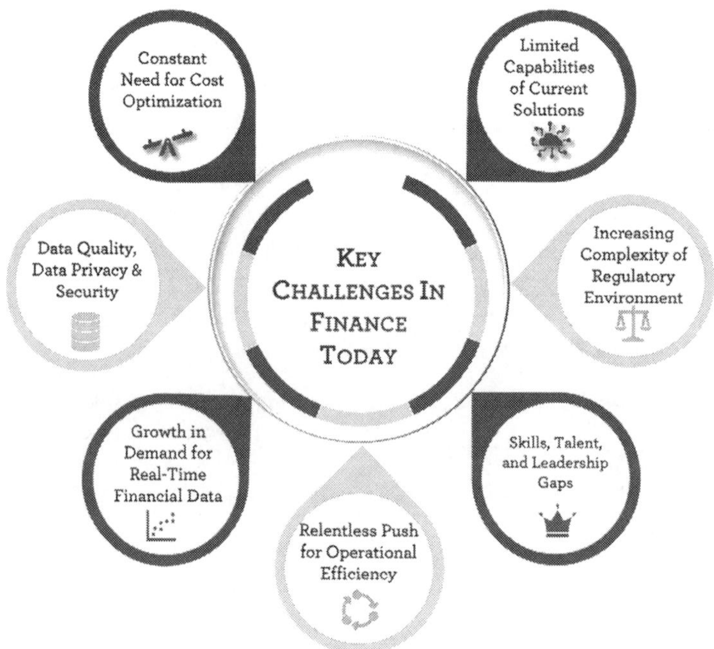

FIGURE 2.1: Key Challenges in Finance Today

These challenges are not merely obstacles but a call to action, signaling the need for Finance teams to evolve and adapt. They underscore the urgency to break free from rigid and outdated Finance practices and adopt more flexible, responsive, and forward-looking approaches. This section reflects the current state, vividly portraying the trials and tribulations that beckon a new era of financial agility.

The following section delves deeper into this need for agility in Finance, exploring strategies and frameworks that translate challenges into opportunities and crafting a financial function that is not just responsive but visionary.

2.3 Unleashing the Power of Agility in Finance

Winston Churchill's maxim, *"To improve is to change; to be perfect is to change often,"* is more than a statement; it is a philosophy that encapsulates the essence of agility (Churchill, 1955). As the world sails into the uncharted waters of the twenty-first century, finance teams find themselves at the helm, grappling with relentless waves of change. The previous section unearthed these challenges, ranging from technological complexities to evolving regulations. As Churchill's wisdom suggests, the path forward lies in embracing change and becoming agile.

Agility in finance refers to the dynamic capability that enables financial functions to adapt swiftly and effectively to change. It represents a paradigm shift from traditional, siloed operations to a more flexible, collaborative model that leverages data-driven insights for long-term strategic vision.

Agility, a term that once thrived solely in the corridors of software development, has transcended its origins to become a universal mantra for modern business (Sutherland, 2014). It represents a transformative shift from rigidity to flexibility, a metamorphosis that empowers organizations to adapt swiftly and effectively in the face of change. In finance, agility is not a mere concept but a revolution. It challenges the traditional sequential processes, siloed operations, and short-term focus, replacing them with an orchestrated dance of flexibility, collaboration, data-driven insights, and long-term vision.

The resonance of Agile Finance is felt across various dimensions:

a) **Managing Uncertainty:** Agile teams navigate uncertainty with grace, making faster and more informed decisions.

b) **Process Efficiency:** Agility enhances efficiency by promoting iterative approaches and continuous improvement.

c) **Accurate Forecasting:** With data-driven insights at its core, Agile Finance delivers more precise forecasts (*Crunch Time V: Finance 2025*, n.d.).

d) **Cultural Transformation:** Agility fosters a culture where mistakes are embraced as learning opportunities, igniting a cycle of growth and innovation (Agarwal et al., 2009).

The allure of these benefits has ignited a pursuit of agility among numerous companies to redefine their finance departments. However, this transition is not a linear path. It is a complex journey requiring a substantial metamorphosis in mindsets, skills, processes, and tools. The road to Agile Finance is both an art and a science, a blend of philosophical alignment and pragmatic execution.

Considering the undeniable potential and complexities of agility, the forthcoming sections will serve as a compass, guiding the reader through the fundamental constituents of Agile Finance. Practical insights will be shared, demystifying the process of integrating agility into the finance function, transforming challenges into opportunities, and crafting a future where finance is not just a function but a visionary force.

2.4 Core Elements of Agile Finance

The essence of Agile Finance is the ability to shift from traditional rigid practices to a more fluid, adaptive, and responsive model. The transformative shift in how financial operations are viewed and conducted is marked by adopting certain core elements, which we explore in this section. Figure 2.2 visually represents the transition from traditional finance to Agile Finance.

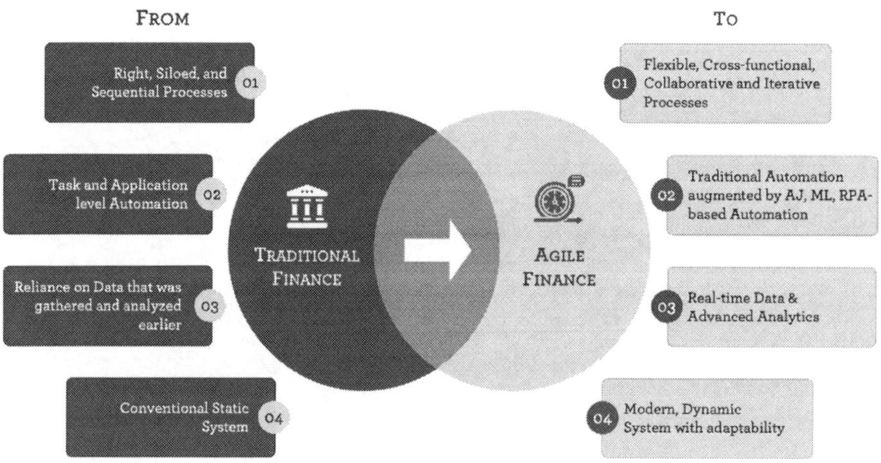

FIGURE 2.2: Transition from Traditional Finance to Agile Finance

Real-Time Data and Analytics

Data is the bedrock of modern businesses. This statement is especially true for Agile Finance, where real-time data and analytics are pivotal. In a traditional finance setup, businesses rely on historical data to make future predictions. This approach, however, had a glaring drawback: By the time data was gathered, analyzed, and decisions made, the market often shifted, rendering these decisions, at best, obsolete or, at worst, harmful.

Agile Finance addresses this problem by leveraging technologies to harness the power of real-time data (Davenport, 2014). Real-time data provides timely, actionable insights, enabling finance teams to respond quickly to changing market trends, customer behaviors, and business performance metrics. Additionally, advanced analytics can assist in identifying patterns, predicting outcomes, and making data-driven decisions. It can enhance the precision of financial forecasting, risk management, and strategic planning (McAfee et al., 2012). The finance function becomes more proactive and efficient by embracing real-time data and analytics.

Technology Integration

Technological integration forms the backbone of Agile Finance. Integrating various technologies, including Enterprise Resource Planning (ERP) systems, RPA, blockchain, and AI, is instrumental in driving efficiency and innovation in financial processes (Loebbecke & Picot, 2015).

For instance, ERP systems can integrate various business functions, providing a single, unified view of the organization. This consolidated view aids in better decision-making and promotes operational efficiency (Rom & Rohde, 2007). RPA can automate repetitive, mundane tasks, freeing human resources (HR) for more strategic roles. Blockchain technology can enhance the security and transparency of financial transactions, reducing fraud and errors (Tapscott & Tapscott, 2016).

At the core, technology integration is about creating an ecosystem where these technologies work synergistically to streamline financial processes, reduce errors, enhance efficiency, and foster innovation. Figure 2.3 showcases the contrast between the technology integration approach commonly observed in traditional finance vs. Agile Finance landscapes.

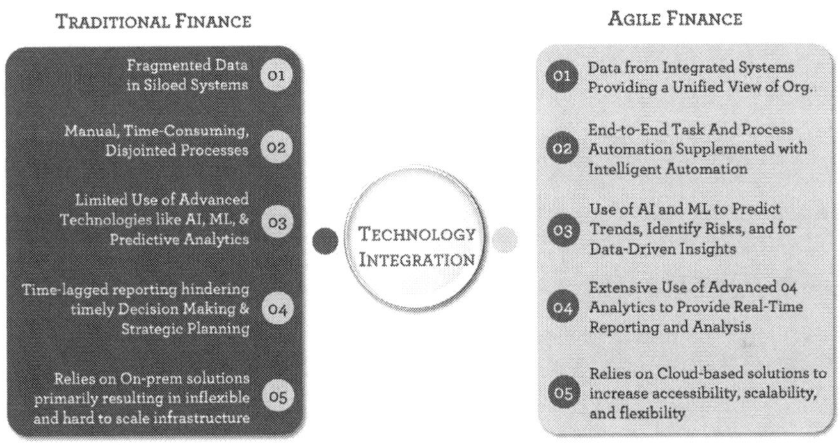

FIGURE 2.3: Technology Integration in Traditional Finance vs. Agile Finance

Cross-functional Collaboration

In traditional finance models, finance departments often operate in isolation, disconnected from the rest of the organization. Agile Finance, in contrast, promotes a culture of cross-functional collaboration. The finance teams work closely with other departments, such as sales, marketing, and operations, creating a comprehensive, integrated approach to business strategy (Appelo, 2016).

This integration leads to more holistic insights and decisions and fosters a sense of shared ownership and accountability across the organization. It breaks down silos, promoting better communication and fostering a more collaborative and efficient working environment.

Continuous Learning and Improvement

The fourth pillar of Agile Finance is the principle of continuous learning and improvement. It represents a paradigm shift from the traditional static model to a dynamic one that thrives on constant evolution (Denning, 2018). Agile Finance teams regularly review and evaluate their processes, identifying improvement areas and learning from successes and failures.

This ongoing learning cycle enhances the efficiency and effectiveness of financial processes and cultivates a culture of innovation. It encourages teams to experiment, take calculated risks, and embrace failure as a part of the learning process. This growth mindset is crucial in maintaining agility and staying competitive in today's fast-paced business environment.

Navigating the Regulatory Landscape

In the financial ecosystem, regulation is not an obdurate barrier but a mutable membrane permeable to innovation and adaptable strategies. Particularly salient in banking, insurance, and publicly traded corporations, regulatory mandates often present themselves as crucibles—tests of an organization's agility and tactical dexterity. Responding swiftly to these changing demands is not merely an act of dutiful compliance; it represents an expression of competitive advantage in a dynamic marketplace (Porter, 2008).

Traditional approaches to compliance—often characterized by their static, reactive nature—are relics of a bygone era, ill-suited for the brisk cadence and high volatility of contemporary financial markets. In contrast, Agile Finance emerges as a paradigm-shifting alternative, infusing the compliance matrix with unparalleled speed, resilience, and anticipatory foresight. Agile financial teams leverage real-time data and analytics to demystify the contours of impending regulatory changes, dissect their implications, and choreograph nuanced strategies for seamless adaptation (Davenport & Harris, 2007).

However, technology serves as more than mere instrumentation; it is the catalytic enabler of this transformation. Regulatory Technology (RegTech) solutions act as automated virtuosos, orchestrating complex compliance tasks with surgical precision, thereby mitigating human error and liberating finance teams to contemplate the strategic dimensions of regulatory change (Arner et al., 2015). Through such technological integration, organizations metamorphose regulatory imperatives into vectors for operational refinement and innovative prowess.

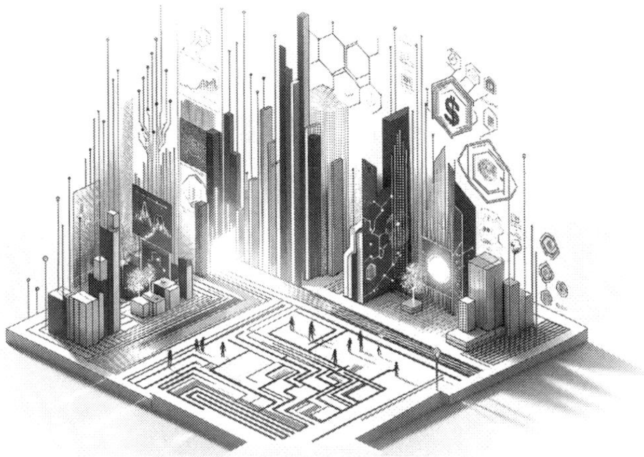

At the nucleus of Agile Finance's regulatory navigation lies the ethic of transparency. Far from being a mere display of procedural candor, transparency cultivates an internal ecology of proactive self-regulation. By illuminating the operational vista, finance teams can identify nascent compliance issues, intercepting and neutralizing them before metastasizing into substantive challenges. This proactive posture facilitates internal governance and cultivates relational capital with external regulatory entities, engendering a more streamlined, efficacious compliance process (Sucher & Preble, 2011).

In summation, navigating the regulatory landscape in the ambit of Agile Finance transcends conventional boundaries. It metamorphoses compliance from a prescriptive checklist into a strategic doctrine, transmuting regulatory challenges into platforms for organizational ingenuity and sustainable competitive advantage.

Measuring Agility: KPIs and Metrics

Peter Drucker, the renowned management consultant, once said, *"You cannot manage what you cannot measure."* The introduction and management of Agile Finance in an organization requires effective tracking and measuring progress. This section will explore the importance of KPIs and metrics in Agile Finance, shedding light on their role in monitoring and enhancing agility.

The selection of appropriate metrics plays a crucial role in gauging agility. It is essential to select metrics that align with the goals of Agile Finance and provide a holistic view of the performance (Kalaignanam et al., 2020). Some relevant metrics for Agile Finance may include:

- **Cycle Time:** The time it takes to complete a financial process, like closing accounts.
- **Financial Forecast Accuracy:** The precision of financial forecasts and budgets.
- **Percentage of Automated Processes:** A measure of how much the financial process is automated, indicating efficiency and innovation.

- **Employee Engagement:** An often overlooked but critical indicator of adopting Agile practices and principles (Journal, 2023).

Once one establishes the appropriate metrics, regular tracking and analysis become crucial. These metrics are crucial in evaluating an organization's Agile Finance practices, identifying areas that need improvement, and making necessary adjustments. The ultimate goal is continuous improvement, where the insights gathered from these metrics drive the iterative learning and enhancement processes integral to Agile Finance (*CFO & Enterprise Value | Accenture*, n.d.).

For instance, the organization can investigate and address bottlenecks if closing accounts takes longer. Similarly, a low percentage of automated processes may indicate a need to invest in more advanced technologies. Additionally, employee engagement levels can provide valuable information about the effectiveness of Agile training programs and the extent to which the team is embracing Agile Finance. Ultimately, these metrics provide valuable insights that guide organizations toward achieving greater financial agility. The focus is not merely on measurement but on deriving actionable insights that can enhance and improve the journey of Agile Finance.

This section painted a comprehensive picture of the core elements that define Agile Finance, illuminating the path from rigid traditional models to a vibrant, adaptable future. Adopting these elements propels financial operations into enhanced responsiveness, efficiency, and innovation, laying the groundwork for the subsequent transformation and practical steps that will be explored in the following sections.

2.5 Implementing Agile Finance: Practical Steps

To implement Agile Finance, we must focus our efforts on the right actions - those that create value and drive us toward our goals. In this context, let us explore the practical steps to implement Agile Finance in an organization. Figure 2.4 illustrates the salient steps in the implementation methodology for an Agile Finance implementation.

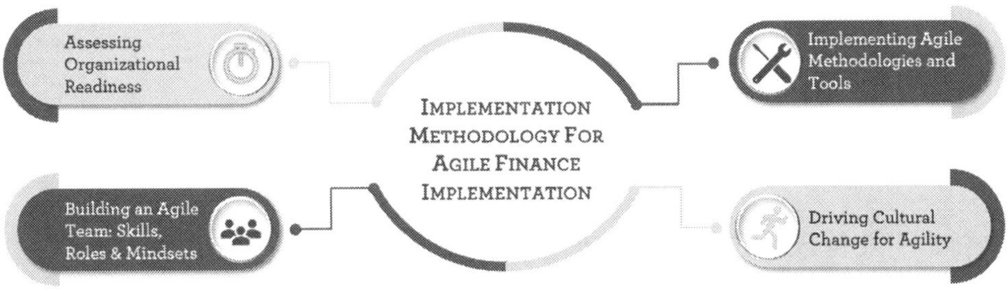

FIGURE 2.4: Implementation Methodology for an Agile Finance Implementation

Assessing Organizational Readiness

The first step in this voyage to Agile Finance involves evaluating the organization's readiness for the change. This assessment is akin to a sailor checking his ship's structural integrity, assessing the stock of provisions, understanding the capabilities of his crew, and meticulously studying the route and potential challenges ahead.

Similarly, finance leaders must thoroughly evaluate their existing financial processes, infrastructure, culture, and team skills (*Agile or Irrelevant: Redefining Resilience*, 2021). This evaluation clearly explains the organization's current state and readiness to embrace Agile Finance. Additionally, it helps identify the gaps and areas of improvement that need to be addressed in the transformation process.

During this assessment, it is essential to remember that every organization has unique strengths, weaknesses, and complexities. Therefore, the approach to Agile Finance should be tailored to fit the organization's specific needs and context (*Finance Digital Transformation: Predictions for 2025*, 2020).

Building an Agile Team: Skills, Roles, and Mindsets

The second phase of the journey is building the right team – the crew that will navigate the organization toward Agile Finance. Building the right team involves developing the required skills and fostering the right mindset among team members. From a skills perspective, Finance professionals must adapt to new roles and responsibilities. They must master new tools and technologies and work collaboratively in cross-functional teams. As data becomes more integral to decision-making, Finance professionals must also develop analytical and data interpretation skills (*The Journey to an Agile Organization*, 2019).

Regarding roles, Agile Finance may necessitate changes in the traditional hierarchical structures. It encourages more flexible, collaborative structures where the task at hand defines roles and not by job titles (*CFO & Enterprise Value | Accenture*, n.d.). Therefore, organizations might need to reconsider their existing structures to facilitate better collaboration and faster decision-making.

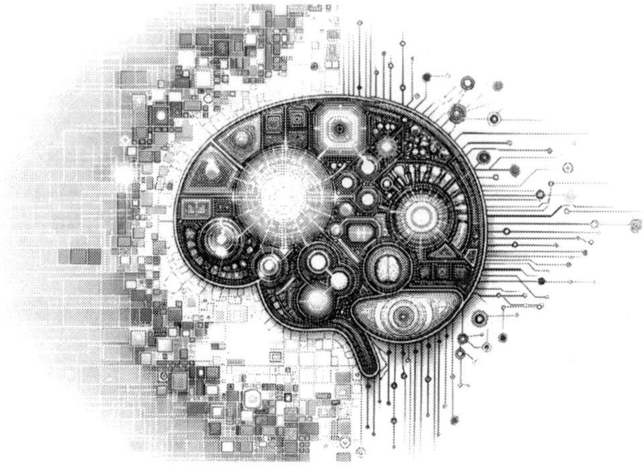

Developing the right mindset is equally, if not more, crucial in building an Agile team. Agile Finance requires a shift from a rigid, rules-based approach to a more flexible, customer-centric approach. It values adaptability over efficiency, customer satisfaction over internal benchmarks, and iterative learning over perfection (Rigby, 2021).

Implementing Agile Methodologies and Tools

While an organizational culture of agility and a team with the right skills form the foundation for Agile Finance, the next crucial step is implementing Agile methodologies and tools. These methods and tools are the ships, sails, and compasses that guide the journey toward financial agility (Leffingwell, 2023).

A rigid and hierarchical structure with a top-down approach to decision-making usually characterizes traditional finance functions. In contrast, Agile methodologies encourage a decentralized structure where decisions are made collaboratively and quickly, enabling a more responsive Finance function (Oren, 2023).

At the heart of these methodologies is the Agile principle of working in rapid, iterative cycles, often called *'sprints.'* For instance, instead of making annual financial forecasts, Agile Finance teams may opt for rolling forecasts updated monthly or quarterly, enabling them to quickly adjust their financial plans in response to changing circumstances (Donnelly, 2023).

Another crucial aspect is using Agile project management tools such as Jira, Asana, or Trello. These tools enable teams to visualize work, limit work-in-progress, and manage flow, facilitating greater efficiency, transparency, and adaptability (*Team Kanban Practitioner (TKP)*, 2023).

However, simply implementing Agile tools and methodologies is not enough. They must align with Agile principles and mindsets to deliver value. The alignment between Agile tools, methodologies, and principles can be driven by prioritizing customer-centricity, embracing change, and promoting transparency (Rigby, 2021).

Driving Cultural Change for Agility

The final, yet arguably most vital step in implementing Agile Finance is driving a cultural change within the organization. Like the winds guiding a ship's journey, an Agile culture acts as a propelling force toward financial agility.

Cultural change is crucial because Agile Finance is not merely a set of new methodologies or tools—but a fundamental shift in how an organization approaches its financial operations (*Doing vs. Being: Practical Lessons on Building an Agile Culture*, 2020). As famed management consultant Peter Drucker once said, *"Culture eats strategy for breakfast"* (Foreman, 2020).

An agile culture is one where change is accepted and embraced. It is a culture that fosters collaboration, encourages continuous learning, and thrives on transparency. However, driving cultural change is no easy task. It requires a concerted effort from all levels of an organization, starting from the top (*Doing vs. Being: Practical Lessons on Building an Agile Culture*, 2020). Leaders have a critical role in driving cultural change. They must embody Agile principles in their actions and decisions, setting an example for the rest of the organization. They must also communicate the importance and benefits of Agile Finance clearly and consistently, fostering understanding and buy-in among all employees (*The Five Trademarks of Agile Organizations*, 2018). Moreover, this cultural change must be driven by actions, not words alone. Implementing training programs to equip employees with Agile skills, setting up platforms for cross-functional collaboration, and redesigning performance metrics to align with Agile principles are often strategies organizations leverage to drive such cultural change (Leffingwell, 2023).

Ultimately, driving cultural change is a continuous process, not a one-time effort. It requires persistence, resilience, and patience. However, the rewards that await at the end of this journey—a finance function that's more efficient, innovative, and adaptable—are undoubtedly worth the effort.

This section serves as a roadmap to Agile Finance, offering actionable insights for every stage of the journey. By following these practical steps, organizations can transform their finance function into a dynamic, responsive, and innovative entity, well-equipped to navigate the challenges and opportunities of the modern financial landscape. Implementing

Agile Finance is not merely a procedural adjustment; it is a profound metamorphosis that reimagines the very essence of the finance function, paving the way for a future where finance becomes a function and a visionary force.

2.6 Future Trends & Opportunities in Agile Finance

As the horizon of Agile Finance expands, it unveils a mosaic of future trends and opportunities that promise to reshape the financial landscape. These emergent patterns provide glimpses of a world where finance is not merely a numbers game but a strategic, innovative, and adaptive force. Driven by technological advancements, regulatory shifts, global challenges, and societal demands, the future of Agile Finance beckons organizations to explore new territories and forge once uncharted paths.

The following sections will unfold a panoramic view of the anticipated trends and opportunities in Agile Finance. These insights reflect the evolution of finance and signify a transformational shift that emphasizes agility, innovation, and strategic alignment. From advanced technology integration to the growing focus on ESG factors, the kaleidoscope of trends paints a picture of a finance function ready to embrace the complexities and opportunities of a rapidly changing world.

Let us journey through these profound shifts, exploring how Agile Finance is poised to become a lighthouse guiding organizations towards resilience, growth, and excellence. Figure 2.5 below provides a visual overview of the future trends and opportunities in Agile Finance.

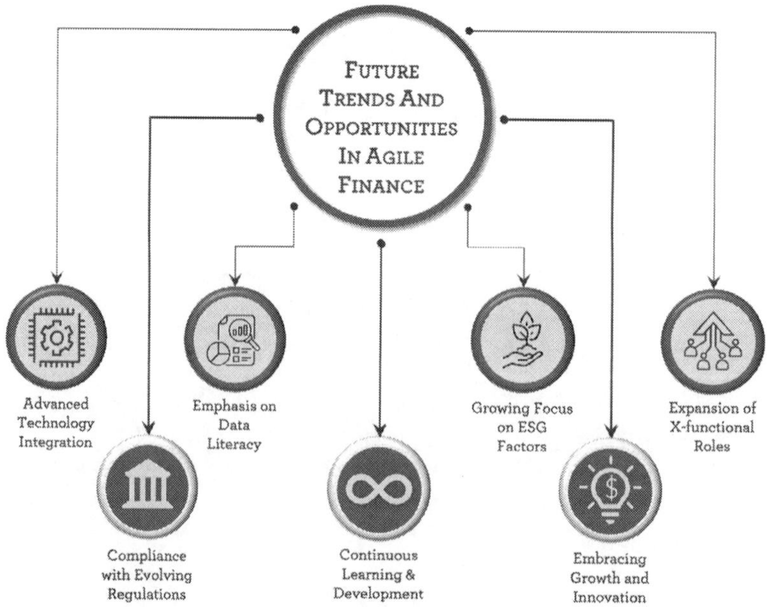

FIGURE 2.5: Future Trends and Opportunities in Agile Finance

Advanced Technology Integration

The wave of technological innovation is continually washing over the finance function, profoundly impacting its operations (*Seven Technologies Shaping the Future of Fintech*, 2021). Advanced technologies like AI, ML, RPA, Big Data, Advanced Analytics, Cloud Computing, SaaS, and APIs are set to redefine the finance landscape, pushing the boundaries of what is possible.

Combining AI and automation, Intelligent Automation promises to streamline financial processes, improve accuracy, and free up time for strategic tasks (*What Is Intelligent Automation? | IBM*, n.d.). Meanwhile, hyper-automation, which involves the integration of multiple automation technologies, offers the potential to automate complex processes that involve decision-making and problem-solving (AhamedMB, n.d.).

These advanced technologies will enhance the efficiency and accuracy of finance operations and provide unprecedented insights, leading to more intelligent and quicker decisions, which, in turn, will elevate the strategic role of the Finance function, transforming it into a value-adding business partner (*Autonomous Finance - Everything a CFO Should Know | Gartner*, 2022).

Emphasis on Data Literacy

With the surge in data generated by advanced technologies, the importance of data literacy is set to rise. Data literacy, the ability to read, understand, and communicate with data, will become a critical skill for finance professionals (Panetta, 2021).

As finance teams evolve into strategic business partners, the ability to interpret and communicate data-driven insights will become increasingly important. Data literacy will enable Finance professionals to make sense of complex data, draw meaningful insights, and communicate these insights effectively to stakeholders to drive data-driven decision-making across the organization (Panetta, 2021).

Growing Focus on ESG Factors

The increasing emphasis on ESG factors is another trend to shape the future of Agile Finance. As stakeholders demand greater transparency and accountability, finance functions integrate ESG factors into their reporting and decision-making processes (Boffo, 2020).

Integrating ESG factors into the financial reporting and decision-making processes is based on collecting, analyzing, and reporting ESG data, which requires agility in adapting to evolving standards and regulations. In addition, finance functions must be able to respond to ESG risks and opportunities, aligning financial strategies with sustainability goals. This growing focus on ESG represents both a challenge and an opportunity for Agile Finance (Boffo, 2020).

Expansion of Cross-Functional Roles

The boundary of the finance function is blurring as finance professionals increasingly engage in cross-functional roles. This trend not only fosters collaboration and breaks silos but also gives finance a seat at the table in strategic decision-making. Agile Finance teams are, therefore, morphing into versatile, multifaceted entities that play a critical role in steering the organization's overall strategy (*How Can Corporate Functions Become More Agile?*, 2022).

As we move forward, these trends present both challenges and opportunities. Navigating these trends will require a commitment to continuous learning and evolution, a cornerstone of Agile Finance. As we delve deeper into Agile Finance, we must embrace these developments not as disruptions but as avenues to forge a more resilient, responsive, and future-ready Finance function.

Compliance with Evolving Regulations

In an ever-changing global landscape, compliance with evolving regulations will continue to challenge Finance teams. Regulatory changes can impact various aspects of finance operations, from financial reporting to tax planning (Greggwirth, 2023).

In this context, Agile Finance can serve as a valuable tool for navigating regulatory changes. An Agile Finance function can swiftly adapt to new regulations, adjusting processes and practices as necessary. Moreover, it can proactively manage regulatory risks, leveraging real-time data and analytics to anticipate potential impacts (Russo, 2023).

Continuous Learning and Development

Another significant trend in Agile Finance is the focus on continuous learning and development. As Finance functions strive to remain agile, they must cultivate a culture of lifelong learning among their teams. Continuous learning in finance may encompass various areas, including new technologies, regulatory changes, emerging business models, and more. By fostering a learning culture, finance teams can ensure their resources are equipped with the knowledge and skills needed to navigate the ever-evolving business landscape (*The Five Trademarks of Agile Organizations*, 2018).

Finance professionals will need to embrace this culture of continuous learning, investing in their professional development to keep pace with the demands of their roles. As such, Agile Finance changes how finance functions operate and the mindset and behaviors of the individuals within them (ACCA, https://www.accaglobal.com, n.d.).

Embracing Growth and Innovation

Lastly, Agile Finance is about more than just adapting to change - it is about leveraging change for growth and innovation. Agile Finance functions are those that not only respond to disruptions but use them as springboards for innovative solutions. These innovative solutions might involve the creation of new business models, the development of novel financial products, or the implementation of groundbreaking technologies. Whatever form it takes, this focus on growth and innovation is a defining feature of Agile Finance that will continue to shape its future (Russo, 2023).

This section painted a vision of the future where Agile Finance is not merely a response to change but an embrace of innovation and growth. These trends represent a call to action for finance teams to stay ahead of the curve and forge a more resilient, responsive, and future-ready function. The onus now lies on finance leaders to embrace these developments as opportunities to redefine their roles, reimagine their processes, and reinforce their strategic influence. In the ensuing dance of change, these trends will serve as the rhythm to which the future of finance will sway, etching a new chapter in the annals of financial management.

2.7 Conclusion

The passage through Chapter 2 has been an enlightening journey through the labyrinthine world of Agile Finance. The chapter unfolded the significance of agility in the modern financial landscape, where change is not merely a phenomenon but a perpetual reality.

The resounding theme of this chapter has been the transformative power of agility. In the face of unrelenting change, driven by technological advancements, evolving customer expectations, and a dynamic regulatory environment, the finance function must be akin to a skilled surfer, masterfully riding the waves of change.

The analysis delved into the core elements of Agile Finance, from the integration of real-time data and technology to cross-functional collaboration and continuous learning. The chapter also ventured into the practical steps for implementing Agile Finance, illuminating the path from assessing organizational readiness to driving cultural change. The exploration of future trends and opportunities depicted a vision of a finance function that is not just adaptive but innovative, responsive, and strategically influential. It painted a picture of a future where finance is a visionary force, steering the organization toward growth, innovation, and sustainability.

In the words of Albert Einstein, *"Life is like riding a bicycle. To keep your balance, you must keep moving"* (Goal, 2019). This sentiment is profoundly resonant in the world of Finance. The imperative to move, adapt, and evolve is not a choice but a necessity.

In conclusion, as the curtain falls on this chapter, the stage is set for the journey into the future. The next chapter awaits, ready to explore the concept of adaptability and its crucial role in equipping finance teams with the challenges and opportunities that the future holds. The expedition through the world of Agile Finance is far from over. The road ahead is filled with promise and potential, and the path is lit with the wisdom gleaned from this chapter. The voyage continues, and the future beckons.

2.8 Key Recommendations

a) **Foster a Unified Vision and Empowered Teams:** Establish a clear and shared purpose aligned with the finance function. Break down silos and create networks of empowered teams that promote collaboration, transparency, and accountability. This combination is the North Star guiding the organization's success.

b) **Implement Agile Methodologies and Advanced Technologies:** Embrace methodologies that foster rapid decision-making, such as Scrum or Kanban, and leverage advanced technologies like AI and machine learning. This integration drives innovation accuracy and elevates the strategic role of the finance function.

c) **Champion Continuous Learning, Development, and Cross-Functional Collaboration:** Cultivate a culture of lifelong learning and cross-functional collaboration within the Finance teams. Equip them with the knowledge, skills, and flexibility needed to navigate the evolving business landscape.

d) **Align with ESG Factors and Navigate Compliance with Agility:** Integrate environmental, social, and governance (ESG) factors into reporting and adapt swiftly to regulatory changes. Utilize real-time data to anticipate impacts and manage risks, aligning financial strategies with sustainability goals.

e) **Embrace Growth, Innovation, and Continuous Improvement:** Leverage change for growth and innovation, continuously reviewing and improving processes. Encourage the development of new business models and financial products and foster a mindset that values adaptability, customer satisfaction, and iterative learning.

2.9 References

2022 global corporate treasury survey. (n.d.). Deloitte Malaysia. https://www2.deloitte.com/my/en/pages/risk/articles/global-corporate-treasury-survey.html

ACCA. (n.d.). Transformational journeys: Finance and the agile organization. *ACCA Global*. https://www.accaglobal.com/gb/en/professional-insights/technology/transformational-journeys.html

Agarwal, R., Angst, C. M., & Magni, M. (2009). The performance effects of coaching: A multilevel analysis using hierarchical linear modeling. *International Journal of Human Resource Management, 20*(10), 2110–2134. https://doi.org/10.1080/09585190903178054

AhamedMB, I. (n.d.). Hyper Automation - The Future of Business Automation. *www.linkedin.com*. https://www.linkedin.com/pulse/hyper-automation-future-business-irshad-ahamed-m-b/

Appelo, J. (2016). *Managing for Happiness: Games, Tools, and Practices to Motivate Any Team*. John Wiley & Sons.

Arner, D. W., Barberis, J., & Buckley, R. P. (2015). *The evolution of Fintech: A new Post-Crisis paradigm?* Social Science Research Network. https://doi.org/10.2139/ssrn.2676553

Autonomous Finance – Everything a CFO should know | Gartner. (2022). Gartner. https://www.gartner.com/en/finance/topics/autonomous-finance.

Boffo, R., and Patalano, R. (2020). ESG Investing: Practices, Process, and Challenges. *OECD Paris*, https://www.oecd.org/finance/ESG-Investing-Practices-Progress-Challenges.pdf

CFO & Enterprise Value | Accenture. (n.d.). https://newsroom.accenture.com/subjects/finance-performance-management/.

CFOs prioritize digital tech investments in finance functions. (2022). https://www.oracle.com/news/announcement/idc-research-road-map-to-finance-transformation-2022-07-19/.

Churchill, W. (1955). *Speech to the Lord Mayor's Luncheon*, Mansion House. London, United Kingdom.

Davenport, T. (2014). *Big data at work: Dispelling the Myths, Uncovering the Opportunities*. Harvard Business Review Press.

Davenport, T. H., & Harris, J. G. (2007). *Competing on analytics: The New Science of Winning*. Harvard Business Press.

Crunch time V: Finance 2025. (n.d.). Deloitte Ireland. https://www2.deloitte.com/ie/en/pages/finance-transformation/articles/crunch-time-v--finance-2025.html

Denning, S. (2018). *The age of agile: How Smart Companies Are Transforming the Way Work Gets Done*. AMACOM.

Doing vs being: Practical lessons on building an agile culture. (2020). McKinsey & Company. https://www.mckinsey.com/capabilities/people-and-organizational-performance/our-insights/doing-vs-being-practical-lessons-on-building-an-agile-culture

Donnelly, S. (2023). Rolling forecast best practices | Finance Alliance. *Finance Alliance*. https://www.financealliance.io/rolling-forecast-best-practices/

Finance Digital Transformation: Predictions for 2025. (2020). Deloitte. https://www.deloitte.com/global/en/services/financial-advisory/perspectives/gx-finance-digital-transformation-for-cfos.html

Foreman, D. (2020). Culture eats strategy for breakfast. *The Management Centre.* https://www.managementcentre.co.uk/management-consultancy/culture-eats-strategy-for-breakfast/

Gartner Top Priorities for HR leaders in 2021. (n.d.). Gartner. https://www.gartner.com/smarterwithgartner/gartner-top-3-priorities-for-hr-leaders-in-2021

Goal, S. (2019). *Life Is Like Riding a Bicycle. to Keep Your Balance, You Must Keep Moving.*

Greggwirth. (2023, January 17). *Financial markets' regulatory outlook for 2023: Resilience, vigilance & positioning for change - Thomson Reuters Institute.* Thomson Reuters Institute. https://www.thomsonreuters.com/en-us/posts/investigation-fraud-and-risk/financial-markets-regulatory-outlook-2023/

How can corporate functions become more agile? (2022, April 1). McKinsey & Company. https://www.mckinsey.com/capabilities/operations/our-insights/how-can-corporate-functions-become-more-agile

Journal, J. (2023). *Agile Innovation and Employee Experience | The Jabian Journal.* The Jabian Journal. https://journal.jabian.com/agile-innovation-and-employee-experience/.

Kafka, F. (1995). The Metamorphosis, In the Penal Colony, and Other Stories (W. Muir & E. Muir, Trans.). Schocken Books. (Original work published 1916)

Kalaignanam, K., Tuli, K. R., Kushwaha, T., Lee, L., & Gal, D. (2020). Marketing Agility: The concept, antecedents, and a research agenda. *Journal of Marketing, 85*(1), 35–58. https://doi.org/10.1177/0022242920952760

Keynes, J. M. (1937). *The general theory of employment, interest, and money.* Springer.

Agile or Irrelevant: Redefining resilience. (2021). KPMG. https://info.kpmg.us/agile-or-irrelevant.html

Leffingwell, D. (2023, June 8). *Organizational agility - Scaled Agile framework.* Scaled Agile Framework. https://scaledagileframework.com/organizational-agility/

Loebbecke, C., & Picot, A. (2015). *Reflections on societal and business model transformation arising from digitization and big data analytics: A research agenda. The Journal of Strategic Information Systems, 24*(3), 149-157.

McAfee, A., Brynjolfsson, E., Davenport, T. H., Patil, D. J., & Barton, D. (2012). Big data: the management revolution. *Harvard Business Review, 90*(10), 60-68.

New technology, new rules: Reimagining the modern finance workforce. (2018). McKinsey & Company. https://www.mckinsey.com/capabilities/operations/our-insights/new-technology-new-rules-reimagining-the-modern-finance-workforce

Oren, I. (2023, August 14). *Principle #9 - Decentralize Decision-Making - Scaled Agile Framework.* Scaled Agile Framework. https://scaledagileframework.com/decentralize-decision-making/

Panetta, K. (2021). *A Data and Analytics Leader's Guide to Data Literacy.* Gartner. https://www.gartner.com/smarterwithgartner/a-data-and-analytics-leaders-guide-to-data-literacy

Porter, M. E. (2008). *The Five Competitive Forces That Shape Strategy.* Harvard Business Review. https://hbr.org/2008/01/the-five-competitive-forces-that-shape-strategy

Prijic, M. (2022). 6 Financial Planning and Analysis Challenges for CFOs. *IT Convergence.* https://www.itconvergence.com/blog/6-financial-planning-and-analysis-challenges-for-cfos/

Rigby, D. (2021). *Embracing agile.* Harvard Business Review. https://hbr.org/2016/05/embracing-agile

Rom, A., & Rohde, C. (2007). Management accounting and integrated information systems: A literature review. *International Journal of Accounting Information Systems, 8*(1), 40-68.

Russo, K. (2023). What is Agile Finance? An expert guide. *Oracle NetSuite.* https://www.netsuite.com/portal/resource/articles/financial-management/agile-finance.shtml

Seven technologies shaping the future of fintech. (2021, November 9). McKinsey & Company. https://www.mckinsey.com/cn/our-insights/our-insights/seven-technologies-shaping-the-future-of-fintech

Sucher, S. J., & Preble, M. (2011). Trust and the global regulation of securities markets. *Journal of Applied Corporate Finance, 23*(3), 29–39.

Sutherland, J. (2014). *Scrum: The Art of Doing Twice the Work in Half the Time.* Crown Business.

Tapscott, D., & Tapscott, A. (2016). *Blockchain revolution: How the technology behind Bitcoin is changing money, business, and the world.* Penguin.

Team Kanban Practitioner (TKP). (2023, October 10). Agile & Scrum Training Courses | Collaborative Leadership Team. https://collaborativeleadershipteam.com/team-kanban-practitioner-tkp

The five trademarks of agile organizations. (2018). McKinsey & Company. https://www.mckinsey.com/capabilities/people-and-organizational-performance/our-insights/the-five-trademarks-of-agile-organizations

The journey to an agile organization. (2019, May 10). McKinsey & Company. https://www.mckinsey.com/capabilities/people-and-organizational-performance/our-insights/the-journey-to-an-agile-organization

The territorial impact of COVID-19: Managing the crisis across levels of government, OECD Policy Responses to Coronavirus (COVID-19). (2020). https://www.oecd.org/coronavirus/policy-responses/the-territorial-impact-of-covid-19-managing-the-crisis-across-levels-of-government-d3e314e1/

What is Intelligent Automation? | IBM. (n.d.). https://www.ibm.com/topics/intelligent-automation

CHAPTER 3

Finance in the Future

A Vision of an Adaptable Financial Framework

"Change is the law of life.
And those who look only to the past or present
are certain to miss the future."

- JOHN F. KENNEDY (KENNEDY, 1963)

3.0 Key Insights

- **Adaptability as the Quintessence of Future-Ready Finance:** In an era of relentless change, adaptability emerges as the guiding force. It integrates resilience and agility, allowing organizations to navigate uncertainty and embrace innovation.
- **Diversity and Inclusion as Catalysts for Innovation:** Beyond demographic representation, diversity fosters creativity and robust decision-making. Inclusion unlocks its potential, weaving various voices into the organization's fabric.
- **Technological Integration and Systemic Alignment:** Adopting AI, cloud computing, and systemic integration transcends traditional boundaries, forging a harmonious ecosystem. This shift necessitates ethical and regulatory considerations.
- **A Holistic Approach to Risk and Reward:** Balancing innovation with prudence requires a sophisticated understanding of risks and rewards. This approach fosters calculated risk-taking within a stable, compliant framework and aligns with global dynamics and social responsibility.
- **The Human Dimension of Finance:** Emphasizing empathy, creativity, and insight, the future of Finance recognizes the invaluable asset of diverse human minds. It reflects a broader societal shift towards ethical business practices and sustainability.
- **A Visionary Roadmap for Transformation:** The insights gleaned present an integrated roadmap for reshaping Finance into a strategic partner. This vision leads toward a sustainable, innovative, and inclusive future, embracing complexity and forging meaningful connections.

3.1 Introduction

In the preceding chapter, a meticulous exploration of the intricacies of modern finance was conducted, laying the groundwork for a profound understanding of the underlying dynamics that govern contemporary financial systems. The marriage of technology with finance, the emergence of ethical considerations, and the evolving role of finance within the organizational ecosystem painted a vivid picture of a field in flux. However, the expedition into the financial labyrinth is far from complete.

The present chapter embarks on a journey into the heart of adaptability within finance. Adaptability, here, is not merely a trait but a cornerstone upon which the edifice of future-ready finance is erected. This concept merits attention, not as an abstract theory but as a tangible force shaping the future, a force resonating with the unprecedented challenges and opportunities of the 21st century.

In a world marked by technological revolutions, shifting economic paradigms, and an unrelenting march toward globalization, adaptability guides financial professionals through the turbulence of change. It embodies a philosophy, a mindset, and a strategic imperative that transcends the traditional boundaries of finance, weaving a cohesive narrative that connects the past, present, and future.

This chapter's primary thesis revolves around adaptability's transformative potential in shaping tomorrow's finance function. As a visionary compass, it navigates through the complexities of an ever-evolving landscape, exploring the multifaceted dimensions that constitute future-ready finance. This exploration is not merely academic but profoundly pragmatic, offering insights, strategies, and frameworks that are both actionable and visionary.

Why should this exploration resonate with the readers? Because it reaches into the very core of what makes organizations thrive in an age marked by uncertainty and innovation. It offers leaders, professionals, and scholars a roadmap, equipping them with the wisdom to survive and flourish in relentless change. It speaks to the zeitgeist of an era where adaptability is not a choice but a necessity.

The ensuing sections will explore building resilience, fostering agility, promoting diversity, achieving systemic integration, and more. Each theme is an essential facet of adaptability, interconnected yet unique, forming a mosaic that defines the future of finance.

This chapter seeks to challenge conventional wisdom, provoke reflection, and inspire action. It invites readers to see finance not merely as numbers and spreadsheets but as a dynamic, living entity that evolves, learns, and grows. It paints a vision of a future where finance is not confined to the back office but stands at the forefront of innovation, strategy, and transformation.

This is not just a chapter; it is a manifesto for the future of finance. It is a call to think differently, embrace complexity, and forge a path toward a sustainable, innovative, and inclusive future. The future of finance awaits, and the journey begins here.

3.2 Adaptability: The Quintessential Virtue

Adaptability transcends mere reaction to change; it embodies proactive anticipation, a commitment to evolution, and a keen foresight that enables financial institutions to shape their future rather than merely respond. This quality resonates with the profound words of Alvin Toffler: *"The illiterate of the 21st century will not be those who cannot read and write, but those who cannot learn, unlearn, and relearn."*

> **Adaptability in finance refers to the proactive, dynamic capability that enables financial functions to anticipate, prepare for, respond to, and leverage change. It is a dynamic force that propels organizations to new heights of innovation and resilience. This capability transcends mere reactivity to embody strategic foresight, resilience, and innovation, transforming challenges into opportunities and uncertainties into advantages.**

To elaborate, adaptability should not be confused with the traditional notion of flexibility. While flexibility implies a reactive capacity to accommodate change as it occurs, adaptability is a far more dynamic quality. It equips financial functions with the tools to weather changes and leverage them strategically, transforming potential threats into opportunities and uncertainties into advantages.

Adaptability transcends a singular temporal framework, embracing short-term responsiveness, aligning with mid-term strategic imperatives, and forecasting long-term visionary paradigms.

a) **Short-Term Adaptations:** These involve immediate responses to changes, such as market fluctuations or regulatory updates. They require agility, rapid decision-making, and tactical execution. As shown in Figure 3.1 below, such challenges typically call for swift actions and agile strategies.

FIGURE 3.1: Short-Term Hurdles Faced by Finance Teams

b) **Mid-Term Adaptations:** As outlined in Figure 3.2, the medium-term hurdles faced by finance teams range from navigating the digital transformation journey to managing workforce upskilling and remote work dynamics, from coping with FinTech disruption to integrating ESG factors in financial decision-making - these transitional challenges demand foresight and adaptability.

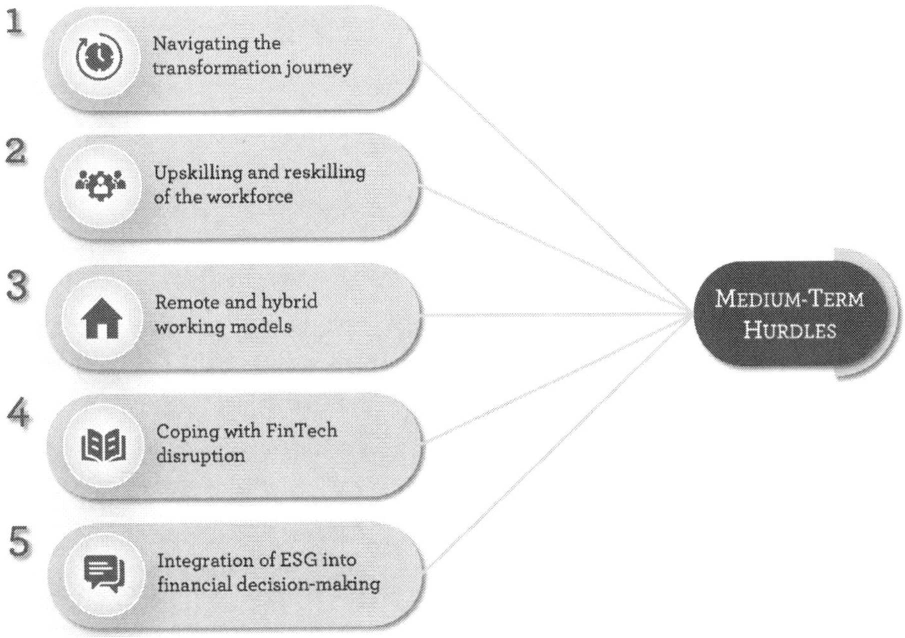

FIGURE 3.2: Medium-Term Hurdles Faced by Finance Teams

c) **Long-Term Adaptations:** Figure 3.3 shows that the interconnection of global financial systems, implications of climate change, demographic and geopolitical shifts, and ethical and regulatory concerns linked with advanced technology present a web of challenges. These invite strategic vision, systemic understanding, and forward-looking financial planning to prepare for the unforeseen.

Adaptability does not merely facilitate smooth sailing; it is a vital tool for weathering storms. In the face of challenges such as economic downturns, regulatory complexities, or technological disruptions, adaptability enables financial functions to recalibrate strategies, optimize resources, and transform challenges into opportunities (Battiston et al., 2012). The real-world applications of adaptability in financial management are abundant and illuminating:

- **IBM's Transformation:** Faced with a near-collapse in the early 1990s, IBM's ability to adapt allowed the company to pivot from a hardware focus to a services and software model, revitalizing the business (Garvin & Levesque, 2006).
- **DBS Bank's Transformation:** DBS Bank embraced the digital future, turning traditional banking paradigms upside down and setting a new standard for digital banking excellence (DBS, 2022).

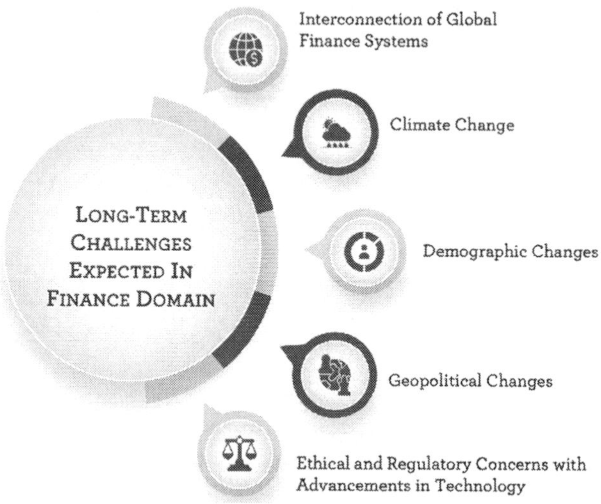

FIGURE 3.3: Long-Term Hurdles Faced by Finance Teams

In conclusion, adaptability is not a skill but a virtue. It is the quintessential quality that defines the future of finance. It is about seeing the bigger picture, connecting the dots, and weaving a story of resilience, innovation, and transformation. The financial landscape is a dynamic, ever-changing canvas, and adaptability is the brush that paints the future.

3.3 Reimagining the Future of Finance

In the face of a rapidly evolving world, the future of finance stands at the threshold of an era defined by a radical shift in roles, responsibilities, and expectations. A transformation, however, demands adaptation and an imaginative reshaping of how finance functions within an organization.

Stepping into this new era necessitates a visionary framework: the **PACE** model—**P**ioneer of Value, **A**dvocate of Ethics and Sustainability, **C**atalyst of Digital Acceleration, and **E**nabler of Change. This model, illustrated in Figure 3.4, is proposed as an insightful guideline to navigate the complexities and dynamics of this transformational phase. These are distinct pivotal roles that finance teams are poised to play in their organizations in the future.

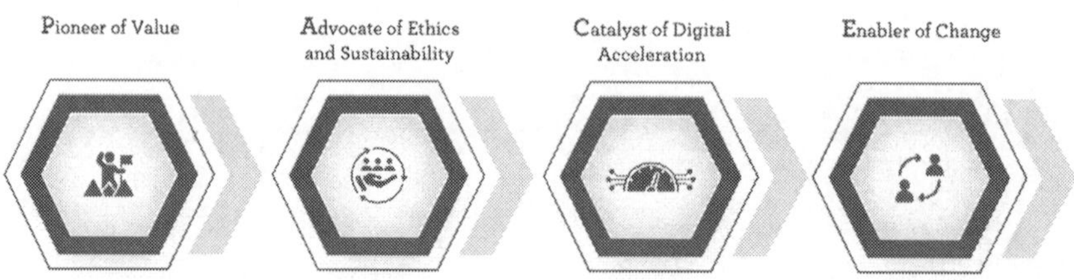

FIGURE 3.4: The Future of Finance - PACE Framework

a) **Pioneer of Value:** In a competitive and evolving business landscape, finance functions must become the Pioneer of Value. This role expands beyond traditional responsibilities to include identifying new business models, revenue streams, and opportunities for value creation. By leveraging a deep understanding of business economics and deploying advanced analytics, finance can provide insightful foresight and drive strategic decision-making processes.

   The Unilever example illuminates this role perfectly. In a strategic initiative led by their CFO, Unilever instituted a value creation project. By rethinking its capital allocation strategy, finance could strategically steer investments toward high-potential segments, fueling growth and maximizing shareholder returns. This role has fundamentally transformed Unilever's business model, emphasizing the finance function's capacity to pioneer value and drive growth (Plc, 2023).

b) **Advocate of Ethics and Sustainability:** The finance function needs to act as an Advocate of Ethics and Sustainability, taking a central role in shaping sustainable business practices and establishing ethical conduct. By spearheading ESG efforts and ensuring ethical compliance, finance can navigate the growing demand for corporate responsibility and sustainable development.

   A shining example is Microsoft. In 2020, Microsoft pledged to be carbon-negative by 2030, and the finance function played a critical role in realizing this vision. Finance led economic modeling and funding mechanisms and monitored the company's progress toward achieving this pledge. This strategic positioning underscores the potential of finance as an advocate of sustainability and ethics (Smith, 2020).

c) **Catalyst of Digital Acceleration:** As a Catalyst of Digital Acceleration, finance must drive the adoption and integration of technology within the organization. This technology adoption and integration requires process automation, AI and ML implementation for predictive insights, and advanced analytics for real-time decision-making. By leading digital transformation, finance can deliver efficient processes, real-time insights, and proactive business strategies.

   DBS Bank provides a compelling case study. Their finance function initiated a radical digital transformation, embracing AI, data analytics, and cloud computing to reinvent processes and customer experiences. This digital acceleration significantly improved efficiency, service quality, and competitive positioning (DBS, 2022).

d) **Enabler of Change:** Finally, finance must become an Enabler of Change, leading the organization in adopting agile methodologies and fostering a culture of adaptability. By encouraging cross-functional collaboration and promoting a continuous learning mindset, finance can help organizations become more resilient and responsive to changes.

   Amazon's finance function encapsulates this role effectively. It has cultivated a culture of innovation, customer-centricity, and agility. By creating a system that quickly responds to customer feedback and market trends, finance has enabled Amazon to maintain its position as a global leader in retail and technology (*Leadership Principles*, n.d.).

The financial landscape is teetering on the brink of a transformation of epic proportions. As we journey into this dynamic future, the finance function must don multiple hats and become adaptable and resilient. Armed with the PACE framework, finance professionals can transform into strategic partners, accelerators of digital transformation, conduits of technological evolution, and guardians of ethical and sustainable practices. Adopting such a comprehensive perspective empowers finance professionals to navigate the complexities of tomorrow, making significant strides in innovation, growth, and sustainability. As we tread this path, the finance function will emerge as the linchpin that connects the dots, fosters collaboration, and shapes the trajectory of organizations in an increasingly digital and interconnected world.

3.4 Transforming Finance for the Future

As we navigate the rapidly evolving business landscape of the 21st century, the role of finance within organizations is undergoing a paradigm shift. In the new era, the finance function is expected to embrace the PACE framework and transform into a Pioneer of Value, an Advocate of Ethics and Sustainability, a Catalyst of Digital Acceleration, and an Enabler of Change. This monumental transformation hinges on a deep and strategic engagement with three core elements: *people, processes*, and *technology*. Their future makeover elements are summarized below in Figure 3.5.

The New People Agenda: Skills, Capabilities, and Mindsets

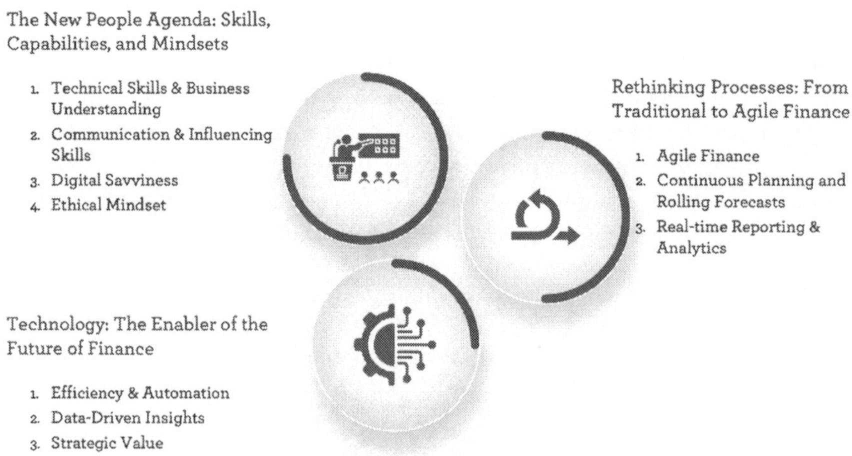

FIGURE 3.5: Makeover Elements of New Finance

At the heart of any organizational transformation are the people—their skills, capabilities, and mindsets. The people's agenda requires a radical makeover to adapt to future demands of the finance function. We are witnessing the need to shift from a transactional, conventional skill set to a dynamic, versatile, and broad-ranging array of competencies.

a) **Technical Skills & Business Understanding:** A recent study (*Finance 2025: Digital Transformation in Finance*, 2022) indicates that traditional finance skills, although still imperative, are being rapidly supplemented and even superseded by other skill sets. Future finance professionals need a keen understanding of data science and a strategic grasp of the business. The ability to comprehend complex business scenarios and respond with relevant, data-backed financial advice would be critical.

b) **Communication & Influencing Skills:** Furthermore, the finance professionals of tomorrow need to be effective communicators. Translating complex financial information into accessible, actionable insights for other departments cannot be overstated. The ability to influence decision-making processes and drive cross-functional collaboration will be pivotal.

c) **Digital Savviness:** Another recent study (*Global Technology Leadership Study*, 2021) further highlights the escalating importance of digital savviness. The future finance function will be heavily entwined with technology, thus making the understanding of emerging technologies, such as AI, ML, and RPA, highly desirable.

d) **Ethical Mindset:** Finally, an ethical mindset is becoming increasingly important. Finance professionals must uphold the highest ethical standards and promote transparency, fairness, and sustainability as finance becomes more integral to strategic decision-making and governance.

A comprehensive and strategic approach to talent development and management is necessary to ensure that the future finance function is staffed with professionals with these skills and mindsets. Companies need to invest in continuous learning and development programs, encourage a culture of curiosity and resilience, and diversify their talent-sourcing methods.

Rethinking Processes: From Traditional to Agile Finance

The second key element in the transformation journey is processes. Standardized steps sequentially progress through traditional finance processes. These processes, including monthly close cycles and annual budgets, were designed for a different era and are becoming increasingly obsolete in today's dynamic and complex business environment.

a) **Agile Finance:** This represents the new frontier in finance process design. Drawing inspiration from agile software development, Agile Finance emphasizes adaptability, iterative improvement, and cross-functional collaboration. Unlike linear processes that rely on large batches of data processed at regular intervals (such as monthly), Agile Finance processes prioritize real-time data processing and continuous improvement.

b) **Continuous Planning & Rolling Forecasts:** In Financial Planning, this shift can manifest in continuous planning and rolling forecasts. Instead of locking in plans and budgets annually, organizations would benefit from frequently updating their plans to reflect the latest business realities.

c) **Real-Time Reporting & Analytics:** Real-time reporting and analytics represent another facet of Agile Finance. By leveraging advanced analytics tools and real-time data, finance can provide stakeholders with timely and actionable insights, enhancing decision-making effectiveness.

An excellent example of a successful transition to Agile Finance comes from IBM. The technology giant's finance function transformed its planning process by implementing a rolling forecasting approach, enabling real-time business strategy adjustments and better responsiveness to changes (Humphries & Humphries, 2022). IBM's success story demonstrates the benefits of Agile Finance, including improved efficiency, better decision-making, and increased adaptability.

Technology: The Enabler of the Future of Finance

The third and arguably the most critical element in the transformation of finance is technology. New technologies like AI, ML, RPA, and cloud computing have already begun transforming the finance landscape. However, the true potential of these technologies can only be realized when they are deeply integrated into the fabric of the finance function and leveraged to redefine roles, processes, and strategies.

a) **Efficiency & Automation:** At the operational level, technologies such as RPA and AI can automate routine, manual tasks, freeing finance professionals to focus on strategic activities.

b) **Data-Driven Insights:** Advanced analytics tools can harness the power of big data to generate deeper insights and enable more informed decision-making.

c) **Strategic Value:** By enabling real-time reporting, predictive modeling, and scenario analysis, technology can help finance the transition from a record-keeping role to a strategic partner.

A report by Accenture (*CFO Strategy & Consulting Services*, 2023) demonstrates the transformative potential of technology. It shows that CFOs who leverage AI and advanced analytics can achieve up to a 35% improvement in forecasting accuracy. A case in point is DBS Bank, which embraced technology to transform its finance function. By leveraging AI and data analytics, DBS enhanced operational efficiency, achieved real-time visibility into financial data, and generated strategic insights, thereby contributing to its award-winning transformation (DBS, 2022).

The future of finance demands a comprehensive and strategic transformation that evolves people, processes, and technology. By focusing on these three elements and guided by the PACE framework, the finance function can successfully navigate the complexities of the 21st century. As pioneers of value, advocates of ethics and sustainability, catalysts of digital acceleration, and enablers of change, finance professionals will play a pivotal role in shaping their organizations and the broader business landscape.

3.5 Gaps in Financial Management Solutions Today

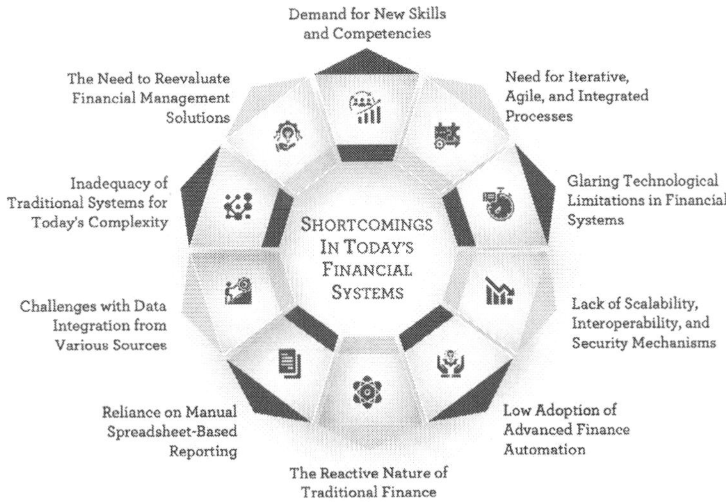

FIGURE 3.6: Shortcomings of Today's Financial Management Systems

The finance function has always been central to organizational success. However, despite its critical role, today's financial management solutions often keep pace with the dynamic and rapidly transforming business environment. As illustrated in Figure 3.6, the limitations of today's financial management solutions primarily stem from three key domains: people, processes, and technologies (Agrawal et al., 2020). Interestingly, these domains were previously identified as the focal points for the future finance makeover in Section 3.4.1 above.

Firstly, the people factor is significant. As technology revolutionizes the finance function, it demands a new set of skills and competencies from the workforce. According to a Deloitte study (2022), 86% of business leaders agree that their organizations must significantly improve their digital skills to adapt to the new financial paradigm. This gap extends beyond technical skills encompassing analytical and strategic thinking, adaptability, and a growth mindset. However, the urgency of day-to-day tasks often leaves little room for the professional development needed to acquire these competencies.

Meanwhile, processes established over the years are also proving to be bottlenecks. Traditionally, financial processes have been linear and sequential, often executed in silos. However, the modern digital landscape requires iterative, agile, and integrated processes. Rigid, legacy processes can slow response times and reduce overall operational efficiency, limiting an organization's ability to adapt to changes swiftly.

Technological limitations are perhaps the most glaring. Many financial systems are anchored in legacy technologies that were not designed for the interconnected, data-intensive operations of the digital age. These systems often lack scalability, interoperability, and robust security mechanisms to handle modern threats.

Traditionally, finance has been bound by its linear processes and dated technology, is mainly reactive, and relies heavily on historical data and hindsight. As businesses strive to become proactive, forward-looking entities, the limitations of traditional finance become increasingly apparent.

One example is the continued reliance on manual spreadsheet-based reporting. Despite technological advancements, 57% of organizations still depend on this method quarterly across every business unit (*Agility in Planning, Budgeting, & Forecasting*, 2021). The consequence? Potential errors, delays in reporting, and missed insights due to limited data-processing capabilities.

Also, finance teams often grapple with integrating data from various sources, especially in organizations with multiple business units or geographic locations. Without a unified data platform, gaining a single source of truth can become a monumental task, impeding swift and informed decision-making.

Traditional financial systems were designed for a different era. In today's world, they are expected to handle a higher degree of complexity and myriad regulatory requirements and do so at breakneck speed. However, a KPMG survey ("Economic Outlook," 2022) reveals that 75% of CEOs doubt their technology investments can meet these demands over the next three years due to the anticipated recession. This stark disconnect underscores the need for organizations to reevaluate the future readiness of their financial systems critically.

Therefore, organizations must address these shortcomings as they strive to adapt and thrive amidst constant change. By recognizing the limitations of their current financial management solutions, they can pave the way for the next era of financial operations.

3.6 Characterizing Future-Ready Finance

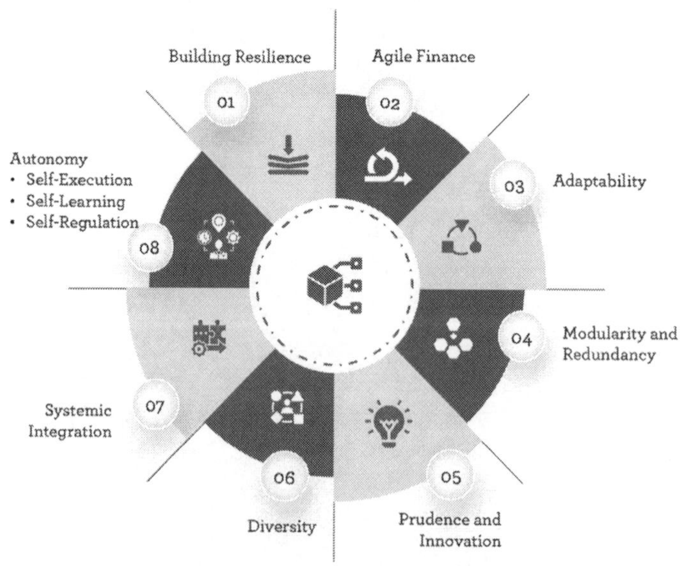

FIGURE 3.7: Future Ready Finance

As we delve into the world of finance of the future, it is essential to discern the defining characteristics that will set future-ready finance functions apart. The ongoing wave of digitalization, the ever-evolving regulatory landscape, increasing volatility in global markets, and the demands of a new generation of stakeholders are all reshaping the finance function as we know it.

In this context, we must ask ourselves: What will characterize future-ready finance? What attributes will successful financial departments possess as they navigate this evolving landscape? What traits will allow them to withstand the disruptions and turn them into opportunities for value creation?

This section explores eight key traits emblematic of future-ready finance functions (see Figure 3.7 below): resilience, agility, adaptability, autonomy, modularity, prudence, diversity, and systemic integration. Each trait represents a fundamental shift from traditional approaches and significantly impacts how finance functions operate and deliver value. Together, they form a roadmap for transforming finance for the future.

Let us dive deeper into each of these characteristics and explore their implications for the future of finance.

Building Resilience: Financial Management's Shield against Volatility

FIGURE 3.8: Impact of Resilience on the Finance Domain

Financial management's resilience will be the prominent line between organizations that flourish and those that falter. However, what does resilience in the finance function mean? It is the ability to bounce back from setbacks, absorb shocks, and adapt to new situations with minimal disruption.

During the 2008 financial crisis, we saw a stark contrast between resilient companies that weathered the storm and others that crumbled under pressure. Companies that had built resilience into their financial strategies could absorb the blow, recalibrate, and emerge stronger. They had the foresight to set aside reserves during good times, diversify their portfolios to mitigate risks, and invest in innovative technologies to drive efficiency and effectiveness (Battiston et al., 2012).

For instance, consider the case of IBM, which undertook a significant business transformation following a near-collapse in the early 1990s. By pivoting from a hardware-focused company to a services and software company, IBM revitalized its business and made it more resilient against future disruptions (Garvin & Levesque, 2006).

Resilience will continue to be a critical trait for financial management in the post-pandemic era. Figure 3.8 above highlights the significance of such resilience in the finance domain. Strengthening resilience in financial management involves adopting more conservative financial strategies, enhancing risk management capabilities, and fostering a culture of agility and adaptability within the finance function.

Agile Finance: The New Normal in a Fast-Paced World

FIGURE 3.9: Impact of Agility on the Finance Domain

Agility is no longer a luxury. It is a necessity. Financial departments need to be able to shift gears quickly in response to emerging opportunities or challenges. This agility extends beyond flexible budgeting and forecasting, regulatory changes, and evolving business models, etc. as depicted in Figure 3.9 below.

Take, for example, the rise of cryptocurrencies like Bitcoin and Ethereum. Traditional finance functions need help integrating these new payment and investment forms into their operations. In contrast, Agile Finance functions see these developments as an opportunity to innovate and gain a competitive edge. They can quickly adapt their systems and processes to accommodate these new financial instruments, mitigating potential risks while capitalizing on potential benefits (Mougayar, 2016).

However, Agile Finance extends beyond cryptocurrencies. It is about having a forward-thinking mindset that embraces change rather than resisting it. It is about leveraging technology to streamline processes, enhance decision-making, and deliver more value to the business. It is about breaking down silos, promoting collaboration, and fostering a culture of continuous learning and improvement.

Adaptability: The Key to Thriving in Uncertainty

While agility and resilience are critical for future-ready finance, they are insufficient. Finance functions must also be adaptable and able to pivot their strategies, processes, and systems in response to changing circumstances. In unprecedented uncertainty, adaptability will be the key to surviving and thriving.

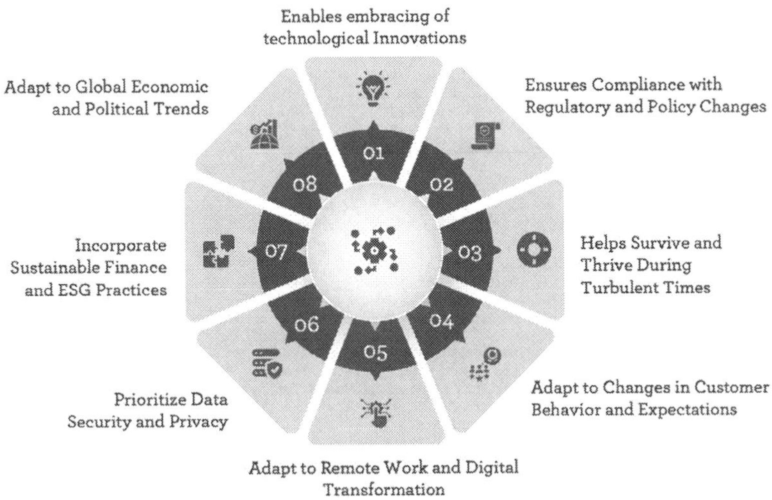

FIGURE 3.10: Impact of Adaptability on the Future of Finance

Consider the implications of Brexit for multinational companies. Suddenly, companies grappled with new trade rules, regulatory requirements, and economic conditions. Companies with adaptable finance functions were able to navigate this new landscape more effectively, adjusting their strategies and operations to mitigate risks and seize new opportunities.

As illustrated in Figure 3.10, adaptability also involves being open to new technologies and working methods. For instance, cloud computing, AI, and advanced analytics are revolutionizing the finance function, automating routine tasks, enhancing decision-making, and providing real-time insights that drive business performance. Adaptable organizations can leverage these technologies to reinvent their finance functions and gain a competitive edge.

Self-Execution, Self-Learning, and Self-Regulation: The Triad of Autonomy

Autonomy in finance is not about eliminating the human element but augmenting human capabilities with intelligent systems that can perform routine tasks, make data-driven decisions, and learn from experience. With the advent of AI, ML, and RPA, the finance function can achieve a high degree of autonomy, freeing finance professionals to focus on more strategic tasks.

Research from the MIT Sloan Management Review and Boston Consulting Group found that companies implementing AI in their finance functions were more likely to achieve cost reductions and revenue increases. They were also more likely to view AI as a tool for creating value rather than just reducing costs (Ransbotham et al., 2017).

However, achieving autonomy in finance is not without its challenges. It requires significant investments in technology and infrastructure, staff upskilling, and organizational culture shift. Moreover, companies must address ethical and regulatory considerations associated with AI and automation.

Modularity and Redundancy: Enhancing Flexibility and Robustness

Modularity—the ability to break down systems and processes into discrete, interchangeable components—provides the flexibility to reconfigure operations quickly to changing needs. Redundancy, on the other hand, offers a safety net against system failures, ensuring that critical operations can continue even when parts of the system fail.

Cloud computing is a prime example of how modularity and redundancy can enhance the flexibility and robustness of the finance function. With cloud-based systems, companies can quickly scale their operations up or down as needed. They can also use built-in redundancy features to prevent data loss and maintain continuity during system failures (Marston et al., 2011).

Moreover, by adopting a modular approach, companies can incrementally upgrade their systems and processes, reducing disruption and facilitating smoother transitions. They can experiment with new technologies and methodologies on a small scale before rolling them out more broadly, thereby mitigating risks and enhancing outcomes.

Prudence and Innovation: Balancing Risk and Reward

Innovation and growth require risk-taking, yet the finance function must balance these risks with prudence. Such an approach comprises three elements: (i) carefully evaluating new initiatives' potential risks and rewards, (ii) making informed decisions based on rigorous analysis, and (iii) establishing robust risk management processes to mitigate potential downsides.

At the same time, prudence should not stifle innovation. The finance function must be willing to embrace new technologies, methodologies, and ideas, even if they involve some risk. By fostering a culture of calculated risk-taking, finance can drive innovation while maintaining financial stability and compliance.

Consider the case of Amazon, which has consistently prioritized long-term growth over short-term profits. Despite criticism for its low profitability, Amazon's calculated risk-taking has paid off handsomely, with the company now dominating multiple industries and achieving a market capitalization of over $1 trillion (Stone, 2013).

Diversity: A Multifaceted Lens for Finance

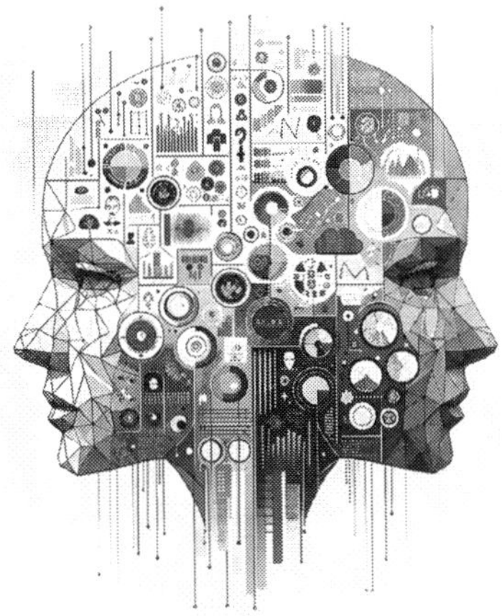

In the complex, interconnected world of finance, a new paradigm is emerging, one that transcends traditional boundaries and challenges conventional wisdom. It is a paradigm shaped not by numbers and algorithms alone but by the rich tapestry of human diversity.

Diversity in finance is no longer confined to mere demographic representation. It has evolved into a multifaceted lens through which organizations are beginning to view their strategies, cultures, and futures. It is a perspective that recognizes the inherent value of different ideas, backgrounds, and ways of thinking.

Imagine a financial team composed not just of number-crunchers and statisticians but of thinkers and innovators from various walks of life. Picture a room where many voices converge, each bringing a unique perspective to the financial puzzles that must be solved. The convergence of many thinkers and innovators from different walks of life is a theoretical construct and a model that some of the most successful finance teams actively pursue.

Why? Research, including a landmark McKinsey study (*Why Diversity Matters*, 2015), has demonstrated that diversity fosters innovation and enhances decision-making. It has shown that when people who think differently come together, the solutions they create are often more robust, creative, and financially sound.

This innovation engine is not fueled by diversity alone but by an inclusive culture that values and nurtures it. Inclusion is the key that unlocks the potential of diversity. It is not enough to have different voices in the room; those voices must be heard, respected, and valued. They must be woven into the very fabric of the organization.

The future of finance demands such a culture. The problems in the financial landscape of an organization today are increasingly complex, and the solutions require a depth and breadth of thinking that can only come from a diverse team. From understanding global market dynamics to navigating intricate regulations, the challenges call for a multifaceted approach.

In global finance, the importance of diversity becomes even more pronounced. Understanding and navigating different cultural landscapes becomes vital as markets become more interconnected. Diversity offers a bridge, connecting different financial ecosystems and creating synergies.

Perhaps most importantly, diversity reflects a broader societal shift towards equality, fairness, and social cohesion. It aligns with a global movement towards corporate social responsibility and ethical business practices.

So, what does the future hold for finance? It is a future where diversity is not just a buzzword but a fundamental business strategy. It is a future where finance is about numbers, people, and ideas. It is a future where the financial landscape is shaped by many voices, all contributing to a more innovative, resilient, and inclusive world.

In this transformative journey, the finance function has the opportunity to lead the way to demonstrate that diversity is not just the right thing to do but the smart thing to do. By embracing diversity in all its forms and creating a culture where everyone is a valued part of the financial tapestry, the world of finance can position itself at the forefront of change, ready to navigate the complexities of the future with creativity, empathy, and insight.

The future of finance, therefore, is not just about algorithms and balance sheets; it is about people. It is about recognizing that the most valuable financial asset is not money but the diverse minds that manage it. This vision challenges us to think differently, see beyond the numbers, and embrace the human element at the heart of finance.

Systemic Integration: Connecting the Dots in an Interconnected World

In the intricate tapestry of modern business, where threads of information weave across continents and digital landscapes, the finance function finds itself at a fascinating crossroad. It is no longer an isolated domain, meticulously crunching numbers behind closed doors. Instead, it has evolved into a dynamic nexus, connecting various aspects of the business, from supply chain logistics to customer relations and HR to global regulatory landscapes.

This transformation is not a mere convenience but necessary in a world where businesses are interconnected. Every decision made within the finance department resonates across the entire organizational structure and beyond. The successful finance function of the future recognizes this interconnectedness and embraces it through systemic integration.

Systemic integration is not a buzzword but a profound shift in how financial operations are conceived and executed. It represents a movement towards a seamless flow of information, a harmonious symphony of coordinated decision-making that transcends departmental boundaries.

Picture a financial function not as an isolated silo but as the keystone uniting diverse internal departments. Envision this core entity as a bastion of indisputable financial data—where visibility, accuracy, and consistency are the norm rather than lofty ideals.

In this transformative vista, contemporary ERP systems serve as paradigmatic embodiments. These are not mere integration mechanisms but transformative engines offering organizations a holistic, 360-degree perspective on their entire business ecosystem. They establish hitherto inconceivable connections between disparate elements and reveal untapped synergistic opportunities, dramatically altering the landscape of what is operationally achievable. However, systemic integration goes beyond the confines of the organization. It reaches into the complex world of globalization and digital business models, where relationships with external stakeholders—suppliers, customers, regulators, and investors—become a critical part of the financial ecosystem.

In this expanded sphere, the finance function must become a master navigator, charting courses through intricate regulatory mazes, managing the ebb and flow of global supply chain risks, and aligning with the values of socially conscious investors. It must become a bridge, connecting internal financial wisdom with external market realities, ensuring the organization is responsive and proactive.

The future-ready finance function, therefore, is not just an aggregator of numbers but a synthesizer of insights. It is characterized by resilience, agility, adaptability, autonomy, modularity, prudence, diversity, and systemic integration. These traits are not isolated virtues but interconnected facets of a holistic vision for finance.

Each trait is unique and essential, forming a comprehensive framework that transcends traditional financial management. They represent a roadmap for transforming finance from a support function to a strategic partner, from a follower of trends to a leader of innovation.

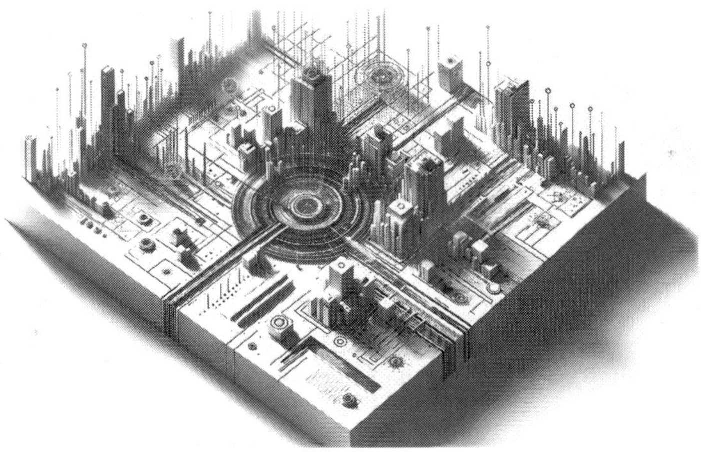

In the final analysis, systemic integration is not just about technology or processes but about a mindset. It is about recognizing that the finance function must become a connective tissue in an interconnected world, linking diverse elements into a coherent, harmonious whole.

It is a vision that challenges conventional wisdom and invites us to see finance not as a mere function but as an essential, integrative force that shapes the future of business. It is a call to embrace complexity, to see connections where others see divisions, and to forge a path toward a future where finance is not just about numbers but about the intricate, beautiful connections that make those numbers meaningful.

3.7 Conclusion

The odyssey through the uncharted terrains of the future of finance, embarked upon in this chapter, has revealed a fascinating confluence of adaptability, technology, human insight, ethical considerations, and systemic integration. It has painted a picture of finance as a guardian of monetary assets and a multifaceted, dynamic, and transformative force shaping the modern business landscape.

The insights gleaned from this exploration have unraveled the complex tapestry of finance in the digital age, highlighting the quintessential virtue of adaptability, the transformative power of technological innovation, and the essential role of human diversity and ethical stewardship. These insights have been illuminated for academic enrichment and to guide, inspire, and challenge financial leaders, professionals, and organizations in a rapidly changing world.

The future of finance is being sculpted here and now through decisions, strategies, and visions that recognize the interconnectedness of diverse elements. It is a future transcending mere numbers and algorithms, embracing complexity, creativity, responsibility, and continuous growth.

However, this chapter is merely one milestone in a broader journey. The path ahead leads to an even more profound exploration of Autonomous Finance, a concept that promises to redefine the very architecture of financial management. The concept beckons with potential, challenges with complexity, and inspires innovation.

In Chapter 4, the lens will focus on this transformative idea of Autonomous Finance. It will delve into its intricacies, probe its potential, and illuminate its implications for building a future that is not only sustainable and innovative but truly autonomous. It will continue the quest to envision a world where finance is not confined to numbers but is a living, evolving entity that shapes businesses, economies, societies, and the essence of our collective future.

The words of Charles Darwin resonate with profound wisdom: *"It is not the strongest of the species that survives, nor the most intelligent; it is the one that is most adaptable to change."* This insight is a beacon in finance, guiding the way to a future filled with promise, resilience, and inclusiveness. The journey continues, and the road ahead is filled with intrigue, inspiration, and infinite possibility.

3.8 Key Recommendations

a) **Cultivate a Culture of Adaptability:** Embrace change and foster an adaptable mindset within the Finance function. Utilize intelligent technologies and embrace adaptability to ensure resilience in ever-changing economic landscapes.

b) **Transform Finance for the Digital World:** Implement robust digital strategies to redefine the role of finance in the digital era. Leverage the PACE Framework and foster new skills and mindsets that align with digital transformation.

c) **Prioritize People, Skills, and Ethical Considerations:** Enhance technical proficiency, communication abilities, and digital savviness, balanced with a strong ethical framework. Recognize the human element at the heart of finance and promote diversity and inclusion.

d) **Rethink Processes and Embrace Agile Finance:** Transition from traditional finance processes to agile methodologies. Encourage flexibility, innovation, and systemic integration to create a future-ready finance function.

e) **Balance Innovation with Prudence and Responsibility:** Encourage calculated risk-taking and innovation with prudence and responsible stewardship. Align financial strategies with global movements towards corporate social responsibility and ethical business practices.

3.9 References

Agility in Planning Budgeting & Forecasting. (2021). FSN - the Modern Finance Forum. https://fsn.co.uk/research-papers/agility-in-planning-budgeting-forecasting/.

Agrawal, A., Eklund, S., Waite, J., & Woodcock, E. (2020). *Transforming to the finance function of the future | McKinsey*. Mckinsey & Company. https://www.mckinsey.com/capabilities/operations/our-insights/finance-2030-four-imperatives-for-the-next-decade.ew

Battiston, S., Gatti, D. D., Gallegati, M., Greenwald, B., & Stiglitz, J. E. (2012). Liaisons dangereuses Increasing connectivity, risk sharing, and systemic risk. *Journal of economic dynamics and control, 36*(8), 1121-1141.

CFO Strategy & Consulting Services. (2023). Accenture. https://www.accenture.com/cz-en/services/cfo-and-enterprise-value-index.

DBS Bank Ltd. (2022). Group Annual Reports. https://www.dbs.com/investors/financials/group-annual-reports.

Economic outlook. (2022). KPMG. https://kpmg.com/xx/en/home/insights/2022/08/kpmg-2022-ceo-outlook/economic-outlook.html.

Finance 2025: Digital transformation in finance. (2022). Deloitte Belgium. https://www2.deloitte.com/be/en/pages/finance-transformation/articles/Finance-2025-Digital-transformation-in-finance.html.

Garvin, D. A., & Levesque, L. C. (2006). Meeting the challenge of corporate entrepreneurship. *Harvard Business Review, 84*(10), 102.

Global Technology Leadership Study. (2021). Deloitte United States. https://www2.deloitte.com/us/en/pages/chief-information-officer/articles/global-technology-leadership-survey.html.

Humphries, D., & Humphries, D. (2022). 2022: Digital transformation becomes a way of life. IBM Blog. https://www.ibm.com/blog/2022-digital-transformation-becomes-a-way-of-life/

Kennedy, J. F. (1963). *Remarks in the Rudolph Wilde Platz, Berlin*. John F. Kennedy Presidential Library and Museum. https://www.jfklibrary.org/archives/other-resources/john-f-kennedy-speeches/berlin-19630626

Leadership principles. (n.d.). Amazon. jobs. https://www.amazon.jobs/content/en/our-workplace/leadership-principles

Marston, S., Li, Z., Bandyopadhyay, S., Zhang, J., & Ghalsasi, A. (2011). Cloud computing—The business perspective. *Decision support systems, 51*(1), 176-189.

Mougayar, W. (2016). *The business blockchain: promise, practice, and application of the next Internet technology*. John Wiley & Sons.

Plc, U. (2023, February 13). *Unilever commits to help build a more inclusive society.* Unilever. https://www.unilever.com/news/press-and-media/press-releases/2021/unilever-commits-to-help-build-a-more-inclusive-society/

Ransbotham, S., Kiron, D., Gerbert, P., & Reeves, M. (2017). Reshaping business with artificial intelligence: Closing the gap between ambition and action. *MIT Sloan Management Review, 59*(1).

Resilience and value: A CFO toolkit for withstanding shocks. (2019). McKinsey & Company. https://www.mckinsey.com/capabilities/strategy-and-corporate-finance/our-insights/resilience-and-value-a-cfo-tool-kit-for-withstanding-shocks

Smith, B. (2020). *Microsoft will be carbon negative by 2030 - The Official Microsoft Blog*. The Official Microsoft Blog. https://blogs.microsoft.com/blog/2020/01/16/microsoft-will-be-carbon-negative-by-2030/

Stone, B. (2013). *The everything store: Jeff Bezos and the age of Amazon*. Random House.

Why diversity matters. (2015, January 1). McKinsey & Company. https://www.mckinsey.com/capabilities/people-and-organizational-performance/our-insights/why-diversity-matters

SECTION II

The Future is Autonomous

The Impact of Artificial Intelligence on Finance

Section 2 of this book delves into the swiftly advancing frontier of Autonomous Finance. The ensuing chapters thoroughly examine the role and influence exerted by AI on the financial domain. This segment aims to comprehensively understand how this groundbreaking technology affects a paradigmatic shift in the financial landscape.

Chapter 4, *"Autonomous Finance: Transforming Finance with Automation and Intelligence,"* forms the entry point into the realm of Autonomous Finance. This chapter introduces how AI drives this intelligent evolution, transitioning organizations from traditional finance practices to AI-powered Autonomous Finance solutions.

Chapter 5, *"Showcasing Success: Autonomous Finance's Breakthroughs,"* aims to consolidate your understanding of the tangible impact of Autonomous Finance. This chapter will highlight various global instances of transformative success driven by Autonomous Finance, shedding light on the potential and promise of this novel concept and technology.

Chapter 6, entitled *"Inside the Engine: The Tech that Powers Autonomous Finance,"* serves as an illuminating exploration beneath the surface of Autonomous Finance. This chapter aims to elucidate the sophisticated technologies that facilitate autonomous operations, enhancing comprehension of the complex interplay among AI, ML, RPA, Big Data, Advanced Analytics, SaaS, Cloud Computing, and APIs within Autonomous Finance.

The conclusion of Section 2 establishes a comprehensive understanding of Autonomous Finance, delineates its real-world applications, and elucidates the advanced technology propelling this evolution. This understanding will lay the groundwork for a detailed blueprint for success in Autonomous Finance, which constitutes the central focus of Section 3 of this book.

CHAPTER 4

Autonomous Finance

Transforming Finance with Automation & Intelligence

"In today's era of volatility, there is no other way but to re-invent. The only sustainable advantage you can have over others is agility, that's it. Because nothing else is sustainable, everything else you create, somebody else will replicate."

- JEFF BEZOS (CRAIGTHOMAS, 2022)

4.0 Key Insights

- **Embracing the Dual Nature of Autonomous Finance:** The advent of Autonomous Finance heralds a new era of efficiency, accuracy, personalized services, and cost reduction. However, it also brings forth challenges, including technological complexities, data privacy, security concerns, regulatory obstacles, cultural resistance, and skill gaps. The balance between leveraging opportunities and mitigating challenges becomes the defining aspect of this revolutionary shift.
- **Strategizing Solutions for a Seamless Transition:** Overcoming the hurdles in Autonomous Finance requires a multifaceted approach that involves phased technology implementation, robust cybersecurity, continuous communication, targeted training, building adaptable systems, and maintaining human oversight in AI-driven processes. A strategic alignment of in-house expertise, vendor support, and collaboration on standards sets the stage for success.
- **Exploring the Multidimensional Applications of Autonomous Finance:** From personalized financial advice, automated trading, fraud detection, and regulatory compliance to treasury management and lending decisions, the applications of Autonomous Finance extend across the financial landscape. These use cases symbolize the transformative potential of Autonomous Finance.
- **Visioning a Strategic and Inclusive Financial Future:** Autonomous Finance is a technological advancement and a future where strategic financial systems predict, strategize, and optimize financial well-being. It promises to promote sustainability and financial inclusion, integrating financial practices with broader societal goals.
- **Crafting the Roadmap to Excellence:** Achieving excellence in Autonomous Finance requires embracing digital transformation, incorporating AI and ML, and fostering new skills and capabilities. Partnerships, collaborations, and risk management become vital components of this journey. The roadmap to excellence is a compass guiding organizations toward an innovative and resilient financial future.

4.1 Introduction

In the wake of the preceding discourse, Chapter 4 embarks on a path less traveled, venturing into the transformative realm of Autonomous Finance. The echoes of history, where adaptability has been the quintessential virtue that has shaped civilizations and nurtured progress, resound in the corridors of modern finance. Charles Darwin's profound wisdom, *"It is not the strongest of the species that survives, nor the most intelligent; it is the one that is most adaptable to change"* (Megginson, 1963), is a guiding beacon in this journey.

Examining Autonomous Finance is not a mere academic endeavor but a visionary pursuit to redefine the essence of financial systems. It is an exploration of a future where finance transcends numbers and algorithms and evolves into a living, breathing entity that shapes businesses, economies, societies, and the collective destiny of humankind. Why should this idea captivate the reader's attention? Because the world stands on the cusp of a technological revolution where AI, ML, RPA, Big Data, Advanced Analytics, SaaS, Cloud

Computing, and APIs are both tools and catalysts of change. They are the brush strokes painting a landscape that is not only sustainable and innovative but genuinely autonomous.

The following pages will delve into Autonomous Finance's intricate examining potential, illuminate its implications, and present a canvas where imagination meets practicality. The journey ahead is filled with intrigue, inspiration, and infinite possibility. It is a journey that invites the reader to shape a financial world that resonates with the values of humanity, embraces the marvels of technology, and celebrates the art of endless innovation.

As a harbinger of this new era, this chapter synthesizes insights, research, and visionary thoughts that will guide, challenge, and inspire. It builds upon the foundational concepts in previous chapters and sets the stage for an exploration that transcends traditional boundaries. The quest for Autonomous Finance continues and beckons all who dare to envision, innovate, and lead.

4.2 The Concept and Origin of Autonomous Finance

A new revolution in financial management is brewing, an evolution named **Autonomous Finance** that combines cutting-edge technology and strategic insight to create a new paradigm for financial management. Autonomous Finance refers to financial systems and processes that can operate with minimal human intervention, using a combination of AI, ML, RPA, Big Data, Advanced Analytics, SaaS, Cloud Computing, and API technologies.

According to Gartner's Hype Cycle for Digital Banking Transformation, 2023, Autonomous Finance is one of the *"key emerging technologies that will drive the Future of Banking."* As Gartner explains, Autonomous Finance is a rapidly evolving field that *"leverages AI and ML to automate personal financial management decisions and optimize consumer finances"* (*Autonomous Finance - Everything a CFO Should Know | Gartner*, 2022).

But what exactly does this entail? Autonomous Finance is about augmenting human capabilities with intelligent systems. Rather than replacing human decision-makers, Autonomous Finance uses technology to take care of routine tasks, make data-driven decisions, and learn from experiences, augmenting our capabilities and enabling us to focus on strategic decisions and creative problem-solving.

This new approach to finance goes beyond just automating processes. It provides unprecedented efficiency, predictive capabilities, continuous optimization, real-time decision-making, and systemic integration, transforming how we manage and think about finance.

The path toward Autonomous Finance has been long and winding, shaped by technological advancements, economic necessities, and evolving consumer expectations. Its roots can be traced back to the early days of digital computing and the emergence of the first automated systems in finance. However, the rapid development of technologies such as AI, ML, RPA, Big Data, Advanced Analytics, SaaS, Cloud Computing, and APIs, coupled with the growing demand for personalized, convenient, and instantaneous financial services, has expedited this journey and brought us to the threshold of a complete autonomous financial ecosystem.

4.3 The Need for Autonomous Finance

Chapter 1 navigated through the disruptions of the past that rippled across the finance domain, driven by technological advancements, shifting consumer expectations, regulatory changes, and the unpredictable impact of global events like the COVID-19 pandemic. These disruptions brought attention to the limitations of traditional financial systems, which struggled to adapt and respond efficiently.

In **Chapter 2**, our focus turned to the present and the importance of agility. We explored how businesses harness innovative technologies to navigate through these turbulent times. The increasing adoption of AI, ML, RPA, Blockchain, Big Data, Advanced Analytics, Cloud Computing, SaaS, and APIs signifies an industry amid rapid evolution, striving to match the pace of change.

In **Chapter 3**, we gaze toward the horizon, envisioning a future that demands adaptability. We discussed transitioning from traditional finance to Autonomous Finance—a transformative journey that addresses current challenges while proactively preparing for future uncertainties.

The overwhelming conclusion of the journey we have taken so far is that we live in a world that is in a near-constant state of disruption, and as a result, Autonomous Finance is a necessity. It paves the way for greater resilience in the face of unpredictability. It embodies agility by leveraging cutting-edge technologies to process vast amounts of data, provide personalized services, and make accurate decisions at a speed and scale beyond human capability.

Ultimately, Autonomous Finance is the compass guiding us toward a future-ready financial ecosystem that is adaptable and prepared for unforeseen challenges and opportunities.

4.4 The Fundamentals of Autonomous Finance

Now that the need for Autonomous Finance has been defined and contextualized, the next priority is to look under the hood to understand its principles and building blocks.

Principles of Autonomous Finance

Within the realm of Autonomous Finance lies a set of guiding principles that serve as the bedrock for its design, development, and operation. As depicted in Figure 4.1, these principles form the essence of a philosophy that prioritizes ethical, user-friendly, and effectual functioning. Let us embark on a journey to explore each principle, unwrapping its implications, challenges, and the promise they hold. Together, they define the vision that Autonomous Finance seeks to manifest - a financial world that is personalized, efficient, responsive, transparent, secure, fair, and inclusive.

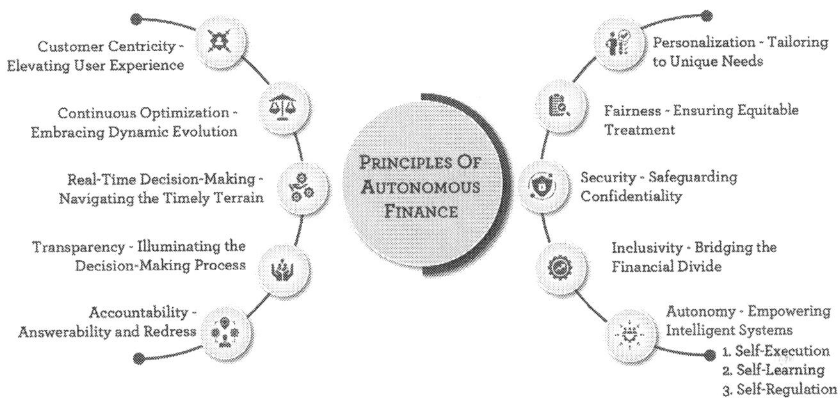

FIGURE 4.1: Principles of Autonomous Finance

a) **Customer Centricity - Elevating User Experience:** In the quest for an autonomous financial ecosystem, customer-centricity stands at the forefront, steering the course of design and implementation. The focus is on providing customers with a seamless, satisfying experience at every touchpoint. Financial solutions must be easy to use, convenient, and tailored to individual financial needs and goals (Shaw, 2016). The goal is to forge lasting relationships and drive loyalty by catering to each customer's unique preferences and demands.

b) **Continuous Optimization - Embracing Dynamic Evolution:** Continuous optimization becomes a vital principle in a landscape of rapid change. Autonomous Finance systems must be adaptable and flexible, continuously learning from data and feedback to improve their performance and decision-making capabilities (Stuart & Peter, 2016). Machine learning algorithms play a central role, refining strategies and adapting as they gather more data, ensuring that financial operations remain at the cutting edge of efficiency and efficacy.

c) **Real-Time Decision-Making - Navigating the Timely Terrain:** Time is of the essence in the financial world, and the principle of real-time decision-making is essential for Autonomous Finance. The systems in an Autonomous Finance landscape must provide instant financial insights and decisions, responding rapidly to market fluctuations and customer needs (*Autonomous Finance - Everything a CFO Should Know/Gartner*, 2022). The ability to make swift decisions can make the difference between capitalizing on opportunities or missing out in a highly dynamic and competitive landscape.

d) **Transparency - Illuminating the Decision-Making Process:** Despite the sophisticated algorithms that power Autonomous Finance, transparency remains paramount. The system must clarify its decision-making processes for financial choices on behalf of users to instill confidence and trust (Mittelstadt et al., 2016). Transparent systems enable users to comprehend the rationale behind recommendations and actions, promoting a harmonious relationship between customers and autonomous financial technologies.

e) **Accountability - Answerability and Redress:** As financial systems become increasingly autonomous, the principle of accountability becomes critical. Autonomous Finance systems must be answerable for their decisions, with clear mechanisms to address errors or disputes (Bryson & Winfield, 2017). As these systems handle sensitive financial matters, providers must be able to explain and justify outcomes, ensuring that users have recourse in case of unexpected outcomes.

f) **Personalization - Tailoring to Unique Needs:** Gone are the days of one-size-fits-all solutions. The principle of personalization drives Autonomous Finance to cater to each customer's unique needs and circumstances (Spring et al., 2018). AI and ML capabilities customize financial advice and products to match individual preferences, goals, and risk profiles. The aim is to create highly tailored financial experiences that resonate with each customer.

g) **Fairness - Ensuring Equitable Treatment:** Fairness lies at the core of Autonomous Finance, rejecting any form of discrimination. The systems must avoid biased decision-making and foster inclusivity (Chouldechova & Roth, 2018). Algorithmic fairness methods and bias testing are essential to promote equitable treatment and financial inclusion.

h) **Security - Safeguarding Confidentiality:** Security is paramount in sensitive financial data. Autonomous Finance must implement robust data security and privacy measures to prevent unauthorized access and maintain customer trust (Cavusoglu et al., 2004). By prioritizing data protection, financial systems ensure the confidentiality and integrity of customer information.

i) **Inclusivity - Bridging the Financial Divide:** Finally, the principle of inclusivity seeks to democratize access to financial services. Autonomous Finance aims to extend its reach to traditionally underserved populations, bridging the gap in financial inclusion (Demirgüç-Kunt et al., 2018). By providing accessible and tailored financial solutions, Autonomous Finance contributes to a more equitable and inclusive financial ecosystem.

j) **Autonomy: The Luminous Beacon in Empowering Intelligent Systems:** Central to Autonomous Finance is the principle of autonomy—a trifecta comprising self-execution, self-learning, and self-regulation. These elements act as the dynamo that powers the machines, imbuing them with unprecedented capabilities that augment human faculties.

k) **Self-Execution, the Embodiment of Automation in Financial Operations:** The concept of self-execution is at the vanguard of this transformation. This notion empowers algorithms and intelligent systems to undertake routine tasks autonomously within the intricate universe of financial transactions. No longer encumbered by the quotidian demands of transaction processing, account reconciliation, or report generation, human capital can ascend to its rightful place, focusing on the strategic, the analytical, and the innovative.

l) **Self-Learning, the Continuous Cycle of Improvement and Adaptation:** Adaptation signifies more than mere survival; it is the quintessence of growth and improvement. In Autonomous Finance, self-learning algorithms harness data's raw, untapped power to evolve ceaselessly. Through complex computations and analyses, these ML algorithms distill patterns and discern correlations from an ocean of data, continually refining strategies, optimizing operations, and sharpening predictive accuracy.

m) **Self-Regulation, the Unseen Guardian of Ethical and Regulatory Compliance:** Yet, this technological behemoth operates not in a vacuum but within the confines of a legal and ethical framework. Self-regulation ensures that these autonomous financial systems operate within the boundaries of pre-established rules, policies, and ethical considerations. Thus, they contribute to an ecosystem where compliance with financial regulations, risk management protocols, and corporate governance standards is not an afterthought but an intrinsic attribute.

Autonomy—encompassing self-execution, self-learning, and self-regulation—heralds the maturation of the finance function. Autonomy instigates a seminal shift by unlocking the doors to data-driven decision-making, immediate responsiveness, and unparalleled efficiency. Such a transformation is not merely technological advancement; it represents the heralding of a new epoch in financial management, one ripe with possibilities and pregnant with promise.

Together, these principles form the philosophical compass that guides Autonomous Finance (Vihtelic, 1996). They represent a vision that transcends the boundaries of traditional finance, beckoning us towards a financial world that is highly personalized, optimally efficient, real-time responsive, transparent, accountable, fair, secure, and inclusive.

Building Blocks of Autonomous Finance

In this intricate landscape, essential building blocks form the bedrock for the cutting-edge financial model of Autonomous Finance. Figure 4.2 demonstrates how these foundational elements weave together seamlessly, shaping an intelligent ecosystem transcending traditional finance. As we embark on a journey to unravel the intricacies of Autonomous Finance, let us recognize the critical components that converge to redefine the financial landscape.

a) **Data: The Pillar of Knowledge:** Data is the foremost building block of Autonomous Finance, a vast repository of information that fuels the system's intelligence (Halevy et al., 2009). Diverse and high-quality data collected from various sources is the lifeblood of AI and ML algorithms, enabling informed financial decisions and predictive capabilities. In Autonomous Finance, data encompasses structured and unstructured data collected from various internal and external sources. The data landscape is expansive, from transactional data within the organization to market data, customer behaviors, social media interactions, and more. Harnessing big data analytics, Autonomous Finance extracts valuable insights, identifies trends, and makes real-time predictions, enabling finance teams to respond proactively to dynamic market conditions and customer demands.

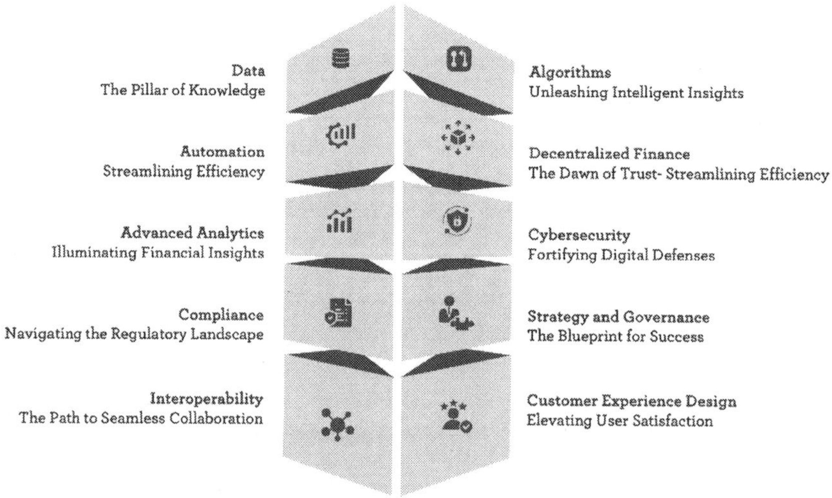

FIGURE 4.2: Building Blocks of Autonomous Finance

b) **Algorithms: Unleashing Intelligent Insights:** AI and ML form the cognitive powerhouse of Autonomous Finance, infusing advanced analytics, decision-making, and learning capabilities (Goodfellow et al., 2016). These algorithms discern patterns and anomalies within data, unlocking the ability to make informed predictions and optimize financial strategies. In Autonomous Finance, AI and ML algorithms are trained on historical data to identify patterns and relationships, enabling them to predict market trends, customer preferences, and potential risks. For instance, these algorithms can forecast stock prices, assess creditworthiness, detect fraud, optimize investment portfolios, and automate customer service interactions. As AI and ML algorithms continuously learn from new data, they improve their accuracy and adaptability, providing finance teams with powerful decision-making tools to navigate an ever-changing financial landscape.

c) **Automation: Streamlining Efficiency:** Automation technologies, such as RPA and Intelligent Automation, revolutionize the efficiency and accuracy of financial processes (Lacity & Willcocks, 2016). By reducing human intervention and minimizing errors, automation unleashes the full potential of finance teams, allowing them to focus on strategic decision-making. The automation in Autonomous Finance is instrumental in streamlining routine tasks such as data entry, reconciliation, and reporting. For instance, RPA can automatically update financial records, process invoices, and execute transactions, freeing finance professionals from mundane tasks. Intelligent Automation extends this efficiency further, incorporating machine learning capabilities to handle more complex processes and adapt to dynamic scenarios. As a result, finance teams can redirect their efforts toward analyzing data, driving innovation, and delivering strategic insights that shape the organization's financial future.

d) **Decentralized Finance (DeFi): The Dawn of Trust:** The DeFi paradigm provides a blockchain-based infrastructure that is transparent, open, and resilient (Mougayar, 2016). DeFi fosters trust and diverse stakeholders' participation in Autonomous Finance by enabling peer-to-peer financial transactions and reducing reliance on traditional intermediaries. Blockchain technology, a distributed ledger system, powers finance decentralization and ensures the transparency and immutability of transactions. In the context of Autonomous Finance, DeFi enables secure and efficient peer-to-peer lending, borrowing, and trading without central authorities. This disintermediation reduces transaction costs, speeds up settlement times, and promotes financial inclusion for individuals and businesses worldwide. DeFi protocols also allow for automated, self-executing financial agreements without the need for intermediaries' programmable smart contracts. With DeFi as a building block, Autonomous Finance blazes a trail toward a more democratic and accessible financial ecosystem.

e) **Advanced Analytics: Illuminating Financial Insights:** Advanced analytics, powered by sophisticated statistical methods and tools, extracts invaluable insights from data (Provost & Fawcett, 2013). By providing a deeper understanding of market trends and customer behavior, advanced analytics empowers stakeholders to make strategic financial decisions. Within the context of Autonomous Finance, advanced analytics transcends traditional business intelligence, encompassing data mining, ML, and predictive modeling. These techniques enable finance teams to uncover hidden patterns, identify emerging risks, and anticipate customer needs. Advanced analytics informs financial planning, investment strategies, risk management, and customer experience optimization by analyzing vast amounts of structured and unstructured data. With such insights, finance teams can proactively respond to market dynamics, deliver personalized financial services, and drive the organization toward sustainable growth.

f) **Cybersecurity: Fortifying Digital Defenses:** In the digital realm, cybersecurity is a formidable shield against data breaches, fraud, and other threats (Shackleford, 2016). Robust encryption, secure authentication, and continuous system monitoring are key protective measures ensuring safe transactions in an Autonomous Finance landscape. As Autonomous Finance relies on digital platforms and cloud-based services to a large extent, safeguarding sensitive financial data becomes paramount. Cybersecurity safeguards encompass multi-factor authentication, data encryption at rest and in transit, robust firewalls, intrusion detection systems, and constant monitoring for potential threats. In Autonomous Finance, the Finance teams work diligently to fortify digital defenses, ensuring compliance with industry standards and regulations while building customer trust. By mitigating cyber risks, Autonomous Finance assures stakeholders that their financial transactions and data remain secure, fostering an environment of trust and confidence.

g) **Compliance: Navigating the Regulatory Landscape:** Regulation Technology (RegTech) solutions facilitate compliance and risk management within Autonomous

Finance (Arner et al., 2016). These solutions safeguard institutions from legal penalties and uphold their reputation by ensuring adherence to complex financial regulations. RegTech solutions deployed in Autonomous Finance monitor regulatory changes, automate compliance reporting, and assess risk exposure in real-time. Through data analytics and AI-driven compliance monitoring, RegTech solutions enable the Finance teams to navigate the intricate landscape of financial regulations efficiently. Integrating RegTech solutions in Autonomous Finance ensures that financial activities remain transparent, accountable, and aligned with the prevailing regulatory environment.

h) **Strategy and Governance: The Blueprint for Success:** Embarking on the path of Autonomous Finance requires a clear strategy and robust governance framework (*Introducing the Next-generation Operating Model*, 2017). These guide the ethical and responsible development and implementation of Autonomous Finance, aligning its objectives with the broader goals and values of the organization. A well-defined strategy should chart the course for adopting Autonomous Finance, outlining the transformation's objectives, scope, and milestones. Finance teams collaborate with stakeholders to develop a governance framework that addresses ethical considerations, risk management, and accountability. Such a framework helps establish the roles and responsibilities of stakeholders, sets benchmarks for performance and compliance, and ensures transparency in decision-making processes. Strategy and governance are the compass that steers Autonomous Finance toward its intended outcomes while upholding principles of integrity, trust, and societal welfare.

i) **Interoperability: The Path to Seamless Collaboration:** The ability of Autonomous Finance systems to interface seamlessly with many other financial technologies and platforms is critical for optimized operations. This frictionless data exchange and transactional capabilities allow for high collaboration among various platforms and services. As a critical cornerstone in a vast financial ecosystem, this attribute connects diverse financial applications, platforms, and service providers. Adopting universally recognized protocols and APIs ensures uninterrupted data flows and transactional engagements between Autonomous Finance mechanisms and outside systems. This connectivity results in streamlined cross-border monetary transfers, efficient payment processing, and fluid interaction among DeFi protocols, thereby elevating opportunities for value generation and customer-focused financial services.

j) **Customer Experience Design: Elevating User Satisfaction:** A pivotal element in successfully implementing Autonomous Finance is constructing an exemplary user experience. Platforms prioritizing easy-to-navigate and engaging interfaces tend to secure user satisfaction, encouraging sustained adoption and brand loyalty. Within Autonomous Finance, the focus rests on decidedly human-centered principles in design. The aim is to understand users' requirements and unique inclinations and challenges deeply. By doing so, one achieves a milieu of intuitive interactions and dashboards that simplify the user's financial journey. Designing a system's architecture to offer personalized financial insights and effortlessly understandable navigation cultivates a lasting relationship with its user base and further solidifies its integral role in their financial life cycle.

In conclusion, the principles and building blocks of Autonomous Finance provide a foundation for its ethical, technological, and strategic execution. With careful consideration of these principles and effective use of the building blocks, finance teams can effectively navigate the shift toward Autonomous Finance.

4.5 Understanding the Shift from Automation to Autonomous Finance

The journey to Autonomous Finance began with the automation of manual tasks in the last century, ushering in improved efficiency and streamlined processes. As technology advanced, automation expanded its reach to encompass more complex financial functions, resulting in faster decision-making and reduced errors. However, as discussed earlier in this chapter, the finance domain is at the cusp of a fundamental paradigm shift—moving beyond automation in finance towards a new era defined by Autonomous Finance. The convergence of guiding principles and building blocks discussed in the previous section drives this paradigm shift. This paradigm shift redefines finance with the following key distinctions:

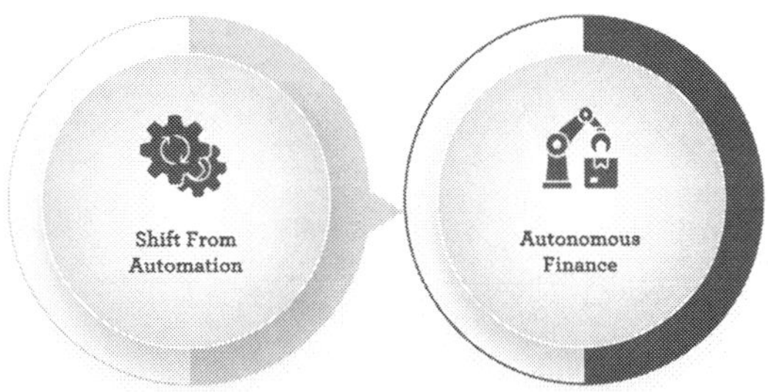

FIGURE 4.3: The Shift from Automation to Autonomous

a) **Scope of Automation:** Automation in finance is often limited to specific tasks or processes, targeting repetitive and rule-based activities. In contrast, Autonomous Finance expands the scope of automation to encompass a broader array of financial functions, including decision-making, risk analysis, customer engagement, and compliance. It operates across the entire financial ecosystem, optimizing processes holistically.

b) **Complexity of Automation:** Automation handles relatively straightforward and standardized tasks. In contrast, Autonomous Finance deals with complex financial scenarios, leveraging AI and ML to analyze vast datasets, identify patterns, and make sophisticated decisions. The level of complexity in Autonomous Finance surpasses traditional automation, allowing for more accurate predictions and adaptability to dynamic market conditions.

c) **Human Intervention vs. Machine Decision-Making:** Human intervention often sets rules and parameters for the system in traditional automation. In contrast, Autonomous Finance exhibits machine decision-making capabilities. AI algorithms process data and generate insights autonomously, reducing the reliance on human intervention for routine decisions.

d) **Efficiency vs. Optimization:** Automation aims to streamline processes and improve operational efficiency. While it undoubtedly enhances productivity, Autonomous Finance takes it further by focusing on optimization. The solutions in Autonomous Finance landscape continuously learn, evolve, and optimize their performance, leading to better strategic decision-making and enhanced outcomes.

e) **Risks vs. Rewards:** While automation in finance mitigates certain risks, such as human errors and process delays, it may introduce new risks without adequate monitoring and controls. On the other hand, Autonomous Finance addresses these challenges by incorporating self-regulation and risk management mechanisms, mitigating potential risks while maximizing rewards.

Adopting Autonomous Finance requires a delicate balance between automation and autonomy. Striking the right balance between the two is crucial to maximizing the benefits of both approaches while ensuring prudent risk management.

4.6 Autonomous Finance: A Comparative Advantage

The impact of Autonomous Finance reaches beyond incremental improvements. As illustrated in Figure 4.4 below, Autonomous Finance drives a revolutionary era beyond being efficient and accurate, completely reshaping the financial landscape in the following ways (Thorpe & Rawlinson, 2013):

a) **Efficiency and Accuracy:** Autonomous Finance has completely transformed the landscape of financial processes. With financial operations becoming increasingly data-driven and intricate, autonomous systems have risen to the challenge, offering unmatched speed and precision. For example, RPA has seen application in loan processing, significantly reducing errors and cutting processing time from weeks to hours (Jędrzejka, 2019).

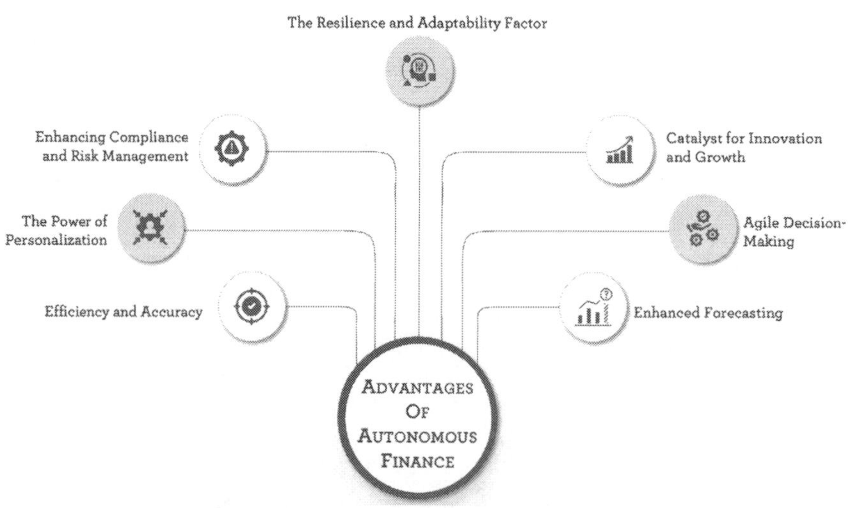

FIGURE 4.4: Advantages of Autonomous Finance

b) **The Power of Personalization:** Personalization is another critical advantage of Autonomous Finance. By leveraging data analytics and AI, Autonomous Finance enables hyper-personalization, empowering Finance teams to offer financial advice and products tailored to each individual's unique circumstances and preferences. Robo-advisors are a case in point, using complex algorithms to design investment portfolios that align with a client's risk tolerance and financial objectives (Zhang et al., 2023).

c) **Enhancing Compliance and Risk Management:** Autonomous Finance revolutionizes handling compliance and risk management. Continuously monitoring transactions and ensuring strict adherence to the ever-changing regulations significantly reduces the risk of human error (Sum et al., 2017).

d) **The Resilience and Adaptability Factor:** Beyond these, Autonomous Finance systems display remarkable resilience and adaptability. They can self-diagnose and rectify disruptions and are inherently self-optimizing, continually enhancing their performance. Furthermore, they are self-protecting and capable of anticipating, detecting, and safeguarding against external threats (Alam et al., 2023).

e) **Catalyst for Innovation and Growth:** Autonomous Finance also catalyzes innovation and growth. Automating routine tasks and freeing up resources enables the reallocation of these resources toward strategic growth initiatives, fostering a more agile and innovative business culture (Thorpe & Rawlinson, 2013).

f) **Agile Decision-Making:** The ability of Autonomous Finance to process and analyze vast quantities of data in real-time equips it for quick and informed decision-making, leading to more effective resource allocation (Merendino et al., 2018).

g) **Enhanced Forecasting:** Autonomous Finance bolsters forecasting and scenario planning. These systems can generate more accurate financial forecasts and robust scenario analyses using advanced algorithms and ML, equipping organizations with the tools to navigate the future more effectively and instilling confidence in their decision-making.(Bounfour & Bounfour, 2016).

In conclusion, the paradigm shift from automation to Autonomous Finance marks a watershed moment in the financial industry. By harnessing the power of AI, ML, and advanced analytics, financial institutions can propel themselves into a future defined by efficiency, accuracy, resilience, agility, adaptability, innovation, growth, empowerment, and customer-centricity. Embracing Autonomous Finance is not merely an option but an imperative for financial organizations seeking to thrive in the digital age.

4.7 Overcoming the Barriers to Autonomous Finance

Despite the many advantages of Autonomous Finance, certain obstacles hinder its full potential. The top 10 barriers to success in shifting to the Autonomous Finance paradigm are shown in Figure 4.4 below.

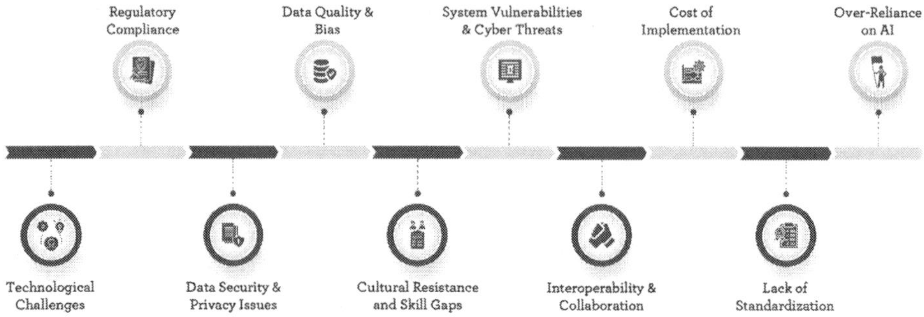

FIGURE 4.5: Barriers to Autonomous Finance

These barriers and the potential solutions for addressing such barriers are described in Table 4.1 below.

TABLE 4.1: Potential Solutions to Barriers of Autonomous Finance

NO.	BARRIERS TO AUTONOMOUS FINANCE	POTENTIAL SOLUTIONS
1	**Technological Challenge:** Implementing Autonomous Finance requires advanced technologies, such as AI, ML, Big Data Analytics, and RPA, to work harmoniously and integrate seamlessly into existing systems. Financial institutions may encounter the following technological challenges: • **Data Integration:** Autonomous Finance relies on vast and diverse datasets. Integrating and harmonizing these datasets can be complex and time-consuming. • **Infrastructure Readiness:** Adopting advanced technologies may require significant hardware and software infrastructure investments. • **System Compatibility:** Existing legacy systems may not be compatible with Autonomous Finance technologies and necessitate upgrades or replacements. • **Scalability:** Autonomous Finance systems must be scalable to handle growing volumes of data and transactions without compromising performance.	Addressing these technological challenges requires meticulous planning, collaboration with technology partners, and a clear roadmap for integration and upgradation. (Abubakar & Handayani, 2018).
2	**Regulatory Compliance and Data Privacy and Security Hurdles:** The Finance domain is subject to stringent requirements, ensuring consumer protection, data privacy, and financial stability. Adopting Autonomous Finance may raise regulatory and compliance concerns: • **Data Privacy and Security:** Autonomous Finance involves the processing and analysis of vast amounts of sensitive data, necessitating robust cybersecurity measures to safeguard against data breaches and unauthorized access. • **Regulatory Compliance:** Financial institutions must ensure that Autonomous Finance systems comply with relevant laws and regulations, including data protection laws, financial reporting standards, and AML regulations. • **Explainability and Transparency:** AI and ML algorithms in Autonomous Finance may be complex and challenging to interpret. Ensuring transparency and explainability of decision-making is essential for regulatory compliance and building trust with customers. • **Regulatory Uncertainty:** The evolving nature of AI and autonomous technologies may lead to regulatory uncertainty, requiring financial institutions to stay abreast of changing regulations and adapt their strategies accordingly.	Addressing regulatory and compliance challenges necessitates close collaboration with regulatory authorities, investing in robust data privacy measures, and implementing explainable AI models. (Sum et al., 2017).

NO.	BARRIERS TO AUTONOMOUS FINANCE	POTENTIAL SOLUTIONS
3	**Cultural Resistance and Skill Gaps:** Introducing Autonomous Finance may encounter resistance from employees who fear job displacement or feel uncomfortable with technology-driven changes. Cultural resistance and skill gaps may manifest in the following ways: • **Workforce Training:** Employees may require training to adapt to the changing work environment, acquire new skills and embrace technology-driven solutions. • **Change Management:** Successfully implementing Autonomous Finance requires effective change management strategies to manage cultural resistance and ensure a smooth transition. • **Trust in AI:** Building trust in AI-driven decision-making among employees and customers is crucial for successful adoption. • **Upskilling and Reskilling:** Financial institutions must invest in upskilling and reskilling programs to equip employees with the necessary skills to work alongside AI and autonomous systems.	Addressing cultural resistance and skill gaps requires proactive communication, employee engagement, and a commitment to workforce development. (Hurn, 2012).
4	**Data Quality and Bias:** The success of Autonomous Finance depends on the quality and accuracy of the data used to train AI and ML algorithms. Poor data quality and biases may lead to inaccurate or unfair decision-making: • **Data Bias:** Biases in historical data may lead to biased AI-driven decisions, potentially resulting in unfair treatment of certain customer groups. • **Data Accuracy:** Financial institutions must ensure that the data used to train AI algorithms is accurate and representative of the target population to avoid erroneous conclusions. • **Model Interpretability:** Ensuring the interpretability of AI models is essential for detecting and addressing potential biases in decision-making.	Improving data quality and addressing biases requires data validation, diversity in data sources, and constant monitoring of AI algorithms for fairness and accuracy.
5	**System Vulnerabilities and Cyber Threats:** Autonomous Finance systems are susceptible to cybersecurity threats, including hacking, data breaches, and malware attacks. Protecting sensitive financial information and ensuring system integrity are top priorities. • **Cybersecurity Measures:** Financial institutions must implement robust cybersecurity measures, including encryption, access controls, and intrusion detection systems. • **Ethical Hacking and Vulnerability Testing:** Regular ethical hacking and vulnerability testing help identify system weaknesses and address potential security loopholes. • **Continuity and Disaster Recovery:** Implementing disaster recovery plans ensures system continuity and resilience in the face of cyber incidents.	Maintaining system Security and resilience requires continuous monitoring, periodic security assessments, and a proactive approach to cybersecurity.
6	**Interoperability and Collaboration:** As the financial ecosystem evolves, financial institutions must collaborate and share data to ensure seamless integration of Autonomous Finance systems. • **Industry Standards:** Collaborating on industry standards ensures interoperability between financial systems and facilitates data exchange. • **Data Sharing Agreements:** Finance teams must establish agreements prioritizing data privacy and security. • **Fostering Collaboration:** Collaboration between financial institutions, technology providers, and regulators promotes the development of a robust Autonomous Finance ecosystem.	Navigating the challenges and barriers to Autonomous Finance implementation requires a comprehensive and integrated approach. By addressing technological complexities, regulatory compliance, cultural considerations and cybersecurity threats, financial institutions can successfully embrace the transformative potential of Autonomous Finance.

NO.	BARRIERS TO AUTONOMOUS FINANCE	POTENTIAL SOLUTIONS
7	**Cost of Implementation:** The deployment of Autonomous Finance systems represents a financial exigency that often seems prohibitive, especially for organizations operating on a constrained budget. One of the principal cost drivers is the integration of sophisticated technologies like AI, ML, and RPA. These technologies require specialized skill sets, which are scarce and expensive. The upfront investment also extends to replacing or retrofitting existing systems and the unquantifiable cost of organizational change management. Regulatory compliance, too, necessitates continuous monitoring and frequent updates, adding another layer to the total cost of ownership. Lastly, the imperative for continual adaptation to rapidly evolving technological landscapes necessitates a consistent investment in system upgrades and employee training. Hence, while the promise of Autonomous Finance is tantalizing, its implementation remains a complex and costly endeavor.	A strong business case highlighting potential savings and efficiency gains can justify the investment. The cost of transitioning to Autonomous Finance is intrinsically linked to an organization's technological sophistication. The more advanced the starting point, the less prohibitive the financial outlay for implementation.
8	**Lack of Standardization:** The absence of industry-wide standards can create integration challenges (Abubakar & Handayani, 2018).	Collaborating with industry bodies to create and adopt standards can help.
9	**Over-Reliance on AI:** While AI is a crucial component of Autonomous Finance, over-reliance can lead to complacency and increased risk (Ponnam, 2009).	Regular audits and maintaining human oversight can prevent this.

4.8 Envisioning the Future of Autonomous Finance

The advent of Autonomous Finance promises to herald a paradigm shift in financial services, transforming how savings, investments, and financial management are conducted. This section offers a peek into this future, illuminating the potential changes and developments that Autonomous Finance may bring (Noble, 2018).

a) **A Data-Driven Ecosystem:** As big data analytics continues to mature, the reliance on machine learning algorithms to process substantial amounts of data and derive actionable insights will increase. These developments will lead to more precise and personalization of customer experience, allowing finance teams to manage risks proactively, make informed decisions, and provide real-time tailored financial products and advice (Marr, 2018).

b) **The Emergence of Self-Regulating Financial Networks:** In a future dominated by Autonomous Finance, we might witness the rise of self-regulating financial networks. Blockchain and similar distributed ledger technologies enable these networks to ensure decentralized and autonomous transactions, enforce financial contracts, mitigate fraud, and maintain transparency. This technological evolution would eliminate intermediaries, thereby enhancing the efficiency and reliability of financial systems (Mougayar, 2016).

c) **The Evolution of Regulatory Compliance:** Designing autonomous systems to enforce and ensure compliance can significantly transform regulatory compliance. By continuously monitoring and analyzing financial transactions, these systems enable real-time reporting and ensure swift adaptation to regulatory changes, reducing non-compliance risk (Risius & Spohrer, 2017).

d) **Reshaping the Role of Human Talent:** As Autonomous Finance systems handle more routine tasks, the role of humans in finance will evolve. Financial professionals will shift their focus toward strategic decision-making, customer service, and innovation. This shift toward more strategic roles could potentially enhance job satisfaction and productivity within the finance domain (National Academies of Sciences, Engineering, and Medicine, 2016).

e) **The Future of Customer Experience:** Customers can anticipate a future of personalized, seamless, and responsive financial services. Advanced AI and ML algorithms will provide highly personalized financial advice, instant customer service, and real-time financial management. Autonomous Finance integrates further into consumers' daily lives, empowering them to make more informed financial decisions and manage their finances more effectively (Gleasure & Feller, 2016).

f) **The Rise of Responsible Autonomous Finance:** Lastly, an increased focus on ethical considerations will mark the future of Autonomous Finance. As autonomous systems gain more decision-making authority, there will be a growing emphasis on ensuring fairness, transparency, and accountability. This development will necessitate significant strides in responsible AI, coupled with robust regulatory frameworks to ensure that Autonomous Finance operates in the best interests of all stakeholders (Bryson et al., 2017).

The future of Autonomous Finance, as portrayed in Figure 4.6 below, unfolds a spectrum of exciting prospects that could redefine our association with finance teams at its core. While the journey to this future presents its share of challenges and uncertainties, the potential benefits of a more efficient, personalized, and equitable financial system make it worth undertaking. As we progress on this path, it will be crucial to ensure that the evolution of Autonomous Finance aligns with the broader objective of enhancing financial well-being for all (Choudhary & Jain, 2022).

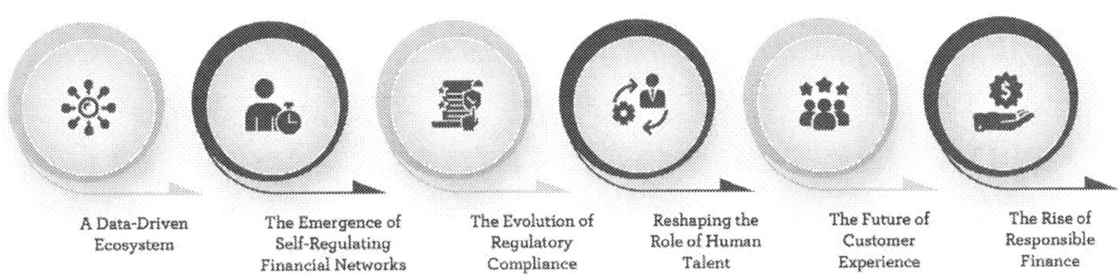

FIGURE 4.6: The Future of Autonomous Finance

4.9 Conclusion

The dawn of Autonomous Finance marks more than a technological advancement; it heralds a philosophical and strategic revolution encompassing the core of the financial industry. This paradigm shift transcends the realm of incremental improvements and enters the domain of transformative innovation.

a) **A Future Defined by Intelligence and Empathy:** Autonomous Finance is not merely about algorithms and automation; it is about creating a financial ecosystem that understands, learns, adapts, and aligns with human needs. It is about crafting personalized financial experiences that are not only efficient but compassionate.

b) **The Rise of Ethical Machines:** The era of Autonomous Finance will witness machines that calculate and contemplate. Ethics, fairness, and transparency will be incorporated into AI and ML, ensuring that financial decisions are made with a conscience.

c) **Democratizing Finance:** By leveraging AI and big data, Autonomous Finance will make high-quality financial advice and products accessible to all, irrespective of economic status. It is a step towards a more inclusive and egalitarian financial world.

d) **Resilience and Adaptation:** Autonomous Finance systems will become living, evolving entities. Their ability to self-diagnose, rectify disruptions, anticipate threats, and continually enhance performance will set new standards for resilience and adaptability.

e) **Catalyzing Innovation and Growth:** Autonomous Finance will ignite creativity and strategic thinking by freeing human intellect from routine tasks. It will be the bedrock of a culture that thrives on innovation, agility, and growth.

f) **A Harmonious Symphony of Man and Machine:** The future is not about machines replacing humans but a harmonious collaboration where machines augment human intelligence, creativity, and empathy. It is a vision of a world where technology serves humanity, not competes with it.

The visionary exploration of Autonomous Finance is not about predicting the future but about shaping it. It calls financial institutions, technology companies, regulators, and individuals to come together to think, innovate, and build a future that reflects our shared values and aspirations.

As this chapter concludes, it becomes evident that the threshold crossed is not merely one of technological change but rather heralds a new age in finance. This new era is characterized by intelligence, ethics, inclusivity, resilience, innovation, and a profound human touch. Embarking on this journey demands wisdom, courage, collaboration, and foresight. It is a journey that invites participation from all. The subsequent chapter, Chapter 5, will continue this exploration, delving into the dynamic interplay between theory and practice. There, the promise of Autonomous Finance will transition from being a visionary concept to becoming an operational reality.

4.10 Key Recommendations

a) **Embrace Digital Transformation:** Incorporation of AI, ML, and other digital technologies into financial operations can propel an organization toward adopting Autonomous Finance and reaping its benefits.

b) **Invest in Skills and Talent:** Upskilling current staff and hiring new talents proficient in areas like data analytics, AI, and strategic decision-making can ensure the successful implementation and optimization of Autonomous Finance.

c) **Establish Strategic Partnerships:** Collaboration with tech companies, fintech startups, and other Finance teams accelerates the journey toward Autonomous Finance and fosters a more integrated and robust financial ecosystem.

d) **Promote Cultural Change and Acceptance:** Drive continuous communication about the benefits of Autonomous Finance and invest in training to overcome cultural resistance. Encouraging a culture that is open to change and values learning will promote the successful adoption of Autonomous Finance.

e) **Pursue Ethical and Responsible Development:** The Future of Autonomous Finance demands a focus on ethical considerations, ensuring fairness, transparency, and accountability, which calls for strides in responsible AI and regulatory measures to ensure alignment with the collective good.

4.11 References

Abubakar, L., & Handayani, T. (2018). *Financial technology: Legal challenges for Indonesia financial sector*. In IOP Conference Series: Earth and Environmental Science (Vol. 175, No. 1, p. 012204). IOP Publishing.

Alam, M. F. B., Tushar, S. R., Zaman, S. M., Gonzalez, E. D. S., Bari, A. M., & Karmaker, C. L. (2023). Analysis of the drivers of Agriculture 4.0 implementation in the emerging economies: Implications towards sustainability and food security. *Green Technologies and Sustainability, 1*(2), 100021.

Arner, D. W., Barberis, J., & Buckley, R. P. (2015). The evolution of Fintech: A new post-crisis paradigm. *Geo. J. Int'l L., 47*, 1271.

Autonomous Finance - Everything a CFO should know | Gartner. (2022). Gartner. https://www.gartner.com/en/finance/topics/autonomous-finance

Bounfour, A., & Bounfour, A. (2016). Beyond 2020: Network Abundance, Data, and the Future of Organizing. Digital Futures, Digital Transformation: From Lean Production to Acceluction, 113-117.

Bryson, J., & Winfield, A. (2017). Standardizing ethical design for artificial intelligence and autonomous systems. *Computer, 50*(5), 116-119.

Choudhary, M. A., & Jain, A. (2022). Finance and inequality: The distributional impacts of bank credit rationing. *Journal of Financial Intermediation, 52*, 100997.

Chouldechova, A., & Roth, A. (2018). The frontiers of fairness in machine learning. *arXiv (Cornell University)*. https://arxiv.org/pdf/1810.08810.pdf.

Craig Thomas. (2022, January 7). Digital transformation and becoming an agile business. *Institute of Directors*. https://www.iod.com/resources/blog/science-innovation-and-tech/digital-transformation-and-becoming-an-agile-business/

Demirguc-Kunt, A., Klapper, L., Singer, D., Ansar, S., & Hess, J. (2018). The Global FinDex Database 2017: Measuring Financial Inclusion and the Fintech Revolution. In *Washington, DC: World Bank eBooks*. https://doi.org/10.1596/978-1-4648-1259-0.

Gleasure, R., & Feller, J. (2016). Emerging technologies and the democratization of financial services: A metatriangulation of crowdfunding research. *Information and Organization, 26*(4), 101-115.

Goodfellow, I., Bengio, Y., & Courville, A. (2016). *Deep learning*. MIT Press.

Halevy, A., Norvig, P., & Pereira, F. (2009). The unreasonable effectiveness of data. *IEEE Intelligent Systems, 24*(2), 8-12.

Hurn, B. J. (2012). Management of change in a multinational company. *Industrial and commercial training, 44*(1), 41-46.

Hype Cycle for Digital Banking Transformation, 2023 | Gartner. (2023). Gartner, Inc. https://www.gartner.com/en/documents/4549699

Introducing the next-generation operating model. (2017, March 1). McKinsey & Company. https://www.mckinsey.com/capabilities/mckinsey-digital/our-insights/introducing-the-next-generation-operating-model

Jędrzejka, D. (2019). Robotic process automation and its impact on accounting. *Zeszyty Teoretyczne Rachunkowości*, (105), 137-166.

Lacity, M. C., & Willcocks, L. P. (2016). *A new approach to automating services*. MIT Sloan Management Review.

Marr, B. (2018). *Data-driven HR: how to use analytics and metrics to drive performance*. Kogan Page Publishers.

Megginson, L. C. (1963). Lessons from Europe for American Business. *Southwestern Social Science Quarterly, 44*(1), 3-13.

Merendino, A., Dibb, S., Meadows, M., Quinn, L., Wilson, D., Simkin, L., & Canhoto, A. (2018). Big data, big decisions: The impact of big data on board level decision-making. *Journal of Business Research, 93*, 67-78.
Mittelstadt, B. D., Allo, P., Taddeo, M., Wachter, S., & Floridi, L. (2016). The ethics of algorithms: Mapping the debate. *Big Data & Society, 3*(2), 2053951716679679.
Mougayar, W. (2016). *The Business Blockchain: Promise, Practice, and Application of the Next Internet Technology.* Wiley.
National Academies of Sciences, Engineering, and Medicine. 2017. *Information Technology and the U.S. Workforce: Where Are We and Where Do We Go from Here?* Washington, DC: The National Academies Press. https://doi.org/10.17226/24649.
Noble, S. U. (2018). *Algorithms of oppression. In Algorithms of oppression.* New York University Press.
Ponnam, A. (2009). Information systems risk management: An audit and control approach. In *Handbook of Research on Information Security and Assurance* (pp. 68-84).
Provost, F., & Fawcett, T. (2013). Data science and its relationship to big data and data-driven decision making. *Big data, 1*(1), 51-59.
Risius, M., & Spohrer, K. (2017). A blockchain research framework: What we (don't) know, where we go from here, and how we will get there. *Business & information systems engineering, 59*, 385-409.
Shackelford, S. (2016). *2016 Cybersecurity Symposium (CYBERSEC)*. https://doi.org/10.1109/cybersec201640843.2016
Shaw, R. (2016). *Customer-centricity: The future of banking.* Market Force Information, Inc.
Spring, K. J., Graham, C. R., & Ikahihifo, T. B. (2018). Learner engagement in blended learning. *Encyclopedia of Information Science and Technology, Fourth Edition*, 1487–1498. https://doi.org/10.4018/978-1-5225-2255-3.ch128
Stuart, R., & Peter, N. (2016). *Artificial intelligence, A modern approach.* Pearson.
Sum, R. M., & Saad, Z. M. (2017). Risk management in universities. In 3rd International Conference on Qalb-Guided Leadership in Higher Education Institutions (iQALB, 2017, pp. 128-142).
Thorpe, R., and Rawlinson, R. 2013. (2013). *The Role of UK Business Schools in Driving Innovation and Growth in the Domestic Economy*, London: ABS.
Vihtelic, J. L. (1996). Personal finance: An alternative approach to teaching undergraduate finance. *Financial Services Review, 5*(2), 119-131.
Zhang, Y., & Chen, G. (2023). Research on the Personalized Teaching Model of College English Based on Big Data: Taking Guangdong University of Finance and Economics (GDUFE) as an Example. In 2023 The 8th International Conference on Information and Education Innovations (pp. 48-54).

CHAPTER 5

Showcasing Success

Autonomous Finance's Breakthroughs

"The future is already here —
it's just not very evenly distributed."

- WILLIAM GIBSON (CHATTERTON & NEWMARCH, 2017)

5.0 Key Insights

- **The Expansive Horizon of Autonomous Finance:** Autonomous Finance is not confined to a monolithic application but spreads its transformative tendrils across an expansive spectrum. Its influence is poised to redefine various financial functional areas, transcending the boundaries of both the private and public sectors. This broad scope stands as a testament to the universal appeal and adaptability of Autonomous Finance.
- **A Tapestry of Use Cases within the Private Sector:** The private sector offers a fertile ground for applying Autonomous Finance. Its top use cases weave a rich tapestry, encompassing diverse functional realms such as Financial Accounting, Management Accounting, FP&AS, Treasury, Risk Management, and Compliance and Regulatory Reporting. Each use case illustrates the profound impact and innovation Autonomous Finance brings to the private sector landscape.
- **Elevating Organizational Capabilities through Autonomous Finance:** Implementing Autonomous Finance is a technological advancement and an organizational metamorphosis. By harnessing the power of automation and AI, organizations can significantly elevate their efficiency, accuracy, and strategic decision-making prowess. It is a leap towards a future where finance is not just a function but a strategic partner in business growth.
- **Unearthing Distinct Potential within the Public Sector:** Though marked by unique regulatory and structural intricacies, the Public Sector is not immune to the allure of Autonomous Finance. The technology also finds resonance here with use cases that parallel yet distinctively differ from those in the private sector. It is a compelling narrative of how innovation adapts and thrives even within the complex web of public governance.
- **Envisioning a Fluid and Dynamic Future for Autonomous Finance:** The future of Autonomous Finance is not cast in stone; it is fluid, vibrant, and ever-evolving. The anticipation of new use cases, the burgeoning complexity in financial predictions, the growing appetite for personalized solutions, and the opportunity for cross-domain integration paint a promising and exhilarating future. It is a glimpse into a world where finance becomes an agile, intelligent, and responsive entity.

5.1 Introduction

As we continue our exploration into the intriguing realm of Autonomous Finance, we find ourselves standing at the intersection of the present and future, where the emergent reality of automation meets the tantalizing possibilities of advanced AI. Here, the inspiring quote from (Chatterton & Newmarch, 2017) takes on a poignant relevance. The future of finance, shaped by the transformative power of automation and AI, is not a distant dream but an unfolding reality. However, this future is not uniformly distributed and is evolving at varying paces across industries, organizations, and functions.

In the previous chapter, we delved into the concept, need, principles, and comparative advantages of Autonomous Finance. We unpacked its building blocks, broached upon some exciting use cases, contemplated its promising future, and mapped out a roadmap to achieving excellence in Autonomous Finance. The narrative, however, is far from complete. The intricacies of Autonomous Finance manifest differently across industry sectors, posing unique challenges and unearthing exclusive opportunities. As the landscape of Autonomous Finance evolves, so does the suite of use cases that demonstrate its practical applications and transformative potential.

This chapter, therefore, pivots toward a deeper exploration of these use cases. It aims to paint a vivid picture of today's autonomous financial landscape while peering into the exciting horizon of emerging possibilities. We start by delving into the top use cases of Autonomous Finance in the private sector, each carefully chosen to illustrate the profound impact of automation and AI across various financial functions and domains. We then turn our lens toward the public sector to explore how this promising technology finds resonance in this distinct landscape.

However, the future of Autonomous Finance is not static, nor is it confined within the boundaries of today's capabilities or use cases. We, therefore, take a step further to envision how these use cases might evolve, morph, and multiply as technology advances, regulatory environments change, and organizational needs grow. We contemplate the challenges and opportunities inherent in this exciting evolution, providing a glimpse into a future where Autonomous Finance could transcend beyond finance to revolutionize other domains.

Embark on this enlightening journey through the multifaceted spectrum of Autonomous Finance. Traverse the contours of its present state, foresee its promising future, and discover the opportunities therein. By the conclusion of this chapter, the intent is to impart a nuanced understanding of Autonomous Finance's current state and to kindle the imagination with a visionary glimpse into its boundless potential.

5.2 Top Use Cases in the Private Sector

Finance, as a domain, is a complex mosaic interwoven with various components that play a vital role in shaping the fiscal health of an organization. Within this intricate framework, Autonomous Finance emerges as a catalyst, driving efficiencies, insights, and strategic agility.

This section will present various use cases of Autonomous Finance, grouped by functional areas and processes within each area. Please note that the 'top' use cases in each functional area-process indicate and broadly represent common scenarios where Autonomous Finance can deliver significant value. These top use cases are subject to vary based on various factors, such as industry vertical, geographic location, organizational structure, the maturity level of existing processes and technology, and regulatory requirements that the organization needs to comply with, among others.

For each use case, a brief description and a list of the critical features of the Autonomous Finance application involved, the potential business impact, and recommended success metrics are provided. These examples should help organizations envision the possibilities of Autonomous Finance in their respective contexts and inspire them to explore these technologies further.

Acknowledging an inherent overlap across many of these use cases is essential as they deal with the same financial data, but from different perspectives or objectives. Consequently, a use case may straddle multiple functional areas and processes. This chapter provides a comprehensive overview of how Autonomous Finance can revolutionize financial operations, management, and decision-making.

Functional areas in finance and the processes within each functional area are shown in Figure 5.1 below.

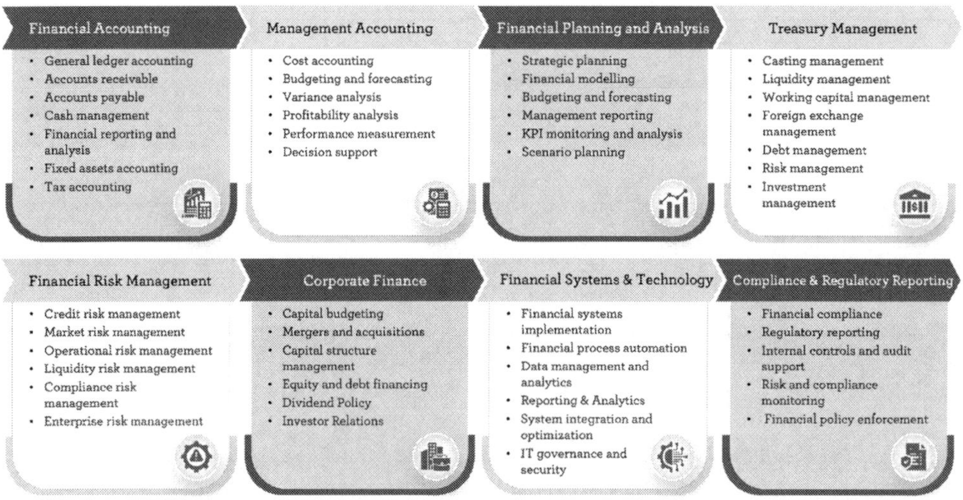

FIGURE 5.1: Functional Areas and the Processes/Functions Within Finance

Now, let us delve into the top 3 use cases of Autonomous Finance for each functional area and process.

Financial Accounting

Financial Accounting, a bulwark of financial integrity and transparency, holds a vital position in sculpting the financial narratives of organizations. With its emphasis on accuracy, timeliness, and adherence to regulations, this domain often grapples with complexities and manual intricacies. Thus, a closer examination of the revolution that Autonomous Finance brings to various processes within Financial Accounting (as illustrated in Figure 5.2) is warranted.

a) **General Ledger (GL) Accounting:** GL Accounting, the central repository for accounting data, grapples with manual entries, reconciliation errors, time-consuming reporting, and lack of real-time insights (What Is a General Ledger and How to Use It, n.d.). With its intelligent automation, the application of Autonomous Finance eradicates human errors, elevates efficiency, and introduces real-time reporting. It transcends the traditional role of GL Accounting, transforming it from an administrative task into a strategic function that offers actionable insights. The fusion of AI-driven analytics and automation underscores the profound impact on accuracy, efficiency, and decision-making.

b) **Accounts Receivable (A/R):** A/R, a critical aspect of liquidity management, faces hurdles in credit risk management, timely collections, and customer relationship nurturing. Autonomous Finance redefines this sphere by leveraging AI-driven predictive analytics and automation (Samaria, n.d.). It enhances efficiency, reduces risk, fosters customer satisfaction, and optimizes cash flow. The transition from traditional methods to futuristic, data-driven management showcases how technology can turn challenges into growth, innovation, and competitive advantage opportunities.

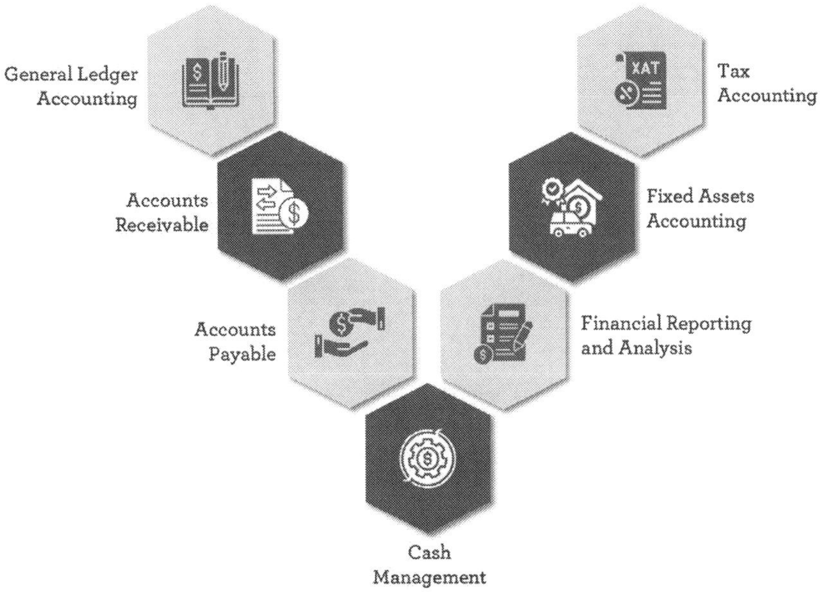

FIGURE 5.2: The Bedrock of Financial Accounting

c) **Accounts Payable (A/P):** A/P, traditionally hampered by manual processes, delayed payments, and vendor dissatisfaction, undergoes a metamorphosis with Autonomous Finance. Sophisticated automation, AI-driven invoice processing, and intelligent payment scheduling transform this function (*Autonomous Accounts Payable Solution | AppZen*, n.d.). The result is not merely a clerical improvement but a complete overhaul of vendor relationships, cash flow management, and strategic alignment. The transformative power of Autonomous Finance reshapes the entire landscape, offering unprecedented efficiency and strategic depth.

d) **Cash Management (CM):** CM, integral to liquidity and investment decision-making, finds profound solutions through Autonomous Finance. Employing predictive analytics, real-time monitoring, and automation assures optimal cash utilization, safeguards against liquidity risks, and aligns investments with strategic goals (HighRadius, 2022). The enhancement in cash visibility, investment returns, and risk mitigation illustrates the vast potential of Autonomous Finance in transforming this crucial function into a strategic asset.

e) **Financial Reporting and Analysis:** Financial Reporting and Analysis, essential for transparency, compliance, and strategic alignment, faces challenges in accuracy, timeliness, and interactivity. Autonomous Finance infuses this space with automation, real-time analysis, and interactive dashboards (Hada, n.d.). It enriches decision-making, aligns reporting with strategic objectives, and fosters a culture of data-driven intelligence. The transformation augments efficiency and opens the doors to a new era of financial intelligence, where insights lead to innovation.

f) **Fixed Assets (FA) Accounting:** FA, responsible for tracking, valuation, depreciation calculation, and compliance, often stumbles on accuracy and real-time insights. Autonomous Finance revolutionizes this process by employing smart algorithms, real-time tracking, and AI-driven valuation models (Global, n.d.). The profound enhancement in accuracy, compliance, strategic asset utilization, and real-time reporting reflects the holistic and transformative influence of Autonomous Finance in this critical domain.

g) **Tax Accounting:** This is a highly complex domain fraught with changing regulations, intricate calculations, and compliance risks, and is seamlessly addressed by Autonomous Finance. Automation in tax computations, real-time regulatory updates, predictive planning, and AI-driven scenario analysis align the process with organizational aspirations and legal mandates (Dorland, 2023). This transformation exemplifies how Autonomous Finance transcends conventional boundaries to redefine financial landscapes, offering agility, compliance, and strategic alignment in an ever-evolving regulatory environment.

This exploration through the multifaceted landscape of Financial Accounting in the private sector is a testament to the transformative power of Autonomous Finance. Its principles and building blocks align with the unique challenges, crafting solutions that are more than mere technological advancements; they reimagine finance. The journey continues in Appendix A, where Table 5.1 provides a comprehensive view of this fascinating world.

Management Accounting

Management Accounting, a strategic partner in organizational decision-making, interfaces with several critical processes. This domain's complexity lies in the need for precise analysis, proactive planning, and real-time insights. With its AI-driven tools and intelligent automation, Autonomous Finance brings forth transformative solutions for each sub-process area (as presented in Figure 5.3 below) within the Management Accounting process.

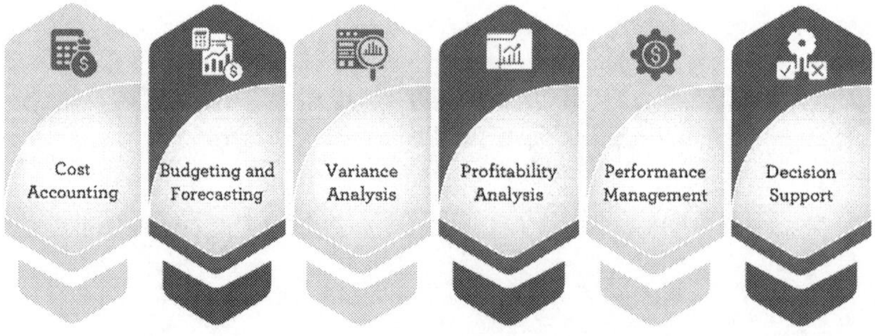

FIGURE 5.3: Navigating Managerial Accounting

a) **Cost Accounting,** essential for product pricing, profitability analysis, and cost control, often grapples with data accuracy, timeliness, and complexity. Autonomous Finance introduces precision and real-time insights through AI-powered cost models, automation in data collection, and predictive cost behavior analysis (Ceylan, 2023). The transformation extends beyond efficiency to strategic alignment, enhanced profitability, and agile response to market dynamics.

b) **Budgeting and Forecasting,** crucial for financial planning, face challenges in aligning with organizational goals, adapting to market changes, and ensuring stakeholder collaboration. Autonomous Finance revolutionizes this process by leveraging predictive analytics, real-time data integration, and intelligent scenario planning (Macoveiciuc, 2023). The alignment of budgets with strategic objectives, the agility to adapt, and the collaborative approach redefine planning and ensure that organizations are future-ready.

c) **Variance Analysis,** vital for performance management, struggles with timely detection, root cause analysis, and corrective action planning. Autonomous Finance introduces automated real-time monitoring, intelligent variance detection, and AI-driven root cause analysis (Acca, n.d.). The transformation transcends traditional analysis, offering actionable insights, enabling swift corrective actions, and aligning performance with strategic goals.

d) **Profitability Analysis,** central to growth and sustainability, often faces hurdles in data granularity, product/customer segmentation, and predictive insights. Autonomous Finance elevates this process by employing sophisticated algorithms, real-time data, and predictive modeling (Macoveiciuc, 2023). The profound insights into profitability drivers, alignment with organizational strategies, and predictive intelligence set the stage for informed decision-making and sustainable growth.

e) **Performance Measurement,** essential for organizational alignment and growth, encounters challenges in real-time tracking, alignment with strategic goals, and actionable insights. Autonomous Finance infuses this sphere with real-time dashboards, AI-driven performance models, and intelligent alert systems (*CFOs Will Move to Autonomous Finance Processes, Says Gartner*, n.d.). The transformation ensures that performance measurement is not a mere evaluative exercise but a strategic tool that fosters alignment, agility, and excellence.

f) **Decision Support,** a linchpin in strategic management, often stumbles on data overload, lack of real-time insights, and predictive intelligence. Autonomous Finance transcends these barriers through AI-driven decision models, real-time data integration, and intelligent scenario planning (HighRadius, 2022a). This metamorphosis turns decision support from a data-driven exercise into a strategic asset, where insights lead to innovation, alignment, and growth.

This exploration through the diverse spectrum of Management Accounting showcases the profound influence of Autonomous Finance. Its alignment with the inherent challenges and innovative solutions transcends traditional practices, symbolizing a new era of insightful and strategic management accounting. Table 5.2 in Appendix A provides further details that enrich this fascinating and comprehensive discourse.

Financial Planning & Analysis

FP&A, the strategic compass guiding an organization's financial journey, embodies the art and science of financial decision-making. This intricate and vital domain faces challenges in predictive accuracy, alignment with strategic objectives, agility in response to market shifts, and synthesis of multifaceted data. With its fusion of AI, data analytics, and automation, Autonomous Finance can breathe new life into each of the sub-process area within the FP&A process (as displayed in Figure 5.4).

FIGURE 5.4: The Financial Planning & Analysis Odyssey

a) **Strategic Planning** harmonizes long-term visions with actionable financial roadmaps. Challenges in this area include aligning financial strategies with organizational objectives, ensuring adaptability to market dynamics, and maintaining a forward-looking approach. Autonomous Finance intervenes with AI-driven scenario analysis, real-time strategic alignment, and adaptive roadmaps, transforming static planning into a dynamic, resilient, and future-oriented strategy (*Autonomous Finance - Everything a CFO Should Know | Gartner*, 2022).

b) **Financial Modeling** crafts the financial foresight necessary for decision-making. The complexities lie in predictive accuracy, adaptability to change, and alignment with business goals. Autonomous Finance revolutionizes this sphere by employing intelligent modeling algorithms, integrating real-time data, and offering adaptive scenario planning (Pro Business Plans, 2023). This transformation bestows models with unprecedented accuracy, strategic alignment, and market responsiveness.

c) **Budgeting and Forecasting** are central to financial control but often face challenges in real-time accuracy, strategic alignment, and adaptability. Autonomous Finance addresses these challenges through AI-driven forecasting, automated budget alignment, and adaptive controls (*A Path to Automated Financial Forecasting*, n.d.). The transformation transcends budgeting and forecasting from static exercises to dynamic, strategic functions, enhancing financial control and aligning with organizational aspirations.

d) **Management Reporting** communicates financial insights and faces challenges in real-time delivery, insightful analytics, and customization to management needs.

Autonomous Finance introduces intelligent reporting algorithms, real-time analytics, and adaptable reporting formats (PricewaterhouseCoopers, n.d.). This introduction ensures that financial reporting is timely and rich in insights, tailored to decision-making needs, and aligned with management's strategic perspectives..

e) **KPI Monitoring and Analysis** gauge organizational performance, grappling with real-time monitoring, insightful analysis, and alignment with strategic goals. Autonomous Finance revitalizes this domain with AI-driven KPI analysis, real-time performance tracking, and strategic alignment checks (*Finance Dashboards Need a Revamp to Drive Performance*, n.d.). This transformation ensures that KPIs are not mere numerical indicators but strategic beacons, guiding the organization toward success and growth.

f) **Scenario Planning** readies an organization for the future but faces complexities in predictive accuracy, adaptability, and alignment with strategic visions. Autonomous Finance leverages intelligent scenario algorithms, real-time market data, and adaptive planning strategies to elevate this vital domain (Tyson, 2022). The alignment with strategic objectives, predictive accuracy, and adaptability turn scenario planning into an essential tool for future readiness and strategic foresight.

Navigating the intricate pathways of FP&A reveals how Autonomous Finance catalyzes future-oriented financial wisdom. Weaving cutting-edge technologies with a deep understanding of financial complexities shapes a new FP&A paradigm. Delve deeper into this captivating realm in Appendix A, with Table 5.3 illuminating the complete spectrum.

Treasury Management

The custodian of an organization's financial vitality, Treasury Management, is entrusted with safeguarding liquidity, optimizing capital, and managing financial risks. This critical realm, however, is fraught with complexities and demands agility, precision, and strategic foresight. With its confluence of automation, AI, and data analytics, Autonomous Finance emerges as a transformative ally for each sub-process area within the Treasury Management process (as exhibited in Figure 5.5).

a) **Cash Management,** the lifeblood of financial operations, often wrestles with real-time visibility, optimization, and control. Autonomous Finance introduces AI-driven cash forecasting, automated reconciliation, and intelligent cash positioning (*Autonomous Finance - Everything a CFO should know | Gartner*, 2022). The transformation extends beyond efficiency to strategic liquidity management, enhancing financial stability and enabling agile responses to market dynamics.

b) **Liquidity Management,** crucial for sustaining operations, faces challenges in alignment with market conditions, regulatory compliance, and predictive planning. Autonomous Finance revitalizes this domain by leveraging real-time data, predictive analytics, and intelligent scenario planning (HighRadius, 2023). The alignment with market trends, adherence to regulations, and adaptation agility set the stage for resilient and responsive liquidity management.

c) **Working Capital Management,** vital for operational efficiency, grapples with a fine balance between liquidity and profitability. Autonomous Finance introduces intelligent algorithms, real-time analytics, and adaptive strategies (HighRadius, 2022b). The transformation ensures optimal working capital levels, aligning with business cycles, enhancing profitability, and fostering financial health.

d) **Foreign Exchange Management,** central to global operations, often faces volatility, compliance challenges, and risk mitigation complexities. Autonomous Finance employs sophisticated risk models, real-time market data, and intelligent hedging strategies (Council, 2022). The profound insights, adaptive policies, and risk-aligned decisions redefine forex management, ensuring alignment with global strategies and risk appetite.

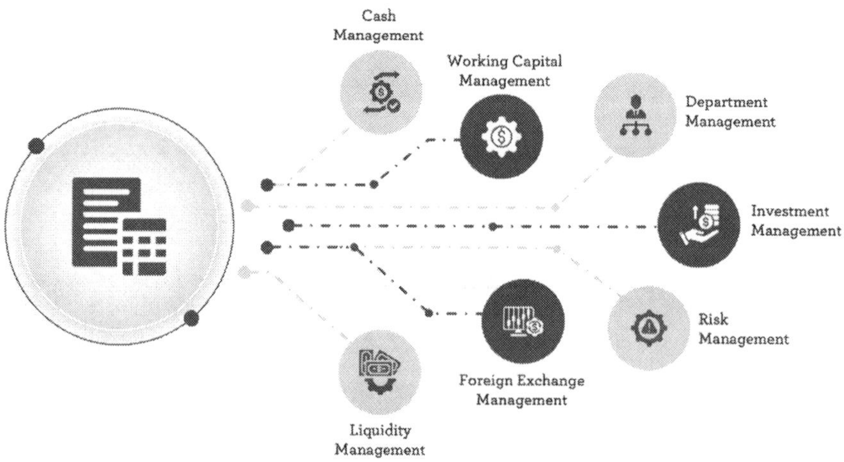

FIGURE 5.5: Treasury Management Tapestry

e) **Debt Management,** essential for financial sustainability, encounters challenges in alignment with organizational goals, risk management, and cost optimization. Autonomous Finance infuses this sphere with AI-driven debt structuring, real-time monitoring, and intelligent refinancing strategies (Strange & Strange, 2023). The transformation ensures that debt management aligns with strategic goals, optimizes costs, and ensures financial resilience.

f) **Risk Management,** a linchpin in treasury functions, often stumbles on predictive insights, real-time monitoring, and adaptive risk mitigation. Autonomous Finance transcends these barriers through intelligent risk models, real-time alerts, and adaptive risk policies (Basrai, 2021). This metamorphosis turns risk management from a compliance-driven exercise into a strategic function, ensuring financial stability and growth.

g) **Investment Management,** the steward of capital growth, grapples with alignment with organizational goals, risk-adjusted returns, and regulatory adherence. Autonomous Finance, employing AI-driven investment strategies, real-time market insights, and intelligent compliance checks, elevates this domain (*Artificial Intelligence*, n.d.). The alignment with organizational strategies, risk-balanced growth, and regulatory compliance set the stage for sustainable capital growth.

The journey through Treasury Management opens up a world where Autonomous Finance merges with financial stewardship, creating a synergy that enhances efficiency, strategy, and adaptability. This union paves the way for a treasury function that is not merely operational but visionary. Embark on a further exploration of these intricate dynamics in Appendix A, where Table 5.4 provides a thorough insight.

Financial Risk Management

Financial Risk Management (FRM) is the bulwark safeguarding an organization's financial well-being. It is a multifaceted domain, characterized by its intricate challenges and the transformative potential of Autonomous Finance in each of its sub-process areas (as revealed in Figure 5.6):

FIGURE 5.6: The Fortress of FRM

a) **Credit Risk Management** assesses and mitigates the risk of default on credit obligations. Challenges include real-time credit risk assessment, predictive modeling of default probabilities, and dynamic credit risk controls. Autonomous Finance transforms this domain by integrating AI-driven credit risk algorithms, real-time risk monitoring, and adaptive risk mitigation strategies (Joseph, 2023). This paradigm shift enhances credit risk control, predictive accuracy, and strategic alignment with lending objectives.

b) **Market Risk Management** controls the potential losses arising from market dynamics. The complexities lie in real-time market risk monitoring, insightful market trend analysis, and adaptable risk controls. Autonomous Finance can introduce intelligent market risk algorithms, real-time market data integration, and adaptive risk management strategies, redefining market risk management as an insightful, real-time, and strategic function.

c) **Operational Risk Management** deals with the risks stemming from internal processes and systems. Challenges in this area encompass real-time operational risk monitoring, insightful risk analysis, and strategic risk controls. Autonomous Finance revolutionizes operational risk management through AI-driven risk analytics, real-time risk monitoring, and strategic risk alignment. The transformation bestows operational risk management with unprecedented efficiency, insight, and strategic relevance.

d) **Liquidity Risk Management** assures the ability to meet short-term obligations, grappling with real-time liquidity monitoring, predictive liquidity planning, and adaptive liquidity controls. Autonomous Finance redefines this sphere by introducing intelligent liquidity algorithms, real-time liquidity tracking, and adaptive liquidity strategies, ensuring liquidity risk management is not just a function but a strategic enabler of financial stability. (Refer to Table 4 in Appendix A for detailed use cases.)

e) **Compliance Risk Management** ensures adherence to legal and regulatory requirements, facing challenges in real-time compliance monitoring, insightful compliance analysis, and adaptable compliance controls. Autonomous Finance transforms compliance risk management using AI-driven compliance algorithms, real-time compliance checks, and adaptive compliance strategies. This metamorphosis elevates compliance from mere adherence to a strategic alignment with regulatory landscapes.

f) **Enterprise Risk Management** orchestrates the overall risk profile, dealing with comprehensive risk assessment, strategic risk alignment, and enterprise-wide risk controls. Autonomous Finance revitalizes enterprise risk management with intelligent algorithms, real-time risk analytics, and strategic risk alignment. This transformation ensures that enterprise risk management transcends from a functional necessity to a strategic organizational resilience and growth orchestrator.

Unraveling the world of FRM, this examination unveils how Autonomous Finance's innovative approach mitigates risks and crafts a robust financial shield. It is a transformation that goes beyond risk aversion to strategic risk intelligence. Continue this intriguing exploration in Appendix A, where Table 5.5 offers a detailed perspective.

Corporate Finance

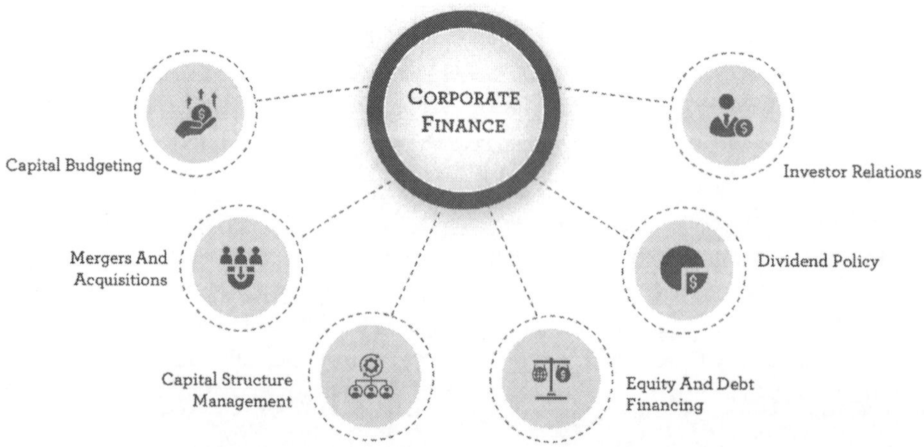

FIGURE 5.7: The Centers of Excellence in Corporate Finance

Corporate Finance is the bedrock of an organization's financial strategy. It orchestrates allocating financial resources, investment decisions, and capital structure, all vital to an organization's growth and sustainability. Let's delve into the intricate processes within Corporate Finance (as rendered in Figure 5.7) and witness the transformative influence of Autonomous Finance.

a) **Capital Budgeting** is the strategic capital allocation for long-term investments. Challenges in this area include precise investment valuation, risk assessment, and alignment with strategic objectives. Autonomous Finance introduces AI-driven investment analytics, real-time risk assessment, and adaptive investment strategies, ensuring that capital budgeting is not merely a financial function but a strategic enabler of growth (HighRadius, 2022a).

b) **Mergers and Acquisitions** (M&A) orchestrate the consolidation or acquisition of entities. Complexities in M&A encompass accurate valuation, risk management, and strategic alignment. Autonomous Finance revolutionizes M&A through intelligent valuation algorithms, real-time risk analytics, and strategic alignment tools (Lauracatalan, 2021). The transformation renders M&A a precise, insightful, and strategically aligned process.

c) **Capital Structure Management** governs the blend of debt and equity financing. Challenges include optimal capital structure modeling, risk management, and alignment with financial objectives. Autonomous Finance redefines capital structure management through AI-driven optimization, real-time risk monitoring, and adaptive financial strategies. This metamorphosis ensures capital structure management transcends from a mathematical exercise to a strategic lever of financial stability.

d) **Equity and Debt Financing** deals with raising capital through equity or debt. Complexities lie in real-time market analysis, risk management, and strategic financial alignment. Autonomous Finance introduces intelligent financing algorithms, real-time market analytics, and strategic financing tools, transforming equity and debt financing into an insightful, real-time, and strategic function.

e) **Dividend Policy** governs the distribution of profits to shareholders. Challenges encompass dividend modeling, risk assessment, and alignment with shareholder expectations. Autonomous Finance revolutionizes dividend policy through intelligent dividend algorithms, real-time risk assessment, and adaptive shareholder strategies, ensuring that dividend policy is a strategic manifestation of organizational values and shareholder alignment.

f) **Investor Relations** is the nexus between the organization and its investors. Complexities in this sphere include real-time communication, insightful investor analytics, and strategic investor alignment. Autonomous Finance transforms investor relations through AI-driven investor analytics, real-time communication platforms, and strategic investor engagement tools. This transformation elevates investor relations from mere communication to strategic partnership and alignment.

Exploring Corporate Finance through the lens of Autonomous Finance reveals a landscape where strategic financial decisions are guided by intelligence, innovation, and insight. It is an alignment that transcends conventional corporate finance, ushering in a new era of strategic financial leadership. Further insights into this transformative world can be found in Appendix A, where Table 5.6 provides a comprehensive overview.

Financial Systems and Technology

In the digitized world, Financial Systems and Technology form the backbone of modern financial operations. This domain is imbued with technological innovation and automation, where the application of Autonomous Finance is both profound and transformative. Let us navigate through the complex pathways of this domain (as manifested in Figure 5.8 below).

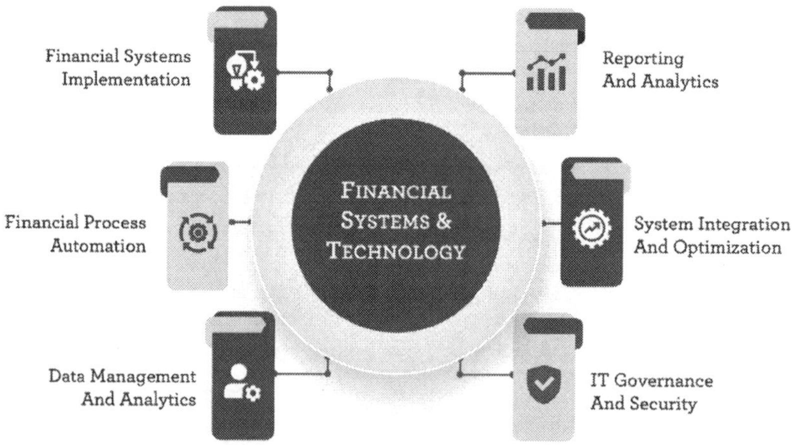

FIGURE 5.8: Financial Systems and Technology Landscape

a) **Financial Systems Implementation** involves deploying financial technologies that support various financial functions. Challenges in this area include system compatibility, cost optimization, and alignment with financial objectives. Autonomous Finance introduces AI-driven implementation strategies, adaptive technology alignment, and cost-effective deployment solutions (*Finance Transformation: Roadmap, Strategy, Trends & Framework | Gartner*, n.d.). These advancements ensure that financial systems implementation transcends mere technology deployment and becomes a strategic enabler. (Refer to Table 6 in Appendix A for detailed use cases.)

b) **Financial Process Automation** focuses on automating repetitive and time-consuming financial tasks. The complexity lies in understanding process intricacies, technology integration, and ensuring compliance. Autonomous Finance revolutionizes this through intelligent automation algorithms, seamless technology integration, and adaptive compliance monitoring. The result is a transformative approach that streamlines processes and elevates them to strategic functionality.

c) **Data Management and Analytics** provide insights into financial data. Challenges encompass data quality, insightful analytics, and strategic alignment. Autonomous Finance reshapes data management through AI-driven data quality tools, intelligent analytics algorithms, and strategic data alignment (Hagerup, 2023). This transformation ensures that data management transcends mere analysis and becomes an insightful guide to strategic decision-making.

d) **Reporting and analytics** enable organizations to understand and communicate their financial performance. Complexities include real-time reporting, insightful analytics, and stakeholder alignment. Autonomous Finance introduces real-time reporting platforms, intelligent analytics tools, and adaptive stakeholder communication strategies. These innovations elevate reporting and analytics from information dissemination to strategic communication and alignment.

e) **System Integration and Optimization** focus on harmonizing various financial systems. Challenges lie in system compatibility, performance optimization, and alignment with financial strategy. Autonomous Finance revolutionizes integration through AI-driven compatibility tools, adaptive performance optimization, and strategic system alignment (Martinis, 2023). This transformation ensures that system integration transcends mere technological alignment and becomes a strategic enabler of financial efficiency.

f) **IT Governance and Security** oversee the policies, procedures, and controls related to financial systems. Complexities include adaptive governance, real-time security monitoring, and alignment with regulatory requirements. Autonomous Finance redefines governance through intelligent governance algorithms, real-time security platforms, and adaptive regulatory alignment (HighRadius, 2022c). This metamorphosis ensures that governance and security transcend mere control mechanisms and become strategic protectors of financial integrity.

The tour of Financial Systems and Technology under the influence of Autonomous Finance uncovers a fusion of technological brilliance with financial mastery. It is a synergy that redefines the very fabric of financial systems, making them more adaptive, intelligent, and strategic. Explore this fascinating world in greater depth in Appendix A, where Table 5.7 provides a detailed panorama.

Compliance and Regulatory Reporting

In the labyrinth of modern financial systems, Compliance and Regulatory Reporting are the guardians of integrity, transparency, and accountability. Although rigorous and structured, this domain lacks complexities and challenges. The introduction of Autonomous Finance ushers in a new era of efficiency, accuracy, and innovation in this vital aspect of the financial world. In this context, it's worth exploring the transformative potential of Autonomous Finance, particularly in the realm of Compliance and Regulatory Reporting (as depicted in Figure 5.9 below).

a) **Financial Compliance** ensures adherence to financial laws, regulations, and standards. The challenges include maintaining up-to-date knowledge, ensuring seamless alignment, and responding to regulatory changes. Autonomous Finance addresses these through AI-driven compliance databases, adaptive alignment algorithms, and real-time response mechanisms (Danwillis, 2023). The transformation transcends mere adherence, fostering a culture of proactive compliance that resonates with organizational integrity.

b) **Regulatory Reporting** involves the timely submission of required financial documents to governmental bodies. The intricacies lie in understanding reporting requirements, maintaining accuracy, and ensuring timeliness. Autonomous Finance revolutionizes this with intelligent reporting algorithms, automated accuracy checks, and adaptive scheduling tools (*The Case for Placing AI at the Heart of Digitally Robust Financial Regulation | Brookings*, 2022). This metamorphosis ensures that regulatory reporting transcends mere statutory obligation, becoming a symbol of transparency and responsibility.

c) **Internal Controls and Audit Support** form the framework for financial oversight within an organization. Challenges encompass designing effective controls, ensuring seamless audits, and aligning with organizational objectives. Autonomous Finance redefines this through AI-driven control design, intelligent audit algorithms, and adaptive alignment strategies (Dokka, n.d.). The transformation ensures that internal controls and audits transcend mere oversight and become strategic enablers of financial excellence.

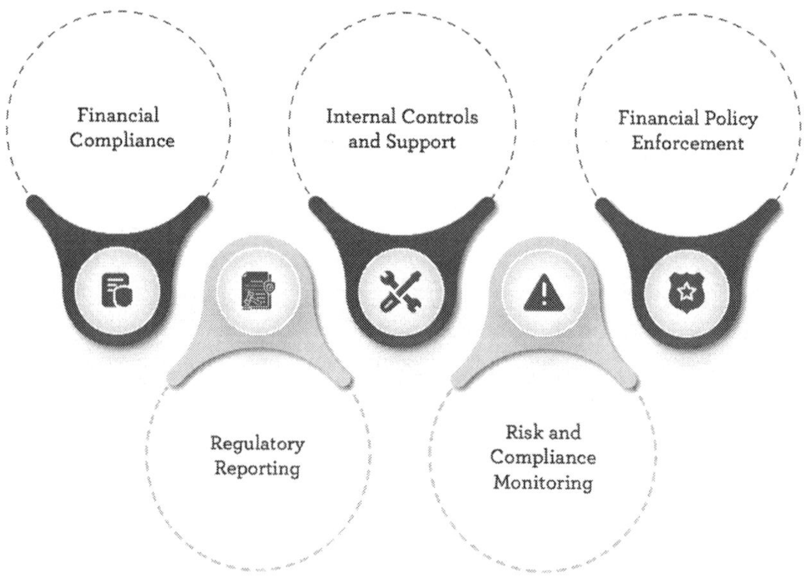

FIGURE 5.9: The Key Cornerstones of Compliance and Regulatory Reporting

d) **Risk and Compliance Monitoring** ensures continuous vigilance over financial risks and compliance adherence. Complexities include real-time monitoring, insightful risk assessment, and adaptive compliance strategies. Autonomous Finance reshapes this with real-time monitoring platforms, intelligent risk assessment tools, and adaptive compliance methodologies (Zhykhareva, n.d.). This reshaping ensures that risk and compliance monitoring transcend mere vigilance, becoming strategic guardians of financial stability.

e) **Financial Policy Enforcement** ensures that financial policies are followed within an organization. Challenges lie in understanding policy intricacies, ensuring alignment, and fostering a culture of adherence. Autonomous Finance transforms this through intelligent policy databases, alignment algorithms, and cultural adaptation tools (*The Case for Placing AI at the Heart of Digitally Robust Financial Regulation | Brookings*,

2022). This transformation ensures that financial policy enforcement transcends mere rule-following and becomes a strategic instrument of financial integrity.

The exploration of Compliance and Regulatory Reporting unravels a realm marked by integrity, transparency and the transformative influence of Autonomous Finance. It unveils a future where compliance and reporting are not mere statutory obligations but strategic manifestations of financial responsibility and excellence. The intertwining of compliance, reporting, and Autonomous Finance is an eloquent testament to the ever-evolving narrative of modern financial management. A more detailed examination is available in Appendix A, where Table 5.8 offers a complete view.

5.3 Top Use Cases in the Public Sector

The use cases described above for Autonomous Finance apply broadly to companies operating in the private sector. Although the underlying principles and potential applications of Autonomous Finance remain consistent, the contexts and specific use cases in the public sector can differ due to distinct operational, regulatory, and societal factors. This section briefly outlines some of these differences.

a) **Nature of Services:** Unlike private sector entities primarily focusing on profitability and shareholder value, public sector entities are driven by the need to deliver public goods and services efficiently and equitably. Thus, Autonomous Finance use cases in the public sector often revolve around improving public service delivery, resource allocation, and compliance with complex regulatory frameworks.

b) **Transparency and Accountability:** Public sector entities are subject to more rigorous standards of transparency and accountability. Autonomous Finance solutions can help streamline and automate compliance, reporting, and auditing processes, making it easier to demonstrate accountability and adhere to transparency norms.

c) **Funding Mechanisms:** Public sector entities have unique funding mechanisms, including taxes, grants, and budgetary allocations. Therefore, Autonomous Finance use cases in the public sector often involve budgeting, forecasting, and financial planning in a politically influenced and often uncertain fiscal environment.

d) **Regulatory Constraints:** Public sector entities often operate under unique and complex regulatory constraints that dictate how they can use and manage funds. Autonomous Finance can automate compliance and ensure that all financial decisions and actions adhere to these regulations.

e) **Scale and Complexity:** Many public sector entities operate on a much larger scale than private companies, handling enormous budgets and serving large, diverse populations. The scale and complexity of public sector operations provide ample opportunities for Autonomous Finance to optimize processes, enhance efficiency, and generate substantial cost savings.

These points merely scratch the surface of how Autonomous Finance can revolutionize financial operations and management in the public sector. Just like in the private sector, the actual 'top' use cases will depend on the specific circumstances of each public sector entity. Further research is encouraged to delve deeper into the unique applications and implications of Autonomous Finance in the public sector.

The following section examines the future of Autonomous Finance use cases regardless of the industry sector.

5.4 The Future of Autonomous Finance Use Cases

The canvas of Autonomous Finance, characterized by a profound interplay of algorithms, data, and financial acumen, resembles an artwork in progress. This fluid, dynamic, and exciting landscape demonstrates that today's defined use cases merely represent the beginning, the initial strokes on a canvas that is destined to fill with increasing color, complexity, and sophistication over time. In this context, the future of Autonomous Finance use cases should be explored, as explored in Figure 5.10 below.

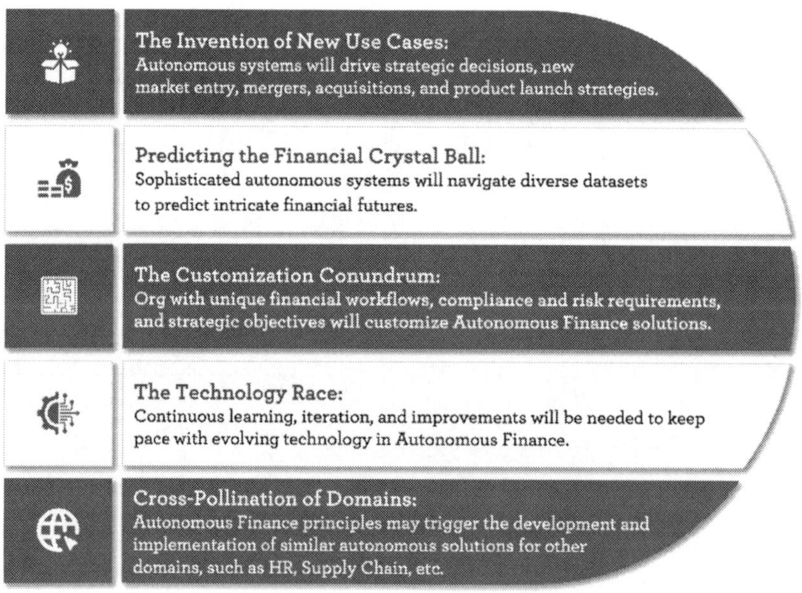

FIGURE 5.10: The Future of Autonomous Finance Use Cases

a) **Invention of New Use Cases:** Imagine Autonomous systems matured enough to autonomously steer the strategic decisions of organizations - gauging the viability of new market entry, evaluating potential M&As, or even chalking out a new product launch strategy. The propulsion in AI and ML capabilities unlocks a horizon brimming with potential, where newer, more intricate, and high-impact use cases of Autonomous Finance can be envisioned and realized.

b) **Predicting the Financial Crystal Ball:** As the narrative of Autonomous Finance evolves, so does the intricacy of predicting financial futures. Harnessing the sheer diversity and volume of data points and factoring in many possible outcomes will demand advanced sophistication from autonomous systems. This complexity represents a significant challenge yet also a stimulating test of the capabilities of future autonomous solutions.

c) **The Customization Conundrum:** With the increasing embracement of Autonomous Finance, organizations will crave solutions that don their unique colors - bespoke applications that address their distinctive financial workflows, compliance needs, risk thresholds, and strategic objectives. Crafting these tailored solutions represents a complex but enriching opportunity in Autonomous Finance.

d) **The Technology Race:** Staying abreast of the whirlwind pace of technology evolution presents an ongoing challenge. However, this dynamism serves as an impetus for continuous learning, iteration, and improvement, ensuring that the narrative of Autonomous Finance remains vibrant and engaging.

e) **Cross-Pollination of Processes:** The principles and promise of Autonomous Finance are not confined within the borders of finance alone. Such principles can revolutionize finance and adjacent landscapes like supply chain management, HR, and operations. The opportunity for such cross-disciplinary application presents itself as a riveting avenue for seeding innovation and efficiency on a broader scale.

The journey through the dynamic landscape of Autonomous Finance reveals that the exploration is just as thrilling as reaching the destination itself. This incessant evolution of Autonomous Finance use cases makes it critical to periodically take stock, evaluate our progress, and deliberate on the course ahead. In the following Conclusion section, we will draw together our discussion threads on the key use cases of Autonomous Finance, reflect on the ground we have covered, and look ahead to the promising horizon of possibilities.

5.5 Conclusion

This chapter has been a captivating journey through the multifaceted landscape of finance, showcasing how Autonomous Finance is not merely an abstract idea but a transformative force with real-world applications.

From Financial Accounting to Compliance and Regulatory Reporting, readers have witnessed the diverse and profound ways Autonomous Finance can redefine financial operations. This exploration has been a testament to the power of automation and AI to transcend traditional barriers, offering solutions as innovative as they are practical.

Nevertheless, what lies beneath these transformative capabilities? How does Autonomous Finance attain the intelligence, precision, adaptability, and efficiency that sets it apart? The answer to these questions lies not just in understanding the "what" and "why" but in delving into the "how."

As this chapter concludes, a new horizon beckons, inviting readers to venture inside the engine that powers Autonomous Finance. The next chapter, *"Inside the Engine - The Technologies that Power Autonomous Finance,"* will peel back the layers, unveiling the technological marvels that make Autonomous Finance possible. It will explore the algorithms, ML models, data analytics, and cutting-edge technologies that translate visionary ideas into tangible realities.

The tables in Appendix A extend the narrative of this chapter, offering a detailed exploration of specific use cases, further enriching the understanding of how Autonomous Finance is shaping the present and future of finance.

The journey thus far has been both enlightening and inspiring, offering a glimpse into a future where finance is not confined to numbers but emerges as a strategic partner in organizational success. As readers turn the page, they are invited to continue this exploration, discovering the technological marvels that turn the dream of Autonomous Finance into an unfolding reality.

5.6 Key Recommendations

a) **Explore the Potential of Autonomous Finance:** Businesses should proactively explore how Autonomous Finance can be implemented or enhanced within their financial operations.

b) **Tailor Autonomous Finance Solutions to Specific Needs:** Private sector organizations must adapt Autonomous Finance solutions to their specific financial workflows, compliance needs, risk thresholds, and strategic objectives.

c) **Leverage Autonomous Finance in the Public Sector:** Public sector entities should recognize the potential benefits of Autonomous Finance and seek ways to incorporate it within their operations while considering their unique process challenges and requirements.

d) **Stay Updated with the Evolving Landscape:** Organizations should continuously learn, iterate, and improve their Autonomous Finance solutions in line with technological advancements and emerging use cases.

e) **Consider Cross-Process Potential:** Entities, whether private or public, should explore the cross-process potential of Autonomous Finance principles to foster innovation and efficiency on a broader scale.

5.7 References

A path to automated financial forecasting. (n.d.). Deloitte United States. https://www2.deloitte.com/us/en/pages/operations/articles/automated-financial-forecasting.html.

Acca, P. J. F. (n.d.). How AI & ML can streamline variance analysis? *www.linkedin.com*. https://www.linkedin.com/pulse/how-ai-ml-can-streamline-variance-analysis-pankaj/.

Artificial intelligence. (n.d.). Deloitte United States. https://www2.deloitte.com/us/en/pages/financial-services/articles/ai-next-frontier-in-investment-management.html

Autonomous Accounts Payable Solution | AppZen. (n.d.). https://www.appzen.com/autonomous-accounts-payable-solution.

Autonomous Finance – Everything a CFO should know | Gartner. (2022). Gartner. https://www.gartner.com/en/finance/topics/autonomous-finance.

Basrai, A. (2021, September 23). Artificial intelligence in risk management. KPMG. https://kpmg.com/ae/en/home/insights/2021/09/artificial-intelligence-in-risk-management.html

Ceylan, B. (2023). *Autonomous Finance in 2023: Key Points & Benefits*. AIMultiple. https://research.aimultiple.com/autonomous-finance/.

CFOs will move to Autonomous Finance processes, says Gartner. (n.d.). Celonis. https://www.celonis.com/blog/cfos-will-move-to-autonomous-finance-processes-says-gartner/.

Chatterton, T., & Newmarch, G. (2017). The future is already here. *Interactions, 24*(2), 42–45. https://doi.org/10.1145/3041215.

Council, C. (2022). Autonomous Finance defined and how CFOs can achieve it. *Controllers Council*. https://controllerscouncil.org/autonomous-finance-defined-and-how-cfos-can-achieve-it/

Danwillis. (2023). How is financial services leveraging AI for compliance purposes? *FinTech Global*. https://fintech.global/2023/03/30/how-is-financial-services-leveraging-ai-for-compliance-purposes/

Dokka. (n.d.). How will autonomous accounting change the work of financial controllers? *www.linkedin.com*. https://www.linkedin.com/pulse/how-autonomous-accounting-change-work-financial-controllers/

Dorland, A. (2023, August 17). *Staying up to speed with artificial intelligence in accounting*. Tax & Accounting Blog Posts by Thomson Reuters. https://tax.thomsonreuters.com/blog/staying-up-to-speed-with-artificial-intelligence-in-accounting/.

Finance dashboards need a revamp to drive performance. (n.d.). Gartner. https://www.gartner.com/smarterwithgartner/tune-your-finance-dashboard-using-metrics-to-drive-performance.

Finance Transformation: Roadmap, Strategy, Trends & Framework | Gartner. (n.d.). Gartner. https://www.gartner.com/en/finance/topics/finance-transformation

Global, A. (n.d.). Unlocking the potential of AI in fixed asset Management: benefits, challenges, and cost considerations. www.linkedin.com. https://www.linkedin.com/pulse/unlocking-potential-ai-fixed-asset-management-/.

Hada, P. S. (n.d.). Autonomous Finance - How Artificially Intelligent Systems are Disrupting the Industry. *www.linkedin.com*. https://www.linkedin.com/pulse/autonomous-finance-how-artificially-intelligent-systems-hada/.

Hagerup, A. (2023, July 10). Can A Finance Organization Become Truly Autonomous? *Forbes*. https://www.forbes.com/sites/forbestechcouncil/2023/07/10/can-a-finance-organization-become-truly-autonomous/

HighRadius. (2022a). Is Your Business Ready for Autonomous Finance? Why Can't CFOs Afford to Wait? *HighRadius Resource Center*. https://www.highradius.com/resources/Blog/what-is-autonomous-finance/.

HighRadius. (2022b). How AI helps CFOs Optimize working Capital Management. *HighRadius Resource Center*. https://www.highradius.com/resources/Blog/optimize-working-capital-management/.

HighRadius. (2022c). Marching towards an Autonomous Finance function – 201. HighRadius Resource Center. https://www.highradius.com/resources/Blog/marching-towards-an-autonomous-finance-function-part-2/

HighRadius. (2023). What is Cash and Liquidity Management? *HighRadius Resource Center*. https://www.highradius.com/resources/Blog/cash-and-liquidity-management-with-importance-and-types/.

Joseph, G. (2023). AI-powered Credit Risk Management software. Emagia™. https://www.emagia.com/products/credit-risk-management/

Lauracatalan. (2021, March 25). *Artificial Intelligence to enhance M&A - ONEtoONE Corporate Finance*. ONEtoONE Corporate Finance. https://www.onetoonecf.com/embracing-artificial-intelligence-to-enhance-mna/

Macoveiciuc, A. (2023). What is Autonomous Finance? A comprehensive guide. *Yokoy - the AI-powered Spend Management Suite*. https://yokoy.io/blog/what-is-autonomous-finance-a-comprehensive-guide/.

Martinis, P. (2023). 7 Trends that will catapult AP into a fully autonomous future. *DOKKA*. https://dokka.com/future-of-autonomous-accounts-payable/

PricewaterhouseCoopers. (n.d.). *Intelligent automation in financial reporting*. PwC. https://www.pwc.com/sg/en/services/risk/intelligent-automation-in-financial-reporting.html.

Pro Business Plans. (2023, June 15). Unlocking the potential of AI for financial modeling in 2023. *Medium*. https://pro-business-plans.medium.com/unlocking-the-potential-of-ai-for-financial-modeling-in-2023-96669e9c5e95.

Samaria, P. (n.d.). *Autonomous Finance 101: Don't waste any more time*. https://learn.g2.com/autonomous-finance.

Strange, A., & Strange, A. (2023). The 'Google Maps for money.' *Andreessen Horowitz*. https://a16z.com/the-google-maps-for-money/

The case for placing AI at the heart of digitally robust financial regulation | Brookings. (2022). *Brookings*. https://www.brookings.edu/articles/the-case-for-placing-ai-at-the-heart-of-digitally-robust-financial-regulation/

Tyson, J. (2022, June 7). *CFOs should use AI to navigate "dark horizons" in scenario planning: Gartner*. CFO Dive. https://www.cfodive.com/news/cfos-should-use-ai-navigate-dark-horizons-scenario-planning-gartner/625069/.

What Is A General Ledger And How To Use It. (n.d.). https://www.zeni.ai/blog/what-is-a-general-ledger.

Zhykhareva, A. (n.d.). Leveraging AI in financial monitoring. *www.linkedin.com*. https://www.linkedin.com/pulse/leveraging-ai-financial-monitoring-alexandra-zhykhareva/

5.8 Appendix A: Autonomous Finance Use Cases

This section provides three use cases of Autonomous Finance per sub-process area within the following finance process areas:

- Finance Accounting
- Management Accounting
- Financial, Planning, & Analytics
- Treasury Management
- FRM
- Corporate Finance
- Financial Systems and Technology
- Compliance and Regulatory Reporting

TABLE 5.1: Autonomous Financial Accounting Use Cases

NO.	USE CASE	DESCRIPTION	KEY FEATURES	BUSINESS IMPACT	SUCCESS METRICS
GL ACCOUNTING					
1.	Autonomous Journal Entry:	Automate the creation, review, and posting of journal entries based on pre-defined rules, exceptions, and ML models.	Rule-based automation, exception handling, and ML models for anomaly detection.	Reduces manual effort and potential for error, ensures timely posting of transactions, and enhances financial data accuracy.	Number of journal entries processed autonomously, reduction in manual effort, accuracy rate.
2.	Autonomous Reconciliation	Automate the reconciliation of GL accounts, highlighting discrepancies for review and preparing reconciliation reports.	Rule-based matching, ML models for pattern detection, and predictive analytics.	Streamlines the reconciliation process, identifies errors and discrepancies early, and reduces risk.	Number of accounts reconciled autonomously, reduction in manual effort, reduction in discrepancies.
3.	Autonomous Period-End Closing	Automate the various steps in the month-end, quarter-end, and year-end closing process based on a pre-set schedule and checklists.	Task automation, scheduling, workflow management, rule-based validations, and exception handling.	Accelerates the closing process, reduces manual effort, enhances data accuracy, and ensures timely financial reporting.	Reduction in close cycle time, reduction in manual effort, accuracy of postings, and timeliness of financial reports.
ACCOUNTS PAYABLE					
4.	Autonomous Invoice Processing:	Automate the receipt, extraction, verification, and posting of supplier invoices.	OCR and AI for data extraction, rule-based validation, ML for anomaly detection, and predictive analytics.	Reduced manual effort and errors and improved supplier relationships through timely payment.	The number of invoices processed autonomously, reduction in manual effort, accuracy rate, and on-time payment rate.

NO.	USE CASE	DESCRIPTION	KEY FEATURES	BUSINESS IMPACT	SUCCESS METRICS
5.	Autonomous Payment Processing	Automate the scheduling, approval, and execution of supplier payments based on due dates, cash flow forecasts, and other parameters.	Rule-based scheduling and approval and integration with bank systems for payment execution.	Improves cash flow management, reduces payment errors, and enhances supplier relationships.	The number of payments processed autonomously, reduction in manual effort, and on-time payment rate.
6.	Autonomous Vendor Reconciliation	Automate the reconciliation of supplier statements and ledgers, highlighting discrepancies for review.	Rule-based matching and machine learning models for pattern detection and predictive analytics.	Streamlines reconciliation, identifies errors and discrepancies early and improves supplier relationships.	The number of reconciliations performed autonomously reduced manual effort and reduced discrepancies.
ACCOUNTS RECEIVABLE					
7.	Autonomous Billing	Automate the creation and dispatch of customer invoices based on contracted terms and service delivery data.	Rule-based invoice creation, automated dispatch via email or portal.	Reduces billing errors, accelerates invoice dispatch, and improves cash flow.	The number of invoices created and dispatched autonomously, reduction in billing errors, and days sales outstanding.
8.	Autonomous Cash Application	Automate matching customer payments to open invoices, dealing with partial payments and discounts.	Rule-based matching, ML models for pattern detection.	Streamlines cash application reduces unallocated cash and enhances customer relationships.	Amount of cash applied autonomously, reduced unallocated cash, and reduced manual effort.
9.	Autonomous Collections Management	Predict and perform proactive collection activities based on payment history and other customer data.	Predictive analytics, automated reminders, and communication	Reduce overdue receivables, improve cash flow, and enhance customer relationships.	Reduction in overdue receivables, increase in on-time payment rate, and customer satisfaction.
CASH MANAGEMENT					
10.	Autonomous Bank Reconciliation	Automate the reconciliation of bank statements with the GL, identifying discrepancies.	Rule-based matching, machine learning models for pattern detection.	Streamlines bank reconciliation and detects errors and fraud early.	The number of bank reconciliations performed autonomously, reduction in discrepancies, and reduction in manual effort.
11.	Autonomous Cash Forecasting	Generate cash flow forecasts based on historical cash flows, upcoming receivables, payables, and other parameters.	Predictive analytics and scenario analysis.	Enhances cash flow management and supports decision-making.	Accuracy of cash flow forecasts, the number of forecasts generated autonomously.

NO.	USE CASE	DESCRIPTION	KEY FEATURES	BUSINESS IMPACT	SUCCESS METRICS
12.	Autonomous Fund Transfer	Automate funds transfer between bank accounts based on cash flow forecasts and preset rules.	Rule-based transfers, integration with bank systems.	Optimizes cash positioning and reduces idle cash and overdrafts.	The number of fund transfers executed autonomously reduced idle cash and reduced overdrafts.
FINANCIAL REPORTING AND ANALYSIS					
13.	Autonomous Financial Consolidation	Automate consolidating financial data from various sub-ledgers and subsidiary ledgers into the GL for reporting purposes.	Advanced data mapping and real-time consolidation.	Enhances speed and accuracy of financial reporting and ensures compliance.	Reduction in time taken for consolidation, reduction in manual effort, accuracy of reports.
14.	Autonomous Financial Analysis:	Apply AI algorithms to analyze financial data for trends, patterns, and anomalies.	Predictive analytics, trend analysis, anomaly detection.	Provides actionable insights for decision-making and improves financial performance.	The number of analyses conducted autonomously, actionable insights derived.
15.	Autonomous Financial Reporting:	Create financial reports, such as income statements, balance sheets, and cash flow statements, autonomously using real-time financial data.	Real-time reporting, pre-defined reporting templates.	Enhances speed and accuracy of financial reporting and ensures compliance.	Number of reports generated autonomously, reduction in time taken for reporting, accuracy of reports.
FIXED ASSET ACCOUNTING					
16.	Autonomous Depreciation Calculation:	Apply pre-set depreciation rules to calculate depreciation for FA autonomously.	Rule-based depreciation calculation, real-time asset tracking.	Ensures accurate accounting for asset depreciation and supports regulatory compliance.	Accuracy of depreciation calculations, the number of calculations performed autonomously.
17.	Autonomous Asset Tracking	Track FA lifecycle autonomously, including acquisition, depreciation, enhancements, and disposition.	Real-time asset tracking and lifecycle event detection.	Enhances asset management and supports accurate asset valuation and financial reporting.	The number of assets tracked autonomously, the accuracy of asset tracking.
18.	Autonomous Asset Verification	Verify the existence and condition of FA based on pre-set schedules and criteria autonomously. This may involve integrating the Internet of Things (IoT) devices and other data sources.	IoT integration, schedule-based verification, and anomaly detection.	It enhances asset management, prevents asset misappropriation, and supports compliance.	The number of assets verified autonomously and the accuracy of verification.

NO.	USE CASE	DESCRIPTION	KEY FEATURES	BUSINESS IMPACT	SUCCESS METRICS
TAX ACCOUNTING					
19.	Autonomous Tax Calculation	Calculate tax liabilities for different tax types (income tax, sales tax, etc.) based on pre-set rules and real-time transaction data.	Rule-based tax calculation, real-time transaction tracking.	Ensures accurate tax accounting and supports compliance.	Accuracy of tax calculations, the number of calculations performed autonomously.
20.	Autonomous Tax Return Preparation	Prepare tax returns based on real-time financial data and pre-set templates.	Real-time data integration and pre-defined return templates.	Enhances speed and accuracy of tax return preparation and supports compliance.	The number of tax returns prepared autonomously, the accuracy of tax returns.
21.	Autonomous Tax Planning	Analyze tax data to identify tax-saving opportunities based on pre-set rules and criteria.	Predictive analytics, rule-based opportunity identification.	Minimizes tax liabilities and enhances financial performance.	The number of tax-saving opportunities identified autonomously, tax savings realized.

TABLE 5.2: Autonomous Management Accounting Use Cases

NO.	USE CASE	DESCRIPTION	KEY FEATURES	BUSINESS IMPACT	SUCCESS METRICS
COST ACCOUNTING					
1.	Autonomous Cost Allocation	Allocates costs to products, services, departments, etc., based on predefined rules while factoring in real-time operational data.	Real-time data integration, rule-based allocation.	Ensures accurate cost accounting and supports product/service pricing and profitability analysis.	Accuracy of cost allocations and number of allocations performed autonomously.
2.	Autonomous Overhead Cost Distribution	Distributes overhead costs to cost objects (like products, services, departments, etc.) based on predefined distribution keys or activity drivers.	Rule-based distribution and real-time data integration.	Ensures accurate overhead cost accounting supports product/ service pricing and profitability analysis.	Accuracy of overhead cost distributions, the number of distributions performed autonomously.
3.	Autonomous Cost Variance Analysis	Analyzes variances between actual costs and standard or budgeted costs to identify cost overruns or underruns.	Predictive analytics, anomaly detection.	Provides insights for cost control and enhances financial performance.	The number of variances detected autonomously and cost savings achieved from variance analysis.

NO.	USE CASE	DESCRIPTION	KEY FEATURES	BUSINESS IMPACT	SUCCESS METRICS
BUDGETING AND FORECASTING					
4.	Autonomous Budget Preparation	Prepares budgets based on historical financial data, projected operational data, and pre-set budgeting rules.	Predictive analytics and rule-based budgeting.	Enhances speed and accuracy of budget preparation and supports financial planning.	Accuracy of budgets prepared autonomously, reduction in time taken for budget preparation.
5.	Autonomous Revenue Forecasting	Forecasts revenue based on historical sales data, market trends, and other relevant data.	Predictive analytics, trend analysis.	Provides accurate revenue forecasts and supports financial planning and decision-making.	Accuracy of revenue forecasts, number of forecasts performed autonomously.
6.	Autonomous Expense Forecasting	Forecasts expenses based on historical expense data, planned operational activities, and other relevant data.	Predictive analytics, trend analysis.	Provides accurate expense forecasts and supports financial planning and cost control.	Accuracy of expense forecasts, number of forecasts performed autonomously.
VARIANCE ANALYSIS					
7.	Autonomous Budget Variance Analysis	Analyzes variances between actual financial performance and budgeted performance to identify areas of concern.	Predictive analytics, anomaly detection.	Provides insights for budget control and enhances financial performance.	The number of budget variances detected autonomously and financial improvements achieved from variance analysis.
8.	Autonomous Forecast Variance Analysis	Analyzes variances between actual financial performance and forecasted performance to Identify areas of concern.	Predictive analytics, anomaly detection.	Provides insights for forecast accuracy and supports financial planning.	The number of forecast variances detected autonomously, improvements in forecast accuracy.
9.	Autonomous Operational Variance Analysis	Analyzes variances between actual operational performance and budgeted/ forecasted performance to identify operational inefficiencies or opportunities.	Predictive analytics, anomaly detection, and operational data integration.	Provides insights for operational improvement and enhances operational efficiency and financial performance.	The number of operational variances detected autonomously and operational improvements achieved from variance analysis.

NO.	USE CASE	DESCRIPTION	KEY FEATURES	BUSINESS IMPACT	SUCCESS METRICS
PROFITABILITY ANALYSIS					
10.	Autonomous Profitability Analysis by Business Segment	The system automatically performs a profitability analysis for each segment, considering all associated revenues and costs.	ML-based revenue and cost allocation, predictive analytics, and data integration from multiple sources.	Assists in strategic decision-making, supports cost optimization, and helps identify more profitable segments.	Profitability increases in analyzed business segments, improved decision-making based on profitability data.
11.	Autonomous Product Profitability Analysis	Analyzes profitability for each product or service the business offers by assigning revenues and costs.	Revenue and cost attribution to products, data integration, and predictive analytics.	Helps in product strategy, pricing decisions, and cost management.	Profitability increases in analyzed products, changes in product strategy, or pricing based on analysis.
12.	Autonomous Customer Profitability Analysis	Evaluates profitability of individual customers or customer groups considering revenues generated and costs incurred.	Customer lifetime value calculation, predictive analytics, and data integration.	Assists in customer relationship management, sales strategy, and customer retention.	Profitability increases in analyzed customer groups, improved customer retention, and changes in customer strategy.
PERFORMANCE MEASUREMENT					
13.	Autonomous Financial Performance Measurement	Autonomous system measures financial performance using KPIs like Return on Investment (ROI), Return on Equity (ROE), Earnings Per Share (EPS), and more. It adjusts for anomalies and considers only valid transactions.	ML-based performance measurement, predictive analytics for trend analysis, and benchmarking against industry standards.	Enhances understanding of financial health, supports strategic planning, and aids investor relations.	Consistent financial performance improves investor relations and better decision-making.
14.	Autonomous Operational Performance Measurement:	The system tracks and measures operational performance using KPIs relevant to the organization's operations.	ML-based operational performance measurement, predictive analytics, and trend analysis.	Supports operational optimization and cost management and drives operational excellence.	Operational efficiency improvement, cost reduction, and achievement of operational KPIs.
15.	Autonomous Sales Performance Measurement	Evaluates sales performance by tracking metrics such as sales growth, sales target achievement, sales per rep, etc.	ML-based sales performance measurement, predictive analytics for sales forecasting.	Drives sales the strategy helps in salesforce management and supports revenue growth.	Sales growth, improved salesforce productivity, and sales target achievement.

NO.	USE CASE	DESCRIPTION	KEY FEATURES	BUSINESS IMPACT	SUCCESS METRICS
DECISION SUPPORT					
16.	Autonomous Profitability Decision Support	The Autonomous system predicts profitability outcomes of different strategic decisions, leveraging historical data and current market trends.	Predictive analytics, scenario analysis, and ML models to estimate profitability.	Enhances strategic decision-making and supports profitability management.	Profitability improvement and successful strategic decisions.
17.	Autonomous Cost Decision Support	Evaluates the cost implications of different operational decisions and recommends the most cost-effective options.	Cost modeling, predictive analytics, and scenario analysis.	Aids in cost management support operational efficiency.	Cost reduction, operational efficiency improvement.
18.	Autonomous Investment Decision Support	Assesses the financial and strategic implications of different investment options, providing recommendations for decision-making.	Investment modeling, predictive analytics, and scenario analysis.	Enhances investment management and supports capital allocation decisions.	Return on Investment, successful investment decisions.

TABLE 5.3: Autonomous Financial Planning & Analysis Use Cases

NO.	USE CASE	DESCRIPTION	KEY FEATURES	BUSINESS IMPACT	SUCCESS METRICS
STRATEGIC PLANNING					
1.	Autonomous Long-Term Business Planning	The system integrates data from various internal and external sources to project long-term financial results based on different strategic business scenarios. It then recommends optimal strategies to achieve the organization's long-term objectives.	Data integration, financial projections, scenario analysis, and strategy recommendation.	Improved strategic decision-making and achievement of long-term business objectives.	Alignment of long-term financial results with strategic goals and accuracy of financial projections.

NO.	USE CASE	DESCRIPTION	KEY FEATURES	BUSINESS IMPACT	SUCCESS METRICS
2.	Autonomous Capital Allocation:	The system uses predictive algorithms to identify the optimal allocation of capital across various business units, projects, or investments to maximize returns and achieve strategic objectives. The allocation is continuously optimized based on real-time performance data.	Predictive algorithms, capital allocation, and continuous optimization.	Optimal use of capital, improved investment returns, and achievement of strategic objectives.	Improvement in ROI and achievement of strategic objectives.
3.	Autonomous Strategy Execution Monitoring:	The system continuously monitors the execution of strategic plans by tracking KPIs and financial results against the strategic objectives. It autonomously adjusts strategies as required based on real-time data.	Includes continuous monitoring, KPI tracking, and strategy adjustment.	Effective execution of strategies and achievement of strategic objectives.	Improvement in achieving strategic objectives and aligning KPIs with strategic goals.
FINANCIAL MODELING					
4.	Autonomous Financial Modeling	The system uses ML algorithms to autonomously create sophisticated financial models that project the organization's financial performance under various scenarios. The models are continuously updated based on real-time data.	ML algorithms and financial modeling continuously update.	Improved financial planning and better decision-making.	Accuracy of financial models and alignment of financial results with projections.
5.	Autonomous Risk Modeling	The system uses advanced analytics to autonomously develop risk models that assess the organization's exposure to various financial risks. It then recommends risk mitigation strategies.	Advanced analytics, risk modeling, and risk mitigation recommendations.	Effective risk management reduces financial risk.	Accuracy of risk models, reduction in financial risk.

NO.	USE CASE	DESCRIPTION	KEY FEATURES	BUSINESS IMPACT	SUCCESS METRICS
6.	Autonomous Investment Modeling	The system uses predictive analytics to develop investment models that project the returns from various investment options. It then recommends optimal investment strategies based on the organization's risk tolerance and financial objectives.	Predictive analytics, investment modeling, and investment recommendations.	Optimal allocation of investments and improved investment returns.	Accuracy of investment models, improvement in investment returns.
BUDGETING AND FORECASTING					
7.	Autonomous Budgeting	The system automates the entire budgeting process, taking into account historical data, real-time data, and future projections. The budgets are continuously updated as the underlying data changes.	Automated budgeting, continuous updates, historical data, real-time data, and future projections.	Accurate budgets, efficient budgeting process.	Accuracy of budgets, the efficiency of the budgeting process.
8.	Autonomous Forecasting	The system uses advanced analytics to generate accurate and timely forecasts of the organization's financial performance. The forecasts are continuously updated based on real-time data.	Advanced analytics, continuous updates, and real-time data.	Accurate forecasts and better decision-making.	Accuracy of forecasts, alignment of financial results with forecasts.
9.	Autonomous Capital Expenditure Forecasting:	The system uses ML algorithms to forecast the organization's capital expenditure needs. The forecasts are updated in real-time based on changes in the business environment.	ML algorithms, real-time updates, and capital expenditure forecasting.	Optimal allocation of capital and improved financial performance.	Accuracy of capital expenditure forecasts and optimization of capital allocation.

NO.	USE CASE	DESCRIPTION	KEY FEATURES	BUSINESS IMPACT	SUCCESS METRICS
MANAGEMENT REPORTING					
10.	Autonomous Management Reports	The system automates the generation of management reports, which are updated in real-time based on changes in the underlying data.	Automated report generation, real-time updates, customizable report templates.	Accurate and timely management reports, efficient reporting process.	Accuracy of reports, timeliness of reports, and the efficiency of the reporting process.
11.	Autonomous KPI Reporting	The system automatically generates reports of the organization's KPIs, which are updated in real-time. The system also alerts management when a KPI is not aligned with expectations.	Automated KPI reporting, real-time updates, and alerts.	Improved performance management and efficient KPI reporting process.	Accuracy of KPI reports and alignment of actual performance with KPI targets.
12.	Autonomous Regulatory Reporting:	The system automates the generation of regulatory reports, ensuring compliance with all relevant regulations. The reports are continuously updated based on changes in the regulatory environment.	Automated Regulatory reporting, real-time updates, and compliance assurance.	Compliance with regulations and an efficient regulatory reporting process.	Compliance with regulations and efficiency of the regulatory reporting process.
KPI MONITORING AND ANALYSIS					
13.	Autonomous KPI Monitoring	The system continuously monitors the organization's KPIs and notifies relevant stakeholders when a KPI is not in line with expectations.	Continuous monitoring, real-time alerts, and stakeholder notification.	Improved performance management and proactive problem identification.	The number of proactive problem identifications and alignment of actual performance with KPI targets.
14.	KPI Anomaly Detection	The system uses ML algorithms to detect anomalies in KPI data. This helps identify potential issues early, enabling timely corrective actions.	Anomaly detection, ML algorithms, proactive problem identification.	Early problem identification and improved performance management.	Number of anomalies detected, time saved through early problem identification.

NO.	USE CASE	DESCRIPTION	KEY FEATURES	BUSINESS IMPACT	SUCCESS METRICS
15.	Autonomous KPI Benchmarking:	The system automatically compares the organization's KPIs against industry benchmarks or best-in-class performers. This helps identify areas of under-performance and potential improvement.	Automatic benchmarking, identification of improvement areas.	Improved performance management and identification of improvement areas.	Improvement in KPIs post identification and alignment of KPIs with industry benchmarks.
SCENARIO PLANNING					
16.	Autonomous Scenario Modelling	The system automatically creates financial models for different scenarios based on input parameters. This helps management understand the potential impact of business decisions or external factors.	Automatic model creation and scenario analysis.	Improved decision-making and understanding of potential business impacts.	Accuracy of the scenario models, the usefulness of scenarios in decision-making.
17.	Autonomous Stress Testing:	The system automatically tests the organization's financials based on predefined scenarios. This helps understand the resilience of the organization under extreme conditions.	Automatic stress testing and scenario analysis.	Understanding of organizational resilience and improved risk management.	Accuracy of stress tests, alignment of actual performance with stress test scenarios.
18.	Autonomous What-if Analysis:	The system automatically performs what-if analyses based on input parameters. This helps management understand the potential impact of business decisions or external factors.	Automatic what-if analyses and scenario analyses.	Improved decision-making and understanding of potential business impacts.	Accuracy of what-if analyses, the usefulness of analyses in decision-making.

TABLE 5.4: Autonomous Treasury Management Use Cases

NO.	USE CASE	DESCRIPTION	KEY FEATURES	BUSINESS IMPACT	SUCCESS METRICS
CASH MANAGEMENT					
1.	Autonomous Cash Flow Forecasting	The system uses predictive algorithms to forecast future cash inflows and outflows based on historical data and external factors. This helps in managing liquidity and planning for future financial requirements.	Predictive algorithms, forecasting.	Improved liquidity management and better financial planning.	Accuracy of cash flow forecasts, effectiveness in liquidity management.
2.	Autonomous Cash Optimization:	The system uses optimization algorithms to determine the optimal cash distribution across various accounts and currencies. This helps in maximizing returns and minimizing costs.	Optimization algorithms, cash distribution.	Increased returns and reduced costs.	Increase in returns, decrease in costs.
3.	Autonomous Bank Reconciliation:	The system automatically matches bank statement entries with accounting entries, identifying any discrepancies for review.	Automatic reconciliation and discrepancy identification.	Time saved in reconciliation, improved accuracy.	Reduction in reconciliation time and reduction in discrepancies.
LIQUIDITY MANAGEMENT					
4.	Autonomous Liquidity Forecasting	The system uses predictive algorithms to forecast future liquidity requirements based on historical data and external factors. This helps in ensuring solvency and planning for future financial requirements.	Predictive algorithms and forecasting.	Improved solvency and better financial planning.	Accuracy of liquidity forecasts, effectiveness insolvency management.

NO.	USE CASE	DESCRIPTION	KEY FEATURES	BUSINESS IMPACT	SUCCESS METRICS
5.	Autonomous Liquidity Stress Testing:	The system automatically tests the organization's liquidity based on predefined scenarios. This helps understand the resilience of the organization under extreme conditions.	Automatic stress testing and scenario analysis.	Understanding of organizational resilience and improved risk management.	Accuracy of stress tests, alignment of actual performance with stress test scenarios.
6.	Autonomous Contingency Funding Plan:	The system automatically creates contingency funding plans based on different scenarios. This helps ensure the organization has backup options in case of unexpected liquidity requirements.	Automatic plan creation and scenario analysis.	Improved risk management and better financial planning.	Effectiveness of contingency plans, reduction in liquidity risks.
WORKING CAPITAL MANAGEMENT					
7.	Autonomous Working Capital Optimization	The system uses predictive and optimization algorithms to manage the company's working capital (A/R, inventory, and A/P) to ensure sufficient cash flow while minimizing costs.	Predictive and optimization algorithms, working capital components management.	Improved cash flow and minimized costs.	Working capital ratio, cash conversion cycle time.
8.	Autonomous Inventory Management:	The system predicts inventory requirements based on demand forecasting and adjusts inventory levels accordingly. This ensures optimal inventory levels and reduces holding and shortage costs.	Demand forecasting, inventory level adjustment.	Reduced inventory costs and improved customer service.	Inventory turnover rate, days of inventory outstanding.
9.	Autonomous Payable and Receivable Management:	The system optimizes the timing of payables and receivables to maximize cash flow. It can also predict late payments and take preventive actions.	Timing optimization, late payment prediction.	Improved cash flow and reduced risk of late payments.	Days payable outstanding, days sales outstanding.

NO.	USE CASE	DESCRIPTION	KEY FEATURES	BUSINESS IMPACT	SUCCESS METRICS
FOREIGN EXCHANGE MANAGEMENT					
10.	Autonomous Hedging	The system predicts the company's foreign exchange exposure and automatically hedges the risk through various financial instruments.	Foreign exchange exposure prediction, automatic hedging.	Reduced foreign exchange risk, potential cost savings.	Accuracy of hedging, reduction in foreign exchange losses.
11.	Autonomous Currency Arbitrage	The system identifies currency arbitrage opportunities and automatically executes trades to benefit from the price differences.	Arbitrage opportunities identification, automatic trading.	Additional profits from currency arbitrage.	Profits from currency arbitrage.
12.	Autonomous Foreign Exchange Risk Analysis	The system analyzes potential Foreign exchange risks based on various factors and predicts their impact on the company's financials.	Risk analysis and impact prediction.	Better understanding of foreign exchange risks and improved decision-making.	Accuracy of risk analysis and effectiveness of risk management strategies.
DEBT MANAGEMENT					
13.	Autonomous Debt Planning	The system forecasts the company's debt requirements based on its financial plans and optimizes the structure and timing of debt.	Debt requirements forecasting, debt structure, timing optimization.	Reduced cost of debt and improved financial structure.	Cost of debt debt-to-equity ratio.
14.	Autonomous Debt Servicing	The system automates the servicing of debt, including the timely payment of interest and principal, and manages refinancing activities if necessary.	Automated debt servicing and refinancing management.	Reduced risk of default, and improved financial stability.	Timeliness of debt servicing, number of refinancing events.
15.	Autonomous Covenant Monitoring:	The system continuously monitors financial covenants in loan agreements and forecasts potential breaches, enabling proactive remediation actions.	Continuous covenant monitoring and breach forecasting.	Reduced risk of covenant breaches and improved relationship with lenders.	Number of covenant breaches, time to remediate breaches.

NO.	USE CASE	DESCRIPTION	KEY FEATURES	BUSINESS IMPACT	SUCCESS METRICS
RISK MANAGEMENT					
16.	Autonomous Interest Rate Hedging	The system predicts the company's interest rate exposure and automatically hedges the risk through various financial instruments.	Interest rate exposure prediction, automatic hedging.	Reduced interest rate risk and potential cost savings.	Accuracy of hedging, reduction in interest rate losses.
17.	Autonomous Credit Risk Management:	The system assesses the credit risk of customers and suppliers, predicts potential defaults and manages credit limits and terms accordingly.	Credit risk assessment, default prediction, credit limit, and terms management.	Reduced credit risk and improved cash flow.	Bad debt ratio, cash conversion cycle time.
18.	Autonomous Liquidity Risk Management:	The system predicts potential liquidity shortfalls and automatically initiates measures to ensure sufficient liquidity.	Liquidity shortfall prediction, automatic liquidity measures.	Reduced liquidity risk and improved financial stability.	Liquidity ratios, number of liquidity shortfalls.
INVESTMENT MANAGEMENT					
19.	Autonomous Investment Planning	The system predicts the company's investment needs based on strategic plans and forecasts and optimizes the portfolio to balance risk and return.	Investment needs prediction. and portfolio optimization.	Enhanced return on investment and improved financial performance.	Return on investment, risk-adjusted return.
20.	Autonomous Investment Execution:	The system automatically executes investment decisions according to the plan and market conditions, reducing the time and cost involved.	Automatic investment execution.	Reduced investment costs and timely investment execution.	Cost of investment execution, time to execute investments.
21.	Autonomous Investment Monitoring and Reporting:	The system continuously monitors the performance of investments and generates regular reports for management.	Continuous investment monitoring and automated reporting.	Improved investment oversight informed decision-making.	Accuracy and timeliness of reports, number of investment issues identified and resolved.

TABLE 5.5: Autonomous FRM Use Cases

NO.	USE CASE	DESCRIPTION	KEY FEATURES	BUSINESS IMPACT	SUCCESS METRICS
CREDIT RISK MANAGEMENT					
1.	Autonomous Credit Risk Assessment	The system automatically assesses the credit risk of customers, suppliers, and other business partners using various data sources and predictive models.	Automated credit risk assessment and predictive modeling.	Improved credit risk management and reduced bad debts.	Accuracy of credit risk assessments, bad debt ratio.
2.	Autonomous Credit Limit Setting:	The system sets and adjusts credit limits for customers and suppliers based on their credit risk assessment and other relevant factors.	Automated credit limit setting and dynamic adjustment.	Reduced credit risk and optimized credit utilization.	Utilization of credit limits, number of credit limit breaches.
3.	Autonomous Collections Management:	The system uses predictive analytics to prioritize collections activities, automatically generates reminders, escalates late payments, and predicts potential defaults.	Predictive collections management, automated reminders, and escalations.	Improved collections efficiency and reduced bad debts.	Collection efficiency, bad debt ratio.
MARKET RISK MANAGEMENT					
4.	Autonomous Market Risk Assessment	The system continually monitors market risk factors and automatically updates market risk assessments based on changes in these factors.	Continuous market risk monitoring and automated risk assessment updates.	Improved market risk management and reduced losses due to market risk.	The accuracy of market risk assessments and losses due to market risk.
5.	Autonomous Market Risk Mitigation	The system identifies and executes risk mitigation strategies automatically, such as adjusting the investment portfolio or hedging.	Automated risk mitigation strategy identification and execution.	Reduced market risk and enhanced financial performance.	Reduction in market risk and return on investment.

NO.	USE CASE	DESCRIPTION	KEY FEATURES	BUSINESS IMPACT	SUCCESS METRICS
6.	Autonomous Market Risk Reporting	The system generates regular market risk reports for management, highlighting key risks and risk mitigation actions.	Automated market risk reporting.	Improved oversight of market risk and informed decision-making.	Accuracy and timeliness of market risk report, number of risk mitigation actions identified and executed.
OPERATIONAL RISK MANAGEMENT					
7.	Autonomous Operational Risk Assessment	The system continuously assesses operational risks in financial operations using various data sources. and predictive analytics.	Continuous operational risk assessment and predictive analytics.	Enhanced operational risk management and reduced losses due to operational risk.	Accuracy of operational risk assessments, losses due to operational risk.
8.	Autonomous Operational Risk Mitigation	The system identifies risk mitigation strategies for operational risks and implements them automatically.	Automated risk mitigation strategy identification and implementation.	Reduced operational risk and improved operational efficiency.	Reduction in operational risk and operational efficiency.
9.	Autonomous Operational Risk Reporting	The system generates regular operational risk reports for management, highlighting key risks and risk mitigation actions.	Automated operational risk reporting.	Improved oversight of operational risk and informed decision-making.	Accuracy and timeliness of operational risk reports, number of risk mitigation actions identified and implemented.
LIQUIDITY RISK MANAGEMENT					
10.	Autonomous Liquidity Forecasting	The system forecasts liquidity needs based on anticipated cash inflows and outflows and adjusts liquidity management strategies accordingly.	Autonomous liquidity forecasting, dynamic liquidity management strategies.	Improved liquidity management and avoidance of liquidity shortfalls.	Accuracy of liquidity forecasts, number of liquidity shortfalls.
11.	Autonomous Liquidity Stress Testing	The system conducts automated stress tests to assess the firm's ability to manage liquidity under adverse market conditions and recommends adjustments to the liquidity management strategy.	Automated liquidity stress testing and automated recommendations for strategy adjustment.	Enhanced resilience to liquidity shocks and improved regulatory compliance.	Effectiveness of liquidity stress tests, number of successful strategy adjustments.

NO.	USE CASE	DESCRIPTION	KEY FEATURES	BUSINESS IMPACT	SUCCESS METRICS
12.	Autonomous Liquidity Reporting	The system produces automated reports for management and regulators on the firm's liquidity position and risks and the measures taken to manage them.	Autonomous liquidity reporting.	Improved oversight of liquidity risk, informed decision-making, and improved regulatory compliance.	Accuracy and timeliness of liquidity reports, several risk mitigation measures identified and implemented.
COMPLIANCE RISK MANAGEMENT					
13.	Autonomous Compliance Monitoring	The system continuously monitors compliance with financial regulations and internal policies and automatically alerts management to potential compliance issues.	Continuous compliance monitoring and automated alerts.	Improved compliance and reduced risk of regulatory penalties and reputational damage.	Compliance rate, number of compliance issues identified and resolved.
14.	Autonomous Compliance Reporting:	The system automatically generates compliance reports for management and regulators, ensuring the firm is transparent about its compliance status.	Independent compliance reporting.	Improved regulatory compliance, transparency, and trust with regulators and stakeholders.	Accuracy and timeliness of compliance reports, number of regulatory inquiries or actions.
15.	Autonomous Policy Enforcement:	The system automatically enforces compliance with internal financial policies and regulations by blocking non-compliant transactions or requiring additional approvals.	Autonomous policy enforcement.	Improved internal controls, compliance, and risk management.	Number of non-compliant transactions blocked or flagged, compliance rate.
ENTERPRISE RISK MANAGEMENT					
16.	Autonomous Risk Identification and Assessment	The system identifies potential financial risks, assesses their potential impact, and presents these findings to management in real-time.	Real-time risk identification and assessment.	Improved risk awareness and decision-making.	Number of risks identified, the accuracy of risk assessments.

NO.	USE CASE	DESCRIPTION	KEY FEATURES	BUSINESS IMPACT	SUCCESS METRICS
17.	Autonomous Risk Mitigation Planning	The system recommends risk mitigation strategies based on the identified risks and their potential impact.	Automated risk mitigation planning.	Enhanced risk management and reduced potential losses.	Effectiveness of risk mitigation strategies, reduction in potential losses.
18.	Autonomous Risk Monitoring and Reporting:	The system continuously monitors the status of identified risks and the effectiveness of mitigation strategies and reports this information to management.	Continuous risk monitoring and reporting.	Improved risk oversight and management.	Accuracy and timeliness of risk reports and the effectiveness of risk mitigation strategies.

TABLE 5.6: Autonomous Corporate Finance Use Cases

NO.	USE CASE	DESCRIPTION	KEY FEATURES	BUSINESS IMPACT	SUCCESS METRICS
CAPITAL BUDGETING					
1.	Autonomous Capital Budgeting	The system recommends an optimal capital budget based on an analysis of potential investments, expected returns, and the organization's risk appetite.	Automated capital budgeting.	Improved allocation of capital resources, the potential increase in return on investment.	ROI, efficiency of capital allocation.
2.	Autonomous Investment Evaluation:	The system evaluates potential investments regarding their expected returns and associated risks and provides an investment rating.	Automated investment evaluation.	Improved investment decisions, the potential increase in ROI.	Accuracy of investment ratings, ROI
3.	Autonomous Capital Expenditure Tracking	The system tracks capital expenditures against the budget and provides real-time updates and forecasts.	Real-time tracking of capital expenditures.	Improved budget control and financial management.	Budget variance, the accuracy of forecasts.

NO.	USE CASE	DESCRIPTION	KEY FEATURES	BUSINESS IMPACT	SUCCESS METRICS
MERGERS & ACQUISITIONS					
4.	Autonomous M&A Target Identification	The system identifies potential M&A targets based on strategic objectives, financial health, market conditions, and other relevant factors.	Automated M&A target identification.	Improved M&A strategy, the potential increase in business growth.	The number of relevant M&A targets identified, business growth resulting from M&A.
5.	Autonomous Due Diligence	The system conducts financial and strategic due diligence on potential M&A targets, identifying potential risks and opportunities.	Automated due diligence.	Improved M&A decision-making, reduced risk of unsuccessful M&A.	Accuracy of due diligence findings, the success rate of M&A.
6.	Autonomous Post-Merger Integration Planning	The system develops a post-merger integration plan based on an analysis of the merging organizations' operations, systems, and cultures.	Automated post-merger integration planning.	Smooth post-merger integration, a potential increase in M&A success.	The success rate of post-merger integration, M&A success rate.
CAPITAL STRUCTURE MANAGEMENT					
7.	Autonomous Capital Structure Optimization	The system recommends optimal capital structure based on analyzing the organization's financial health, risk appetite, and market conditions.	Automated capital structure optimization.	Improved financial health and stability, a potential increase in shareholder value.	Capital structure ratios and shareholder value.
8.	Autonomous Debt-Equity Management	The system manages the balance between debt and equity based on strategic objectives, financial health, and market conditions.	Automated debt-equity management.	Improved financial health and stability, a potential increase in shareholder value.	Debt-equity ratio and shareholder value.
9.	Autonomous Refinancing Analysis:	The system analyzes refinancing options and provides recommendations for assessing potential cost savings and risks.	Automated refinancing analysis.	Potential cost savings and improved financial management.	Refinancing savings, cost of capital.

NO.	USE CASE	DESCRIPTION	KEY FEATURES	BUSINESS IMPACT	SUCCESS METRICS
EQUITY & DEBT FINANCING					
10.	Autonomous Financing Strategy Development	The system develops a financing strategy based on the organization's financial needs, risk appetite, and market conditions.	Automated financing strategy development.	Improved financial management and potential cost savings.	Cost of capital, financing costs.
11.	Autonomous Debt Financing Analysis	The system analyzes debt financing options and provides recommendations based on cost, risk, and strategic fit assessment.	Automated debt financing analysis.	Improved financial management and potential cost savings.	Financing costs, debt ratios.
12.	Autonomous Equity Financing Analysis:	The system analyzes equity financing options and provides recommendations based on an assessment of cost, dilution risk, and strategic fit.	Automated equity financing analysis.	Improved financial management and potential cost savings.	Financing costs, dilution risk.
DIVIDEND POLICY					
13.	Autonomous Dividend Policy Formulation	The system proposes a dividend policy based on the organization's financial health, future investment needs, and market conditions.	Automated dividend policy formulation.	Balanced shareholder returns and retained earnings for future growth.	Dividend payout ratio retained earnings.
14.	Autonomous Dividend Payout Calculation:	The system calculates optimal dividend payouts considering financial performance, future investment needs, and cash flow management.	Automated dividend payout calculation.	Balanced shareholder returns and retained earnings for future growth.	Dividend payout ratio retained earnings.
15.	Autonomous Dividend Impact Analysis	The system assesses the impact of dividend policy on the organization's financial health, market perception, and strategic goals.	Automated dividend impact analysis.	Enhanced understanding of dividend policy impact and improved decision-making.	Shareholder value and market perception.

NO.	USE CASE	DESCRIPTION	KEY FEATURES	BUSINESS IMPACT	SUCCESS METRICS
INVESTOR RELATIONS					
16.	Autonomous Investor Communication	The system generates updates and communications for investors based on the latest financial performance data and strategic updates.	Automated investor communication.	Improved investor relations and increased transparency.	Investor satisfaction, quality of communication.
17.	Autonomous Earnings Call Preparation	The system prepares material for earnings calls, including performance highlights, financial summaries, and strategic updates.	Automated earnings call preparation.	Improved efficiency in earnings call preparation and consistent messaging.	Quality of earnings call, time savings.
18.	Autonomous Investor Query Management	The system responds to investor queries using a vast organizational and market data database.	Automated investor query management.	Improved investor relations and increased responsiveness.	Investor satisfaction, response time.

TABLE 5.7: Autonomous Financial Systems & Technology Use Cases

NO.	USE CASE	DESCRIPTION	KEY FEATURES	BUSINESS IMPACT	SUCCESS METRICS
FINANCIAL SYSTEMS IMPLEMENTATION					
1.	Autonomous System Setup	The system autonomously configures financial systems based on business requirements, industry best practices, and regulatory compliance.	Automated system configuration, benchmarking against best practices.	Reduced implementation time and improved system setup.	Implementation time, system configuration quality.
2.	Autonomous Data Migration:	The system manages the process of data migration from old to new financial systems, ensuring data integrity and minimal downtime.	Automated data migration and data integrity checks.	Smooth transition to new systems and minimized downtime.	Data migration time, data integrity.

NO.	USE CASE	DESCRIPTION	KEY FEATURES	BUSINESS IMPACT	SUCCESS METRICS
3.	Autonomous System Testing:	The system autonomously performs comprehensive testing of financial systems to ensure accuracy and reliability.	Automated system testing, error detection, and correction.	Improved system reliability and reduced error rates.	Testing time, error rates.
FINANCIAL PROCESS AUTOMATION					
4.	Autonomous System Setup	The system autonomously configures financial systems based on business requirements, industry best practices, and regulatory compliance.	Automated system configuration, benchmarking against best practices.	Reduced implementation time and improved system setup.	Implementation time, system configuration quality.
5.	Autonomous Data Migration:	The system manages the process of data migration from old to new financial systems, ensuring data integrity and minimal downtime.	Automated data migration and data integrity checks.	Smooth transition to new systems and minimized downtime.	Data migration time, data integrity.
6.	Autonomous System Testing:	The system autonomously performs comprehensive testing of financial systems to ensure accuracy and reliability.	Automated system testing, error detection, and correction.	Improved system reliability and reduced error rates.	Testing time, error rates.
DATA MANAGEMENT AND ANALYSIS					
7.	Autonomous Data Quality Management	The system autonomously manages data quality by identifying and correcting errors, filling in missing data, and validating data accuracy.	Automated data quality management.	Improved data quality and better decision-making.	Data quality scores, error correction frequency.
8.	Autonomous Data Analytics	The system performs automated data analytics to generate actionable insights for decision-making.	Automated data analytics and insight generation.	Improved decision-making and strategic advantage.	Insight generation frequency, decision impact.

NO.	USE CASE	DESCRIPTION	KEY FEATURES	BUSINESS IMPACT	SUCCESS METRICS
9.	Autonomous Predictive Analytics:	The system uses historical and real-time data to generate predictive analytics for future financial performance.	Automated predictive analytics and forecasting.	Improved decision-making and strategic planning.	Forecast accuracy, decision impact.
REPORTING AND ANALYTICS					
10.	Autonomous Financial Reporting	The system autonomously generates financial reports based on defined schedules and criteria, ensuring the timely availability of crucial financial information.	Automated report generation and scheduling.	Improved timeliness of financial reporting and improved decision-making.	Report generation time and report accuracy.
11.	Autonomous Management Reporting	The system generates management reports with critical financial and operational metrics, contributing to better decision-making.	Automated report generation and metrics tracking.	Improved decision-making and strategic planning.	Report generation frequency and decision impact.
12.	Autonomous Dashboard Generation and Update:	The system autonomously generates and updates dashboards, providing real-time visibility into financial performance.	Automated dashboard generation and real-time updates.	Improved decision-making and operational visibility.	Dashboard update frequency, decision impact.
SYSTEM INTEGRATION AND OPTIMIZATION					
13.	Autonomous System Integration	The system autonomously integrates various financial systems, ensuring seamless data flow and eliminating data silos.	Automated system integration and data flow management.	Improved data flow and reduced data silos.	Integration time and data flow efficiency.
14.	Autonomous System Optimization:	The system continuously optimizes performance based on usage data, enhancing system efficiency and user experience.	Automated system optimization and usage data analysis.	Improved system performance and better user experience.	System performance, user satisfaction.

NO.	USE CASE	DESCRIPTION	KEY FEATURES	BUSINESS IMPACT	SUCCESS METRICS
15.	Autonomous System Maintenance:	The system performs system maintenance tasks such as updates, patches, and fixes, minimizing system downtime.	Automated maintenance and downtime management.	Improved system reliability and reduced downtime.	Downtime frequency and duration, system reliability.
IT GOVERNANCE AND SECURITY					
16.	Autonomous IT Governance	The system autonomously manages IT governance processes, ensuring compliance with standards and regulations.	Automated governance processes and compliance management.	Improved compliance and reduced risk of non-compliance.	Compliance levels, governance process efficiency.
17.	Autonomous Cybersecurity Management:	The system proactively manages cybersecurity threats, ensuring the integrity and security of financial data.	Automated threat detection and response, data protection.	Improved data security and reduced risk of breaches.	Threat detection and response time, data breach incidents.
18.	Autonomous Access Management:	The system manages user access to financial systems, ensuring appropriate access controls and minimizing the risk of unauthorized access.	Automated access management and access control enforcement.	Improved data security and reduced risk of unauthorized access.	Access management efficiency, unauthorized access incidents.

TABLE 5.8: Autonomous Compliance and Regulatory Reporting Use Cases

NO.	USE CASE	DESCRIPTION	KEY FEATURES	BUSINESS IMPACT	SUCCESS METRICS
FINANCIAL COMPLIANCE					
1.	Autonomous Compliance Monitoring	The system proactively monitors compliance with financial regulations and standards, such as Sarbarnes Oxley (SOX), Generally Accepted Accounting Principles (GAAP), International Financial Reporting Standards (IFRS), etc. alerting relevant stakeholders in case of potential breaches.	Automated compliance monitoring and alert system.	Reduced risk of non-compliance, improved governance.	Compliance levels and breach detection time.
2.	Autonomous Policy Enforcement:	The system automatically enforces financial policies, ensuring uniform organization adherence.	Automated policy enforcement and adherence tracking.	Improved policy adherence and reduced risk of breaches.	Adherence levels, policy violation incidents.
3.	Autonomous Regulatory Change Management	The system automatically updates itself in response to financial regulations and standards changes, minimizing the risk of non-compliance.	Automated regulatory update and change management.	Reduced risk of non-compliance and improved adaptability.	Compliance levels, update speed.
REGULATORY REPORTING					
4.	Autonomous SEC Filings	The system generates and submits SEC filings autonomously, ensuring timely and accurate compliance with regulatory requirements.	Automated filings and scheduling.	Improved compliance and reduced risk of penalties.	Filing accuracy and timeliness, compliance levels.
5.	Autonomous Tax Reporting:	The system generates tax reports and files tax returns autonomously, ensuring compliance with tax regulations.	Automated tax reporting and filing and regulatory compliance.	Improved compliance and reduced risk of penalties.	Reporting accuracy and timeliness, compliance levels.

NO.	USE CASE	DESCRIPTION	KEY FEATURES	BUSINESS IMPACT	SUCCESS METRICS
6.	Autonomous Audit Trail Generation:	The system generates comprehensive audit trails for all financial activities, aiding in audit and regulatory scrutiny.	Automated audit trail generation and comprehensive tracking.	Improved audit readiness and transparency.	Audit trail completeness and audit readiness levels.
INTERNAL CONTROLS AND AUDIT SUPPORT					
7.	Autonomous Internal Control Monitoring	The system autonomously monitors internal financial controls, alerting relevant stakeholders in case of any breaches.	Automated control monitoring alert system.	Improved control effectiveness and reduced risk of breaches.	Control effectiveness and breach detection time.
8.	Autonomous Audit Support	The system provides comprehensive support for internal and external audits, including data provision, audit trail generation, and query management.	Automated audit support and comprehensive tracking.	Improved audit readiness and reduced audit time.	Audit readiness levels and audit completion time.
9.	Autonomous Control Testing	The system tests the effectiveness of internal controls autonomously, ensuring their continued effectiveness.	Automated control testing and effectiveness assessment.	Improved control effectiveness and risk mitigation.	Control effectiveness and testing frequency.
RISK AND COMPLIANCE MONITORING					
10.	Autonomous Risk Monitoring	The system autonomously monitors financial risks, alerting relevant stakeholders in case of potential issues.	Automated risk monitoring and alert system.	Improved risk management and proactive issue resolution.	Risk levels, issue detection time.
11.	Autonomous Compliance Tracking:	The system tracks compliance with financial regulations and standards, providing real-time visibility into the organization's compliance status.	Automated compliance tracking and real-time visibility.	Improved compliance and reduced risk of non-compliance.	Compliance levels, tracking accuracy.

NO.	USE CASE	DESCRIPTION	KEY FEATURES	BUSINESS IMPACT	SUCCESS METRICS
12.	Autonomous Risk Reporting:	The system generates risk reports autonomously, providing crucial insights into the organization's risk profile.	Automated risk reporting and insight generation.	Improved risk management and decision-making.	Reporting frequency and accuracy, decision impact.
FINANCIAL POLICY ENFORCEMENT					
13.	Autonomous Policy Compliance	The system autonomously ensures all financial activities align with the company's financial policies, reducing non-compliance risk.	Automated policy compliance check and risk mitigation.	Reduced compliance risk and improved adherence to financial policies.	Policy compliance rate, non-compliance incidents.
14.	Autonomous Policy Breach Detection:	The system detects policy breaches autonomously, providing real-time alerts to ensure swift corrective actions.	Automated breach detection and real-time alerts.	Improved policy enforcement and reduced non-compliance risk.	Breach detection rate, time to action.
15.	Autonomous Policy Update Implement-ation	When a financial policy is updated, the system autonomously implements the changes across all relevant processes, ensuring compliance with the new policy.	Automated policy update implementation and compliance assurance.	Reduced risk of non-compliance due to policy changes and increased policy enforcement efficiency.	Implementation accuracy, time to implementation.

CHAPTER 6

Inside the Engine

The Technology that Powers Autonomous Finance

"Technology is the wind in the sails of Autonomous Finance, propelling organizations toward new horizons of efficiency and innovation."

- ANONYMOUS

6.0 Key Insights

- **Cloud Computing: The Architect of Scalability.** The cloud provides the underlying infrastructure for scalable, responsive financial operations. It reduces upfront investment and maintenance costs while enhancing accessibility and flexibility. To build a cohesive ecosystem, organizations should consider focusing on user adoption and strategic integration of cloud solutions into their current landscapes.
- **Cybersecurity: The Sentinel of Trust.** Cybersecurity is paramount to protect against cyber threats. Autonomous Finance, rich with sensitive data, requires advanced security measures like encryption, authentication, threat intelligence, and secure cloud infrastructure.
- **Emerging Technologies: The Vanguard of Innovation.** Emerging Technologies such as IoT, AR/VR, and quantum computing promise to further revolutionize the trajectory of Autonomous Finance. IoT provides valuable data inputs for personalized services, AR/VR enhances decision-making and user experience, and quantum computing brings potential leaps in computational power and security.
- **Strategic Integration: Beyond Adoption.** Technology selection and integration should consider scalability, complexity, data sensitivity, regulatory compliance, and cost considerations. The focus should be on adoption and strategic integration to form a cohesive ecosystem.
- **Ethical Deployment: A Non-Negotiable Imperative.** Ethical and Responsible Technology Use is imperative to prevent issues like algorithmic bias, breaches in privacy, and irresponsible use of AI. Transparency and accountability should be fundamental tenets in implementing Autonomous Finance across an organization.

6.1 Introduction

The preceding chapter served as a panoramic lens through which readers were afforded the privilege of discerning the compelling tapestry that is Autonomous Finance. Through a detailed exploration of its real-world applications—from Financial Accounting to Compliance and Regulatory Reporting—a vivid tableau of its transformative capabilities unfolded. This chapter thus asserted Autonomous Finance not as a theoretical construct but as an actionable paradigm with sweeping implications for the financial sphere.

However, one may posit that any machinery, as intricate and promising as it may be, remains an enigma until the observer gains insights into its inner workings. The previous discourse offered a compelling understanding of 'what' Autonomous Finance can achieve and 'why' it is revolutionary. However, it stopped at the antechamber of 'how' these marvels exist. As if standing before an awe-inspiring mechanism, readers were allowed to marvel at the results but were left wondering about the gears and cogs behind the façade.

The ensuing chapter, therefore, sets out to illuminate this hitherto obscured terrain. Here, the discourse shifts from outcomes and implications to the engines that power this awe-inspiring revolution in finance: technology. It is crucial to underscore that Autonomous Finance is not a monolith but a complex assembly of distinct yet interlinked technologies.

Each technology—AI, ML, RPA, Big Data, Advanced Analytics, Cloud Computing, SaaS, or APIs—contributes uniquely to orchestrating this monumental shift in financial operations.

In a world where 94% of enterprises worldwide have embraced cloud technology (Statista, 2022) and where AI is expected to add $15.7 trillion to the global economy by 2030 (Bourne, 2023), the urgency to understand these technologies transcends beyond mere curiosity; it becomes a strategic imperative. This chapter serves as a sophisticated lighthouse, illuminating not just the 'what' but also the 'why' and the 'how' of technology's role in Autonomous Finance. It is a treasure trove for those who aim not just to navigate but to captain their ships through the often tumultuous waters of financial innovation.

The transformation of financial operations through technology is a journey that unfolds in several profound layers. This chapter commences with an in-depth analysis of core technologies, the indispensable engines powering the vehicle of Autonomous Finance. These are the tried and tested pioneers of digital transformation, and understanding them is tantamount to grasping the foundations of this seismic shift.

However, the story does not halt at the familiar; it ventures boldly into the realm of emerging pioneers. These frontier technologies have just begun to imprint on the sands of financial innovation. Although nascent, their potential for disruption is undeniable and warrants visionary foresight.

From conceptual brilliance to practical genius, the narrative seamlessly transitions into real-world applications, offering a panoramic view of how theoretical mastery metamorphoses into practical wisdom. These case studies serve as living testaments to the transformative power of technology in finance, bridging the chasm between abstract potential and tangible impact.

As the chapter progresses, it dons a strategic lens, meticulously examining the critical considerations that govern technology selection and integration. This section acts as a navigational North Star, guiding financial leaders through the labyrinthine decisions that influence the realization of an Autonomous Finance ecosystem.

The narrative culminates in a sagacious synthesis, a visionary retrospective that amalgamates the multifaceted insights explored throughout the chapter into a cogent, enlightening narrative, which is then followed by a curated collection of key recommendations designed not merely as a checklist but as an intellectual arsenal for those trailblazers intent on steering their organizations into this transformative paradigm.

Thus, this chapter aspires to be more than a mere repository of knowledge; it seeks to serve as a compass for the intellectually curious and the strategically ambitious, guiding them through the complex yet exhilarating landscape that is Autonomous Finance.

In the words of the late Stephen Hawking, *"Intelligence is the ability to adapt to change"* (Publishing, 2019). This chapter, then, aims to provide the intellectual arsenal required to adapt to and champion the change that Autonomous Finance brings. As one navigates through the ensuing sections, it becomes clear that the future of finance is not written in ink but coded in algorithms, not confined to ledgers but distributed across blocks, and certainly not shackled by tradition but liberated by innovation. Prepare, therefore, to transcend the conventional as this chapter unveils the extraordinary technological tapestry that is Autonomous Finance.

6.2 Core Technologies in Autonomous Finance

In the intricate tapestry of Autonomous Finance, core technologies serve as the warp and weft that provide strength and pattern to the fabric. These technologies support finance functions and catalysts that propel organizations into unprecedented territories of efficiency and innovation. Each technology performs a unique role, from AI that imbues the system with cognitive capabilities to APIs that serve as the crucial links between disparate applications. This section delves into these critical technologies, elucidating their intricacies, impacts, and indispensable roles in sculpting the landscape of Autonomous Finance.

The Quintessential Role of AI & ML in Autonomous Finance

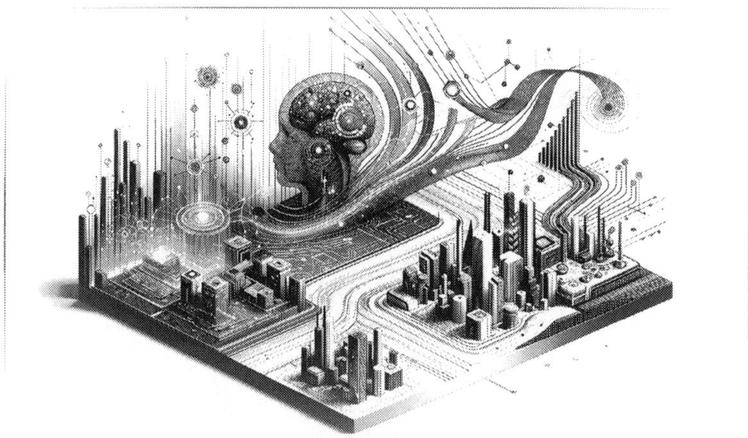

Embarking on the voyage through the technological underpinnings of Autonomous Finance, one encounters the first beacon of transformation: AI and ML. Far from mere buzzwords, these technologies manifest as potent catalysts that invigorate the financial landscape, carving out a realm where efficiency, personalization, and strategic acumen coalesce into an extraordinary symphony.

AI is not merely an adjunct to the financial ecosystem; it is the linchpin. It orchestrates diverse capabilities, from predictive analytics to natural language processing (NLP), thereby sculpting a nuanced environment where optimizing efficiency is not just a goal but an ongoing reality. In this realm, the expansive ocean of data transforms into a navigable reservoir of actionable insights. This metamorphosis enriches customer experience, amplifies revenue streams, mitigates fraud, and ensures regulatory compliance.

ML, a key subset of AI, serves as the oracle that illuminates unseen patterns within data to render informed decisions and predictions. In the volatile arena of stock markets, for instance, ML algorithms sift through the labyrinthine complexities of market movements to foresee future stock prices. Such predictive acumen has given rise to highly efficient algorithmic trading systems, a phenomenon supported by empirical research (Dixon et al., 2020).

However, it is crucial to acknowledge the inherent complexities and challenges that accompany these technologies. The quest for high-quality data and the vigilant avoidance of model overfitting stand as sentinel issues that demand attention and resolution.

Similarly, NLP has emerged as an indispensable asset in customer service. Pioneering examples like Bank of America's chatbot "Erica" have set benchmarks in service efficiency and personalization, serving millions of users within a concise timeframe (Business Insider, 2020).

Computer Vision is the vigilant sentinel, proficient in identifying and interpreting visual information for crucial tasks such as document digitization and fraud detection. Its capabilities are intrinsically tied to the quality of the data upon which it is trained. As highlighted in academic circles, biases in training data can inadvertently result in inequitable or incorrect outcomes (Harvard Business Review, 2020).

In summary, the convergence of AI and ML within the financial ecosystem does not merely signify technological adoption; it marks a paradigmatic shift. It mandates a move from traditional, rule-based systems to intelligent learning systems that adapt and evolve. The implications are profound, extending beyond operational efficiency to ethical responsibility. In steering this technological behemoth, the objective is to employ AI and ML as levers for ethical, equitable, and exceedingly efficient financial services.

RPA: The Evolutionary Leap in Autonomous Finance

In the lexicon of technologies transforming financial ecosystems, RPA emerges as an avant-garde protagonist. It is no longer merely a backstage hand performing tasks in the shadows of grander technologies. No, RPA has ascended to a starring role, its influence both transformational and expansive. It has matured from a tool of operational expediency to a cornerstone in the strategic architecture of Autonomous Finance.

a) **RPA in Symbiotic Harmony with AI and Cloud Computing:** In a world of interconnected technologies, isolation is neither desirable nor feasible. Here, RPA finds its most resonant notes when performing in a technological symphony alongside AI and Cloud Computing. This symbiotic harmony elevates the role of RPA from a mere executor of tasks to an orchestrator of efficiency. Imagine RPA as the conductor, AI as the virtuoso musicians, and the cloud as the grand concert hall. In isolation, each is remarkable but limited. Together, they create a transcendent experience, transforming data into insights, insights into decisions, and decisions into strategic imperatives. They bring the promise of Autonomous Finance closer to realization, constructing a technological edifice greater than the sum of its parts.

b) **The Paradox of Automation: The Complexities and Ethical Quandaries:** Yet, for all its promise, RPA is not without its complexities and ethical quandaries. While RPA bots can perform tasks with mechanical precision, they lack the cognitive abilities to understand ethical or cultural contexts, which can result in unintended consequences. Moreover, the allure of efficiency should not eclipse concerns about job displacement, data governance, and the potential for automated errors at scale.

In this complex landscape, the role of RPA becomes paradoxical. It is a purveyor of efficiency, yet it complicates the ethical milieu of Autonomous Finance. The technology is neutral; its application and impact are shaped by the strategic choices of those who wield it. Therefore, a comprehensive understanding of these complexities is not just beneficial—it is imperative for any institution committed to the responsible and effective implementation of Autonomous Finance.

Blockchain's Transformative Impact on Autonomous Finance

"In blockchain we trust," so goes the mantra of the digital age. This adage is not without merit, for blockchain technology has emerged as a cornerstone in the edifice of Autonomous Finance. Arthur C. Clarke once astutely commented that any sufficiently advanced technology appears as magic to those not versed in its complexities. In a realm as intricate as Autonomous Finance, blockchain serves as both the magician and the wand, defying conventional limitations to rewrite the rules of financial engagement.

Blockchain technology, often lauded as the epitome of security and transparency, is a bulwark against fraud and financial malfeasance. By employing a distributed ledger system, Blockchain has disrupted traditional centralized approaches to data storage and transaction verification (Nakamoto, 2008). The inherent structure of Blockchain, wherein each block is intrinsically linked to its predecessor, fortifies the integrity of the entire chain. These links make unauthorized alterations not only noticeable but virtually impossible.

One of the most transformative effects of blockchain technology is its ability to render intermediaries obsolete. By allowing peer-to-peer transactions to occur in a secure and verifiable manner, Blockchain engenders a level of efficiency previously unattainable (Tapscott & Tapscott, 2016). It has opened the floodgates to an era of reduced transaction fees, expedited transfer times, and elevated financial inclusivity.

Smart contracts represent another revolutionary application of blockchain technology. These self-executing contracts, with the terms of the agreement directly written into code, accelerate transactions while diminishing the potential for disputes or fraud (Mougayar, 2016). Thus, introducing intelligent contracts into the financial ecosystem adds another layer of efficiency and trustworthiness.

However, like any technology, blockchain has its set of challenges. The most notable among these is the scalability issue, wherein the increasing size of the distributed ledger may lead to slower transaction times and higher costs (Wüst & Gervais, 2018). Additionally, the still-evolving regulatory landscape presents both an opportunity and a challenge as compliance becomes a moving target (Zetzsche et al., 2017).

As this segment elucidates, blockchain technology is not merely another cog in the wheel but a veritable linchpin in the machinery of Autonomous Finance. Its influence extends far beyond its technical features, impacting the very ethos of how financial systems operate and evolve. In summary, blockchain is not a mere facilitator but a transformative agent, setting the stage for a more secure, efficient, and transparent financial future.

In navigating the labyrinthine corridors of Blockchain in Autonomous Finance, two compass points beckon with particular urgency.

a) **Prioritize Blockchain Adaptability:** The imperative of adaptability looms large; the volatile landscape of Blockchain technology demands a proactive approach to future-proofing, one that accords with rapid advancements and evolving best practices.

b) **Regulatory Vigilance:** Second, vigilance in regulatory compliance forms the bedrock of optimal blockchain utilization, necessitating continuously monitoring ever-shifting legislative landscapes.

These compass points serve not merely as practical guidelines but as strategic imperatives, underpinning the ascendancy of Blockchain in the new epoch of Autonomous Finance.

Cybersecurity's Role in Financial Autonomy

In the grand tapestry of Autonomous Finance, cybersecurity emerges as an intricate embroidery of protocols, practices, and technologies, effectively serving as the citadel safeguarding the digital realm. The significance of cybersecurity cannot be overstated, as it constructs the bulwark against a torrent of ever-evolving cyber threats. However, what does it indeed entail to construct such a fortress in an age of unbounded digital expansion?

A foray into the world of cybersecurity uncovers a dynamic landscape replete with evolving strategies designed to counteract equally adaptive threats. A pantheon of core technologies, from encryption algorithms to advanced threat intelligence platforms, shapes this unfathomable domain. However, the scope and complexity of cybersecurity's role in Autonomous Finance warrants more than a cursory glance; it calls for an in-depth exploration.

For readers yearning to delve deeper into the intricate mechanics of cybersecurity, including its technologies, strategies, and ethical dimensions, Chapter 14, titled *"The Digital Fortress: Cybersecurity in the Realm of Autonomous Finance,"* awaits to unravel these complexities in detail.

Big Data and Advanced Analytics: The Catalysts of Autonomous Finance

In the grand tapestry of Autonomous Finance, Big Data, and Advanced Analytics emerge as the weft and warp, entwined in a symbiotic relationship that creates a fabric of unparalleled intelligence and foresight. From collecting oceans of data to distilling them into actionable insights, these technologies serve as the compass and the cartographer, mapping out intricate financial landscapes and possible futures.

The role of Big Data can be likened to the very atoms of this universe—indivisible, essential, and profoundly impactful. It is the raw material that, when refined by the alchemy of Advanced Analytics, transforms into the gold of strategic intelligence. This harmony between data collection and interpretation is instrumental in risk management, customer personalization, fraud detection, and various other functionalities at the heart of Autonomous Finance.

For those captivated by the transformative potential of these technologies, a further expedition into their intricate mechanics and applications awaits in Chapter 10: *"The Influence of Insight: Data and Analytics in Autonomous Finance."* There, the curtain will be lifted to reveal not just the 'what' and the 'how' but also the 'why,' furnishing the reader with a panoramic view of a subject as complex as it is fascinating.

The Architectural Backbone: Cloud Computing in Autonomous Finance

In an era where digital landscapes are in perpetual flux, Cloud Computing emerges as the backbone of adaptability, providing the underlying architecture that empowers Autonomous Finance to be agile, scalable, and cost-effective. This technology is not merely an operational necessity; it is the linchpin that unlocks untold efficiencies and enables financial organizations to navigate complex terrains with unparalleled fluidity.

Cloud Computing's scalable nature is a harbinger of operational freedom. By offering computational and storage resources that can be adjusted up or down in real-time, it accommodates the ever-changing demands and complexities of the financial market (Mell & Grance, 2011). A testament to its ubiquity, 94% of enterprises globally had adopted cloud services as of 2022, a statistic that speaks volumes about its transformative impact (Statista, 2022).

Beyond scalability, the economics of Cloud Computing paint a compelling picture. By shifting from capital expenditure models to operational expenditure paradigms, organizations can sidestep the burden of hefty upfront hardware investments, thereby liberating capital for strategic pursuits (Weinhardt et al., 2009). This transition is not merely a cost-saving measure but a strategic maneuver redefining the financial calculus of technology investments.

In the realm of accessibility, Cloud Computing is an emancipator. Its ability to offer financial services unrestricted by geographical barriers democratizes access to financial tools and services, fulfilling the vision of financial inclusivity. Here, Cloud Computing does not merely serve; it liberates.

When examining Cloud Computing's capabilities, one must recognize its role as a security sentinel. Besides being just a repository or a processing powerhouse, cloud platforms are fortified with stringent security measures that safeguard critical financial data and customer information (Ali et al., 2015). In a sector where data breaches can spell catastrophic outcomes, this level of security is not a luxury; it is a mandate.

In conclusion, Cloud Computing transcends its role as a technology enabler to become an essential force shaping the future of Autonomous Finance. It provides the architecture that permits agility, safeguards data, democratizes access, and optimizes costs, acting as a quintessential element in the intricate tapestry of modern financial systems.

Streamlining Operations: The Role of SaaS in Autonomous Finance

In the complex tapestry of technological forces shaping the horizon of Autonomous Finance, SaaS claims a pivotal role, distinguished by its agility, budgetary prudence, and adaptability. This cloud-based software paradigm frees financial enterprises from the rigid bounds traditionally imposed by hardware and localized software configurations.

One of the most salient virtues of SaaS lies in its elasticity, affording financial entities the latitude to adapt to fluctuating market demands in real-time. This dynamic nature is invaluable for modern financial institutions grappling with the uncertainties and turbulences that define contemporary markets.

Equally compelling is the economic sagacity that SaaS cultivates. This technology harmonizes perfectly with cost-containment and operational expediency by circumventing the conventional overheads associated with hardware acquisition and continuous system upkeep. Such alignment places SaaS at the core of Autonomous Finance's operational strategy as a linchpin for cost-effective practices.

SaaS's role is further elevated by its propensity to harmonize and assimilate disparate financial systems. In a financial landscape characterized by a mélange of legacy systems, emerging platforms, and specialized software applications, the capacity for seamless integration is nothing short of a strategic advantage. SaaS is a synergistic nexus, weaving a coherent operational schema from otherwise fragmented elements.

Moreover, the expeditious implementation timelines associated with SaaS are congruent with the brisk tempo that Autonomous Finance aspires to maintain. The acceleration of technological deployment is not merely an operational convenience but an existential requisite, ensuring financial entities remain contemporaneous and competitive.

Adopting SaaS in Autonomous Finance transcends mere technological preference; it embodies a fundamental strategy that introduces robustness, financial prudence, and systemic coherence. Through this lens, SaaS crystallizes as an integral cornerstone, elevating the paradigms that inform and guide the evolution of Autonomous Finance.

Bridging the Digital Divide: APIs in Autonomous Finance

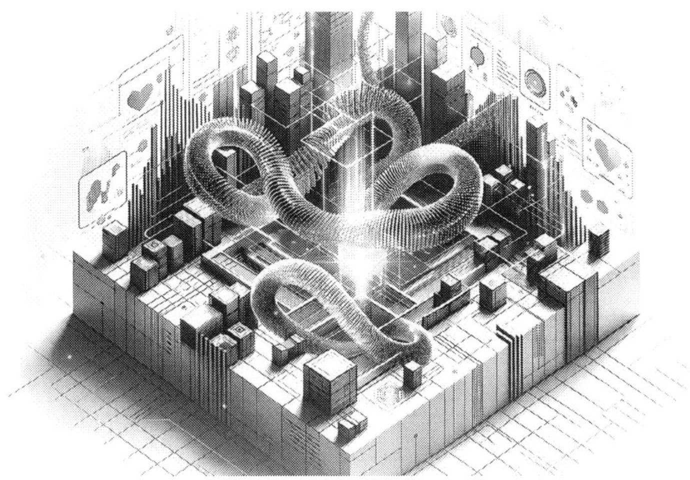

At the crossroads of innovation and integration, APIs are the architects of digital transformation in Autonomous Finance. The role of APIs transcends mere functionality; they are the digital ligaments that bind the body of Autonomous Finance, enabling its various limbs to function in unison (Pautasso et al., 2008).

Imagine an intricate tapestry of financial services and platforms, each woven with distinct threads of code and functionality. APIs are the loom that harmoniously intertwines these diverse elements, crafting a unified and dynamic financial ecosystem (Zachariadis & Ozcan, 2017).

With the ascendancy of open banking paradigms, APIs hold the key to democratizing financial data. By facilitating secure access to banking information, they empower third-party developers to introduce innovative financial products and services, enriching the financial ecosystem and offering consumers greater choice and control (Chishti & Barberis, 2016).

Moreover, the real-time capabilities of APIs are not merely a luxury; they are a necessity in the fast-paced world of Autonomous Finance. With APIs as enablers, instantaneous data sharing and transactions become the new norm, setting the stage for exceptional customer experiences and rapid decision-making (Pautasso et al., 2008).

In essence, APIs serve as the neural network of Autonomous Finance, transmitting signals between different platforms and services to create a cohesive and efficient operational environment.

Concluding Reflections on Technologies Powering Autonomous Finance

As the chapter traverses the intricate avenues of Core Technologies in Autonomous Finance, a panoramic view of their transformative potential emerges. These core technologies—AI & ML, RPA, Blockchain, Big Data & Advanced Analytics, Cybersecurity, Cloud Computing, SaaS, and APIs—serve as the bedrock upon which the edifice of Autonomous Finance is built. Each technology, remarkable in its own scope, contributes to a greater collective impact, akin to the individual instruments in an orchestra harmonizing to create a transcendent melody.

Whether it is the predictive prowess of AI and ML or the operational finesse facilitated by RPA, these technologies are not mere adjuncts to the financial realm. They are the pulse, the lifeblood, the quintessence. Cloud Computing and APIs serve as the architecture and connective tissue, respectively, knitting together disparate systems into a cohesive, agile entity.

It is the elegant choreography of these technologies that propels Autonomous Finance from being a mere concept to an operational reality. Their harmonized application has begun to rewrite the rules of the financial game, laying the groundwork for a future where technology and finance are not just intersecting realms but are deeply integrated facets of a new financial order.

Thus, as this section concludes, the spotlight dims only momentarily, allowing for a reflective pause before plunging into the kaleidoscopic world of emerging disruptors, real-world applications, and strategic considerations that await in the forthcoming sections. The narrative of Autonomous Finance is far from complete; indeed, it is a burgeoning epic in which these core technologies serve as prologue and prophecy.

6.3 Emerging Technological Disruptors

The journey through the annals of Autonomous Finance has arrived at a pivotal juncture, a gateway to a brave new world where the boundaries of possibility are redrawn almost daily. Having explored the cornerstone technologies—AI, ML, RPA, SaaS, and APIs to name a few—that form the sturdy scaffold of this evolving financial architecture, the narrative now ascends to a visionary stratum. This section heralds the disruptive forces poised to unleash transformative shifts, which carry the potential to refine and revolutionize financial paradigms.

Herein lies a meticulously curated compendium of nascent technologies, each representing a frontier of innovation, from Generative AI, which promises to redefine the notion of efficiency, to Quantum Computing, a technology whose computational capabilities defy conventional understanding, these emerging technologies are not mere incremental advancements. Instead, they are quantum leaps, pivots that could redefine the trajectory of Autonomous Finance.

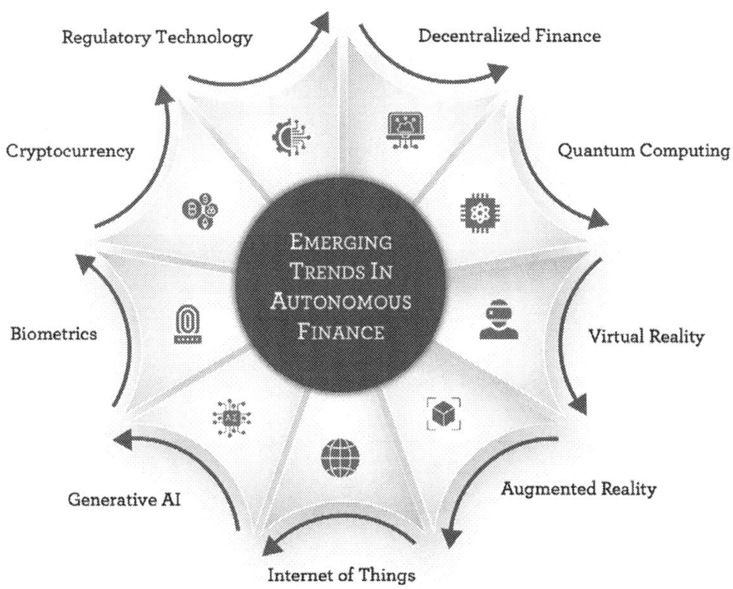

FIGURE 6.1: Emerging Technologies in Autonomous Finance

The objective of this section extends beyond mere acquaintance with these cutting-edge technologies. It aims to provide a nuanced, multi-dimensional understanding of their potential impact, strategic applications, and the complexities they introduce into the financial ecosystem. Each technology is scrutinized through a trifocal lens: its practical applications, its transformative potential, and the challenges it presents. Figure 6.1 is a diagrammatic representation that encapsulates the nine seminal technologies delineated herein, serving as both a navigational aid and a compact synthesis in this unfolding odyssey of financial transformation.

As Autonomous Finance continues to evolve, understanding these emerging technologies becomes not just advisable but imperative. According to Deloitte, 73% of organizations are investing or planning to invest in emerging technologies, underscoring the urgency with which these innovations are being integrated into strategic roadmaps (Deloitte, 2021).

In the eloquent words of Peter Drucker (1974), *"The best way to predict the future is to create it."* This section serves as both a navigational chart and a catalyst, a guide for those committed to adapting to and shaping Autonomous Finance's future. As you traverse this section, may you find the insights to envision new horizons, the wisdom to navigate complex choices, and the inspiration to pioneer the next wave of financial innovation.

Generative AI: The Vanguard of Financial Transformation

Generative AI represents an evolutionary leap for finance, setting the stage for unprecedented levels of automation, efficiency, and innovation. As the sorcerer's apprentice of the 21st century, Generative AI can produce content and insights across myriad modalities – text, code, images, audio, or video. A salient example in this domain is GPT-4, which has demonstrated the ability to generate text indistinguishable from that produced by human intelligence (Brown et al., 2020).

Generative AI is a veritable Swiss army knife for finance functions. Applications range from automating repetitive tasks, such as A/P and reconciliations, to producing preliminary narratives for management reports and board presentations. Additionally, Generative AI can sift through vast datasets to yield insights for forecasting, budgeting, and modeling, thus augmenting the analytical acumen of finance teams.

Integrating Generative AI into finance workflows can catalyze a panoply of benefits. Efficiency gains manifest through the rapid completion of tasks and cost savings. By automating high-volume activities, finance professionals can focus on strategic analysis and decision support, elevating their role as indispensable business partners. Furthermore, Generative AI can uncover novel correlations and insights, empowering finance teams with predictive capabilities hitherto deemed unattainable.

However, this brave new world is not devoid of challenges. Concerns about content authenticity, hidden biases in AI models, and data security necessitate rigorous oversight. Moreover, the enigmatic nature of generative models poses challenges for compliance and explainability. Balancing these challenges with the palpable benefits of Generative AI will dictate how finance teams can metamorphose into agile, data-driven entities.

As Generative AI evolves, finance professionals can anticipate a future punctuated by even greater automation and personalization. One might envision Generative AI capabilities seamlessly integrated within core finance systems, obviating the need for standalone applications. Furthermore, the technology could give rise to hybrid roles focused on training AI systems and translating machine outputs into actionable business insights.

In summary, Generative AI is a sentinel at the frontier of financial transformation. Its adoption demands a strategic and thoughtful approach to technology integration, workforce upskilling, and governance. By navigating this complex landscape with sagacity, finance teams can harness Generative AI as a linchpin for organizational growth and strategic acumen. In this rapidly evolving milieu, Generative AI is not merely an option but an imperative for finance functions that aspire to transcend traditional limitations and achieve unparalleled excellence.

DeFi: The Financial Democratization Movement

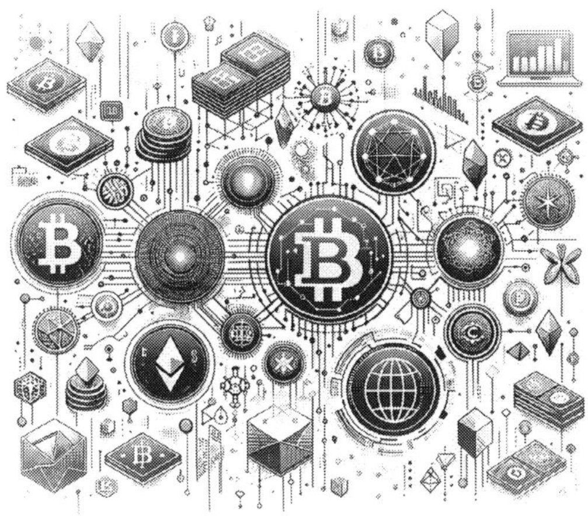

Decentralized Finance, commonly called DeFi, represents a seismic shift in the financial ecosystem. DeFi disintermediates traditional financial services, facilitating peer-to-peer transactions through blockchain technology (Khalil et al., 2022).

In a world dominated by centralized financial institutions, DeFi emerges as the disruptor, offering alternatives to traditional banking services such as lending, insurance, and asset trading. This innovation extends financial inclusion, empowering those without access to conventional banking systems (Zohar, 2015).

The engine propelling DeFi is the use of smart contracts on blockchain networks. These self-executing contracts, with the terms of the agreement directly written into lines of code, enable automated, trustless transactions (Christidis & Devetsikiotis, 2016). However, the nascent nature of DeFi introduces risks, including smart contract vulnerabilities and regulatory ambiguity. The technology's decentralized nature makes it inherently resistant to regulation, posing challenges for financial oversight (Maurer et al., 2013).

DeFi is not merely a trend but a potent movement toward democratizing finance. It challenges the status quo, driving innovation and financial inclusion, but also brings a new set of complexities and risks that require strategic oversight.

Cryptocurrencies and Digital Assets: Unseen Tech in Autonomous Finance

Amidst the tumultuous seas of financial innovation, cryptocurrencies and digital assets chart a course that is both revolutionary and contentious. Their advent challenges traditional paradigms—offering decentralized, transparent, and rapid financial transactions, yet clouded by regulatory ambiguities and market volatility.

Bitcoin, the progenitor of the cryptocurrency movement, disrupted the world stage in 2008, offering an alternative to centralized financial systems (Nakamoto, 2008). Today, cryptocurrency extends far beyond Bitcoin, featuring an array of digital assets, including stablecoins and tokens associated with DeFi. These digital currencies offer unprecedented transaction speed and lower fees, especially for cross-border transactions (Maurer et al., 2013).

Beyond mere currency, blockchain—the underlying technology of cryptocurrencies—enables smart contracts, decentralized applications, and much more, as discussed in Sections 6.2.3 and 6.3.2 above. The implications for Autonomous Finance are profound: instant settlement of asset trades, transparent loan agreements, and automated compliance protocols (Maurer et al., 2013).

While cryptocurrencies promise financial liberation and efficiency, they also bring risks—most notably, regulatory uncertainty and potential for misuse. The lack of oversight can lead to money laundering issues and other financial crimes. Moreover, the volatility of cryptocurrencies adds a layer of investment risk (Vigna & Casey, 2016).

The future of cryptocurrencies and digital assets appears both promising and challenging. While they may never entirely replace traditional currencies, their role in asset diversification, international remittances, and financial inclusion cannot be understated. Regulatory clarity will catalyze broader adoption and integration into conventional financial systems (Böhme et al., 2015).

Cryptocurrencies and digital assets hold the potential to redefine the architecture of financial transactions, offering a decentralized alternative replete with opportunities and risks. As stewards of Autonomous Finance, one must weigh these factors carefully to harness the full potential of this groundbreaking innovation.

Internet of Things (IoT): The Nexus of Interconnected Financial Ecosystems

The tapestry of Autonomous Finance is woven not merely by isolated technologies but through the intricate and nuanced interplay among them. The IoT is a stellar example of this interconnectedness. A digital conduit links various devices, systems, and platforms, aggregating data from disparate sources into cohesive, actionable insights (Atzori et al., 2010). This omnipresent connectivity extends to financial platforms, enabling real-time data collection vital for precise decision-making.

IoT has a broad spectrum of applications in personal and corporate finance, from wearable devices that facilitate contactless payments to smart home systems that manage energy consumption to optimize utility costs (Whitmore et al., 2015).

Nevertheless, the proliferation of interconnected devices expands the attack surface for potential cyber threats, warranting advanced security measures and constant vigilance (Roman et al., 2013). As IoT technology continues its forward march, a transformative shift is inevitable, especially in how real-time data analytics will influence financial decision-making, risk assessment, and customer engagement (Stankovic, 2014).

The IoT is a formidable catalyst in progressing toward a fully realized Autonomous Finance landscape. However, any disruptive technology has challenges and ethical implications requiring balanced, thoughtful integration.

Biometrics: The New Face of Financial Security

As the digital realm burgeons, so does the imperative for robust, reliable, and agile security measures. Biometrics emerges as a cornerstone technology in this respect, uniquely positioning itself as both an enabler and a safeguard within the Autonomous Finance ecosystem.

Biometrics capitalizes on unique physiological and behavioral characteristics — fingerprints, facial recognition, or voice patterns—to authenticate users (Jain et al., 2004). This form of identity verification obviates the need for cumbersome passwords, thus streamlining user experience while augmenting security.

In the financial arena, biometrics has found applications ranging from mobile banking authentication to anti-fraud measures in real-time transactions (Li & Jain, 2009). Its efficacy in combating identity theft and fraudulent activities affirms its pivotal role in Autonomous Finance. However, the integration of biometrics is not without ethical considerations. Issues of privacy invasion and data misuse loom large, demanding rigorous regulations and transparent data-handling policies (Gao, 2010).

As biometric technology continues to evolve, its impact on Autonomous Finance is poised to grow exponentially. Expect more widespread use in customer onboarding, transaction authorizations, and advanced analytics, where behavior biometrics can offer new insights into user behavior (Jain & Ross, 2015).

Biometrics, by its very nature, promises to enhance financial security while improving user experience. Yet, as with any technology, it is the responsibility of the purveyors of Autonomous Finance to implement it in an ethically sound and socially responsible manner.

Regulatory Technology: The Sine Qua Non of Financial Compliance

In a world of ever-evolving regulatory landscapes and complex financial structures, Regulatory Technology—or RegTech—emerges as the compass that guides financial institutions through the labyrinthine maze of compliance. Once considered a mere adjunct to FinTech, RegTech has become a critical enabler of Autonomous Finance.

Fulfilling regulatory requirements is not merely crossing t's and dotting i's; it is a complex undertaking that demands sophisticated solutions. RegTech equips financial organizations with the technological firepower to meet these challenges head-on. From real-time reporting to risk management, RegTech applications are expansive, covering a gamut of needs such as AML protocols, fraud detection, and tax compliance (Arner et al., 2016).

RegTech is about averting penalties, fulfilling regulatory mandates, and elevating the financial ecosystem. By automating compliance tasks, financial institutions can focus on innovation, thus accelerating the pace of financial inclusion and democratization. This capability to act as both a shield and a catalyst makes RegTech indispensable.

However, adopting RegTech is not without its challenges. Issues of data privacy, interoperability with existing systems, and the need for continuous updates to keep pace with regulatory changes necessitate a judicious approach to implementation. RegTech solutions must be integrated with precision, safeguarding against the risk of data breaches and compliance lapses.

As governments and regulators worldwide continue to tighten the noose on compliance requirements, the indispensability of RegTech will only intensify. Future developments may include using AI and machine learning to predict regulatory changes, offering financial institutions a preemptive advantage.

Regulatory Technology is a pillar of strength and innovation for the financial industry. Its transformative potential extends beyond mere compliance to redefine the very ethos of financial operations. For Autonomous Finance to thrive in an increasingly complex regulatory landscape, incorporating RegTech is not just advisable but is imperative.

Quantum Computing: The Paradigm Shift in Computational Power

Quantum computing represents nothing short of a tectonic shift in computational possibilities, enabling a level of problem-solving hitherto considered inconceivable. Traditional bits pale compared to quantum bits or 'qubits', which can exist in multiple states simultaneously, thus exponentially increasing computational power (Nielsen & Chuang, 2000).

In the domain of finance, quantum computing has the potential to redefine risk assessment, optimization, and even fraud detection. Imagine portfolio optimization tasks that take classical computers days to complete being finished in mere seconds or encryption algorithms that would take millennia to break being deciphered almost instantaneously (Orús et al., 2019).

Yet, the advent of quantum computing is not without its paradoxes. The very feature that makes it invaluable—its sheer computational power—also presents a threat to current cryptographic systems that safeguard financial data (Bernstein & Lange, 2017). As quantum technology matures, it promises to unlock new dimensions in financial modeling, risk analysis, and data security. However, the industry must be prepared for the seismic changes it will bring, both in capabilities and security requirements (Varmantchaonala Moudina et al., 2000).

Quantum computing is not merely an incremental advancement; it is a paradigm shift that will redefine the computational boundaries of finance. As such, it merits both awe and a careful, strategic approach to integration within the intricate tapestry of Autonomous Finance.

Augmented Reality & Virtual Reality: The Immersive Financial Experience

The financial realm, traditionally considered a bastion of numbers and spreadsheets, is on the cusp of an experiential revolution, courtesy of Augmented Reality (AR) and Virtual Reality (VR). These technologies augment the user's perception of and interaction with the financial world, providing an immersive, multi-dimensional experience (Milgram & Kishino, 1994).

Imagine a world where potential investors can walk through a virtual portfolio, engaging with data points as interactive elements. In such a dimension, real estate moguls could traverse a digital replica of properties before investing, while insurance companies could assess damages in a virtual environment, thereby expediting claims processing (Mazuryk & Gervautz, 1996).

However, the absorption of AR/VR into the financial fold is not without its caveats. Data privacy and security issues are magnified in an environment where virtual identities might possess substantial financial information (Kourouthanassis et al., 2017). As AR/VR technologies mature, their integration into Autonomous Finance will likely become more nuanced, offering more personalized and secure user experiences (Schuemie et al., 2001).

AR/VR may still be in its infancy in finance, but its potential to revolutionize user engagement and data interaction is colossal. Careful and ethical integration of these technologies could well be the fulcrum upon which the future of immersive financial experiences pivots.

Concluding Reflections on Emerging Technologies

As this section reaches its culmination, the far-reaching implications of the examined emerging technologies on Autonomous Finance crystallize. These disruptors—each groundbreaking in its own right—do not operate in isolation but converge to redefine the financial landscape. Their collective impact is less like individual notes and more akin to a complex symphony of technological advancements that, in concert, have the power to usher in a new era of financial services.

From the advent of Regulatory Tech streamlining compliance to the democratizing forces of Decentralized Finance, each technology harbors the potential for seismic shifts in the financial ecosystem. Indeed, the future of Autonomous Finance will be characterized not merely by the adoption of these individual technologies but by their synergistic integration.

It is this fusion of innovations that holds the promise of a more efficient, more inclusive, and more secure financial future. Therefore, for those at the vanguard of finance and technology, the strategic consideration of these emerging technologies is not just an option but an imperative.

With that, the curtain falls on this section, setting the stage for a deeper dive into real-world applications and case studies. The journey of exploration continues as each subsequent chapter peels away another layer of this complex yet fascinating subject: Autonomous Finance.

6.4 The Panorama of Interconnected Ecosystems

As we venture deeper into the realm of Autonomous Finance, it becomes abundantly clear that this financial revolution does not exist in isolation. It is but one integral part of a far more expansive, interconnected ecosystem of financial models, technologies, and philosophies. This section aims to lift the veil on this complex tapestry, illuminating how Autonomous Finance interacts with, influences, and is influenced by other emerging financial ecosystems.

The unfolding of these interconnected ecosystems—Open Banking, Open Finance, Open Data, Embedded Finance, DeFi, and Autonomic Finance—forms an orchestrated symphony, each contributing its unique melody to the grand composition of future financial services. Together, they offer a panoramic vista that reveals not just the future of finance but a new socio-economic paradigm characterized by democratization, personalization, and intelligent automation.

Upon this conceptual stage, the intricate relationships between Autonomous Finance and its adjacent ecosystems are not merely confined to prose; they find vivid illustration in Figure 6.2, a graphical taxonomy that encapsulates the symbiotic interconnections across Open Banking, Open Finance, Open Data, Embedded Finance, DeFi, and Autonomic Finance.

Each ecosystem is deeply covered, dissecting its core principles, its transformative potential, and, most critically, its symbiotic relationship with Autonomous Finance. This kaleidoscopic view serves as a prism through which the complex, multifaceted nature of Autonomous Finance can be better understood, appreciated, and ultimately, harnessed for societal advancement.

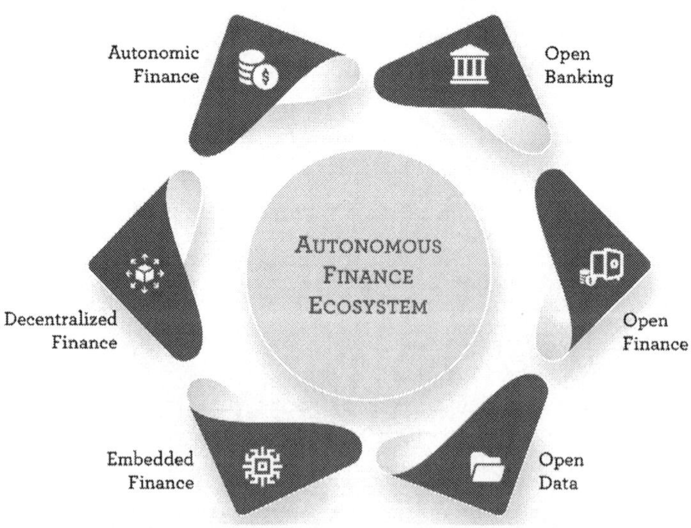

FIGURE 6.2: The Autonomous Finance Ecosystem

Open Banking

Advancing on this trajectory, the crossroads of Autonomous Finance and Open Banking come into view. Parallel to the ethos of Open finance, Open Banking promotes a culture of expansive financial data sharing, albeit with a stronger emphasis on banking data and services (*Financial Services Unchained: The Ongoing Rise of Open Financial Data*, 2021). The convergence of Open Banking and Autonomous Finance unravels a unique tapestry of opportunities and challenges.

In the waters of Open Banking, banking data is made available through APIs, enabling third-party developers to build innovative financial apps and services. When Autonomous Finance sails these waters, it can access a rich trove of banking data to enhance its AI algorithms, offering consumers more personalized and efficient banking services (Samet & Samet, 2023).

Imagine a world where a consumer can use a single app to manage multiple bank accounts, get personalized financial advice, apply for loans, and execute transactions - all autonomously and efficiently. This is the promise of Autonomous Finance in an Open Banking environment.

However, these new waters also bring new challenges to navigate. Data security, privacy, and consent become critical issues. Furthermore, the interoperability of systems, API standardization, and fair access to data are key considerations.

Advancing through the convergence of Autonomous Finance and Open Banking, the trajectory ahead indicates banking services that are remarkably customized, efficient, and pivotally focused on the consumer. Yet, this expedition demands careful steering, harnessing the advantages while effectively diminishing the associated risks.

Open Finance and Autonomous Finance

An integral crossroad on this financial expedition presents itself where Open finance converges with Autonomous Finance. In the expansive sea of finance, Open finance stands as a beacon advocating the free access and sharing of financial data. This transformation signifies a paradigm shift in interactions with financial services, underlined by the principle that data ownership rests with the consumer and should be readily accessible for their advantage and benefit (*Financial Services Unchained: The Ongoing Rise of Open Financial Data*, 2021).

When this philosophy converges with Autonomous Finance, it presents a promising vista. Autonomous Finance relies heavily on data to operate its AI algorithms effectively. In an open finance environment, the availability of freely accessible, rich financial data could greatly enhance the effectiveness and efficiency of these algorithms.

Consider a robo-advisor operating in an open finance environment. With access to a broad spectrum of a client's financial data, it could offer highly personalized and efficient investment advice. For consumers, this could mean more accurate, holistic, and beneficial financial advice, all delivered autonomously.

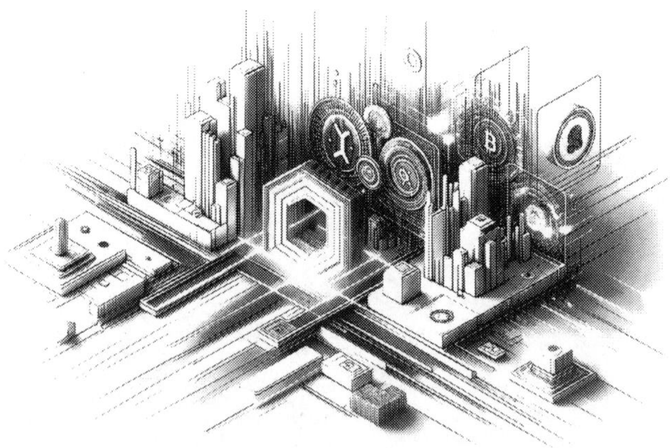

However, the intersection of Autonomous and Open finance is not without challenges. Questions around data privacy and security, data standardization, and interoperability all become more pertinent. Besides, regulatory considerations must evolve to support the data sharing that open finance requires while ensuring adequate consumer protection.

Amid the vibrant confluence of Open and Autonomous Finance, the potential for a more tailored, efficient, and consumer-oriented financial landscape becomes apparent. Nevertheless, steering through these potent currents calls for dexterity and foresight, ensuring the promises of this unison are achieved without compromising consumer rights and privacy.

Open Data and Autonomous Finance

Within the energetic meeting of Open Data and Autonomous Finance, the potential for a custom-tailored, efficient, and consumer-centric financial ecosystem comes to the fore. However, steering through these potent waters demands tactical acumen and anticipatory foresight, ensuring the promises born of this amalgamation are fulfilled without undermining consumer rights and privacy.

The implications are vast. Autonomous Finance is inherently data-driven, leveraging AI to analyze vast amounts of data for decision-making. If that data is open and freely available, the power of Autonomous Finance could be amplified significantly. With more information to feed into its algorithms, the financial advice, risk assessments, and predictive capabilities it could offer would be greatly enhanced.

Imagine a scenario where an AI-driven Autonomous Finance platform, armed with a multitude of open data, could forecast financial market movements, identify investment opportunities, and even predict an individual's financial health with astonishing accuracy.

Sailing the seas of Open Data brings its fair share of stormy weather. Data privacy, consent, and security issues are amplified in an open data environment. Additionally, the quality and reliability of open data and the potential for misuse are significant concerns.

Within the intriguing dance of Autonomous Finance and Open Data, a vision of a future emerges, characterized by heightened efficiency, personalization, and innovation in finance. Yet, navigating the waters toward this future necessitates a keen eye on the compass guiding data ethics, governance, and quality control.

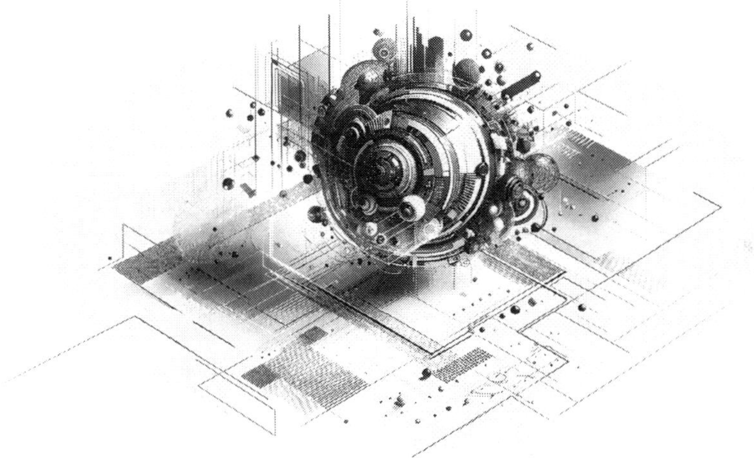

In essence, the interplay of these trends collectively will steer the direction of Autonomous Finance. As these paradigms unfold, they will shape the operational, regulatory, and competitive landscape in which Autonomous Finance evolves, presenting new opportunities and challenges.

Embedded Finance and Autonomous Finance

In the expansive seascape of finance, a novel destination appears on the horizon: the port of Embedded finance. Here, the traditional boundaries between financial and non-financial services fade, creating an integrated digital landscape. Financial services are seamlessly interwoven into other digital platforms, enhancing the user experience. As the ship of Autonomous Finance anchors in this port, many opportunities for collaboration and symbiotic integration come into view, signaling the dawn of an era where Finance and technology are inextricably intertwined.

Embedded finance extends the reach of financial services, bringing them closer to consumers. Similarly, Autonomous Finance enhances these services by making them more efficient and personalized. Thus, when the two combine, consumers can experience highly personalized and efficient financial services within their favorite digital platforms (Embedded or Autonomous Finance? - Chris Skinner's Blog, 2020).

An e-commerce platform, for instance, could leverage Autonomous Finance to offer personalized insurance products or credit options at the point of purchase, creating a seamless and convenient shopping experience. Conversely, a ride-hailing app could offer its drivers real-time loans or savings options, all powered by AI algorithms.

However, integrating Autonomous Finance into Embedded finance requires diligent attention to data privacy and security, ensuring consumers are protected in these more complex digital ecosystems. It also involves addressing regulatory implications, given the more profound involvement of non-financial firms in providing financial services (*Embedded or Autonomous Finance? - Chris Skinner's Blog*, 2020).

As the seascape through the confluence of Autonomous and Embedded finance unfolds, it charts the path to a future defined by greater access, bespoke services, and finance intrinsically woven into daily digital interactions. However, this journey necessitates precise navigation and vigilance to protect consumer rights and interests in this newly merged financial landscape.

DeFi and Autonomous Finance

In our oceanic journey, DeFi emerges like an uncharted territory, a new sea route powered by the blockchain compass. DeFi seeks to democratize financial services by bypassing traditional intermediaries and replacing them with smart contracts on a blockchain (Harvey et al., 2021). This endeavor resonates with Autonomous Finance's goal for efficiency and cost-effectiveness.

The overlapping waters of Autonomous and DeFi present a prospect of democratized, efficient, and transparent financial services. Autonomous Finance's AI-powered decision-making, combined with DeFi's open and transparent financial systems, could potentiate access to finance, provide personalized financial services, and enhance financial inclusion.

For example, while Autonomous Finance can use AI algorithms to offer personalized financial advice or automate investment decisions, DeFi can ensure such services are open and accessible to anyone, anywhere. DeFi could eliminate geographical constraints traditionally associated with financial services by operating on a decentralized network (Harvey et al., 2021). The amalgamation of these two models can bring the best of both worlds to consumers, a blend of personalized, efficient, cost-effective, and globally accessible financial services.

However, navigating these waters also raises new challenges. The transparency and openness of DeFi also invite risks, such as smart contract vulnerabilities, price volatility, and lack of insurance on assets (Harvey et al., 2021). These and the complexities inherent in AI decision-making necessitate robust risk management strategies, regulatory measures, and cybersecurity defenses.

Moreover, given DeFi's nascent nature, consumer understanding and trust in the technology remain underdeveloped; a challenge must be addressed through education and transparent communication.

The journey through the intersection of Autonomous Finance and DeFi sketches a picture of a future financial landscape that is decentralized, autonomous, and equitable. However, bringing this vision to fruition demands careful navigation to mitigate challenges and harness opportunities.

Autonomic Finance and Autonomous Finance

Autonomous and Autonomic Finance are like sibling ships sailing nearby, marked by a shared heritage of automation and AI. Despite their commonality, they offer unique perspectives and strategies for sailing the financial seas.

Autonomous Finance focuses on automating financial decision-making and operations, emphasizing self-sustaining financial systems. It uses AI to navigate complex financial decisions, reducing the need for human intervention. Autonomic Finance, on the other hand, embarks on a more ambitious voyage. It aims to create an all-encompassing, self-regulating financial ecosystem that makes decisions, learns, adapts, and optimizes its strategies based on individual financial behavior and market conditions.

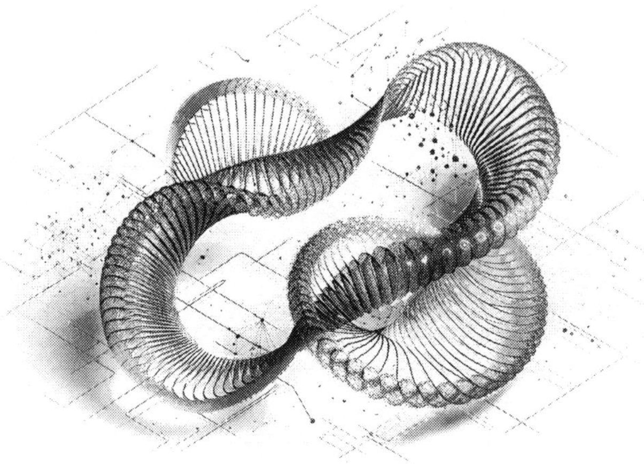

The intersection of these two financial models promises a journey filled with opportunities. The unison of their shared strengths - the ability to learn, adapt, and automate - can create a dynamic and highly responsive financial ecosystem that combines the best of both worlds. This union could lead to financial services that are not only efficient and reliable but also highly personalized, adapting in real-time to meet individual needs and market fluctuations.

However, this voyage is not without its challenges. The confluence of Autonomous Finance and Autonomic Finance will necessitate robust and secure data management practices, clear regulatory guidelines, and transparent communication with customers about using their data and the algorithms that power financial decisions.

Navigating the currents of Autonomous and Autonomic Finance sets a course toward a future in which financial systems exhibit increased intelligence, efficiency, and alignment with individual financial behaviors. This voyage teems with potential and promise, necessitating precise navigation to offset challenges and capitalize on opportunities.

Concluding Reflections on the Symbiosis of Interconnected Ecosystems

The articulation of these interrelated financial ecosystems reveals more than the sum of their parts; it illuminates a grander design of interconnectedness that holds transformative potential. This intricate web of Open Banking, Open Finance, Open Data, Embedded Finance, Decentralized Finance, and Autonomic Finance forms a complex but harmonious symphony, each contributing a unique note yet collectively creating a melody of unparalleled innovation.

Autonomous Finance is the conductor of this dynamic financial symphony, leveraging AI and ML to orchestrate a more efficient, personalized, and ethical financial experience. The role of Autonomous Finance is not simply to automate but to elevate, integrating the strengths of each ecosystem to forge a new financial paradigm. In this confluence of ecosystems, the full potential of Autonomous Finance can be unleashed, crafting a future where finance is not just a service but an experience tailored to individual needs and aspirations.

However, one must recognize that this symphony is still in rehearsal. As these ecosystems evolve, so will the challenges and opportunities they present. Regulatory landscapes will shift; ethical dilemmas will emerge, and technological barriers will be erected and dismantled. The journey towards this futuristic vision will require technological ingenuity, ethical rigor, regulatory foresight, and a steadfast commitment to consumer empowerment.

As the curtain rises on this brave new world of interconnected financial ecosystems, the role of every stakeholder—be it financial teams, public and private sector organizations, technology firms, regulators, or consumers themselves—becomes pivotal. The task ahead is not merely to navigate these waters but to chart a course that reflects shared values, ethical considerations, and a collective vision for a future where finance serves the many, not just the few.

It is not a question of whether these ecosystems will redefine the future of Autonomous Finance; rather, the pivotal issue centers on the direction this transformation will take to manifest the highest collective aspirations. At this crucial juncture, the call to action becomes unequivocal: passive observation must give way to active participation in shaping this impending future. The journey ahead offers an invitation to all.

6.5 Key Criteria for Technology Selection & Integration

In the grand tapestry of Autonomous Finance, selecting and integrating technologies is tantamount to a masterful orchestration, a confluence of precision, foresight, and strategic depth. The objective is no longer merely about adopting technologies but doing so with a finely calibrated strategy that aligns with organizational goals and market dynamics.

a) **Scalability vs Complexity - The Balancing Act:** Scalability and complexity often present a paradox in technology. While technologies that scale efficiently are coveted, their complex nature can pose challenges for seamless integration and user adoption. Achieving a harmonious balance between scalability and complexity becomes a strategic imperative (Wüst & Gervais, 2018). One must ponder: *Does the technology in consideration scale to accommodate growing workloads, and if so, at what cost in terms of complexity and resources?*

b) **Cost-Benefit Analysis - The Economic Calculus:** The financial alchemy of Autonomous Finance involves astute economic calculus. This meticulous cost-benefit analysis transcends mere numbers (Lacity & Willcocks, 2016). This evaluation is not confined to immediate expenditures but extends to long-term operational costs and the projected return on investment. Finance leaders must scrutinize this economic balance sheet closely as the implications reverberate across the enterprise.

c) **Holistic Technology Ecosystems - The Cohesive Matrix:** Integrating disparate technologies into a cohesive matrix requires an architectural vision that appreciates both these tools' utility and collective synergy. It is akin to constructing a complex puzzle, where each piece, while valuable on its own, derives its true worth from being part of a more extensive, interconnected whole (Hevner & Chatterjee, 2010). This holistic approach ensures that technologies do not operate in silos but contribute to a unified, robust, agile financial ecosystem.

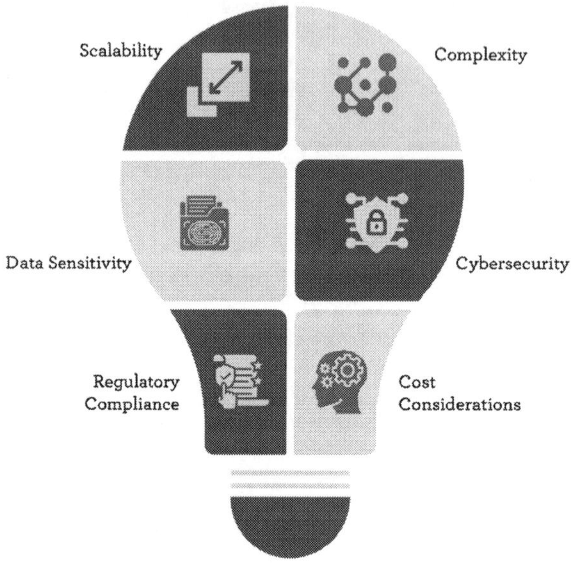

FIGURE 6.3: Technology Selection and Integration

In summary, the strategic considerations for technology selection and integration in Autonomous Finance are multi-faceted and profoundly consequential. They require a blend of technical acuity, economic wisdom, and strategic vision. As organizations embark on this transformative journey, these considerations serve as navigational beacons, guiding them through the labyrinthine complexities of technological adoption and integration.

The insights shared in this section aim to arm finance leaders with the strategic tools needed to traverse this complex terrain, laying a strong foundation for the future of Autonomous Finance. The following section will offer a panoramic view of the future trajectory, illuminating what lies ahead in this exhilarating odyssey.

6.6 Ethical and Responsible Use of Technology

Amid the kaleidoscopic unfolding of technological wonders, pausing and contemplating the ethical imperatives accompanying the journey toward Autonomous Finance becomes essential. These imperatives are not mere footnotes but the bedrock upon which the entire edifice is constructed.

a) **A Glimpse into Ethical AI:** Ethical considerations become paramount as algorithms increasingly dictate financial decision-making. The pitfalls of biased algorithms and the ethical mandates for diverse and representative data sets are not to be taken lightly (For a deeper exploration, refer to "*Chapter 13: Unveiling the Mystique: Ethical and Explainable AI & ML in Autonomous Finance*").

b) **Data Privacy - An Indispensable Concern:** While data grease the algorithmic wheels of Autonomous Finance, the sanctity of individual privacy remains sacrosanct. Robust measures to ensure data privacy are optional and obligatory (Detailed insights are available in Section 10.8 of *"Chapter 10: The Influence of Insight: Data and Analytics in Autonomous Finance"*).

c) **Transparency and Accountability - A Prelude:** Opaque algorithms are anathema to trust. Transparent decision-making processes are not just a 'nice to have' but a foundational element of Autonomous Finance (In-depth discussion to follow in Chapter 13).

d) **The Human-AI Symbiosis - A Teaser:** The romance between AI and human judgment is neither a replacement nor a rivalry but a partnership. Striking this balance is more an art than a science (For an extensive discourse, refer to *"Chapter 15: The Symbiotic Dance: Autonomous Finance and The Future of Work"*).

This section serves as an aperitif, whetting the reader's appetite for the more substantive discussions on these critical ethical dimensions in the coming chapters. The ethical and responsible use of technology remains a cornerstone, guiding the trajectory of Autonomous Finance not just in economic terms but within the ethical and moral compass.

6.7 Conclusion

As this journey through the technological vistas of Autonomous Finance nears its conclusion, a sense of awe at the sheer transformative power of these tools and frameworks becomes palpable. The narrative unfolded in this chapter serves as more than just an intellectual endeavor; it is a clarion call to finance leaders, signaling the imperative to embrace innovation with both arms wide open.

The technologies dissected herein—ranging from AI to Blockchain, from Cloud Computing to DeFi—stand as towering landmarks in the evolving landscape of Autonomous Finance. While each has its own unique set of capabilities and limitations, collectively, they signify an inflection point in the history of finance. At this moment, the frontier of what is possible expands exponentially.

However, it would be remiss to consider this exploration complete without casting an eye toward the horizon. Emerging technologies, those nascent disruptors waiting in the wings, are poised to reshape the contours of the financial industry further. The implications are dizzying and laden with both potential and peril. As this chapter gracefully concludes, it leaves behind a trail of intellectual breadcrumbs, a guide for those venturing into the uncharted territories of Autonomous Finance.

This is not an end but a beginning, a launching pad for further inquiry, exploration, and transformation. The following chapters will delve into the more nuanced aspects of this technological revolution, from ethical considerations to the nitty-gritty of implementation and governance. However, let this chapter serve as a foundational bedrock, a knowledge repository to return to as the journey through Autonomous Finance unfolds.

The future of Autonomous Finance is not a static destination but a dynamic realm, continuously molded and remodeled by technological innovation. The onus, therefore, falls on today's finance leaders to be the visionaries and pioneers of tomorrow, armed with the insights and strategies elucidated herein. As the curtain falls on this chapter, the next act in the grand drama of Autonomous Finance awaits, pregnant with untold possibilities and unprecedented opportunities.

6.8 Key Recommendations

a) **Strategic Technology Integration:** Prioritize technologies that align with business objectives and develop a seamless integration plan. This should include adaptive strategies to remain flexible as technologies and regulations evolve.

b) **Resource Allocation and ROI Assessment:** Prioritize advanced cybersecurity measures and other technologies based on meticulously evaluating implementation costs, projected ROI, and the required human and financial resources.

c) **Ethical and Responsible Technology Use:** Incorporate ethical considerations and risk mitigation strategies in technology selection and usage. Refer to Chapter 13 for an in-depth discussion on ethical AI and to Section 10.8 of Chapter 10 for guidelines on data privacy.

d) **Performance Metrics and Accountability:** Establish KPIs to measure the success of technology implementation. Also, implement quality assurance processes to ensure content authenticity and counter hidden biases in AI algorithms.

e) **Future-Proofing and Continuous Learning:** Given the rapid advancements in technologies like blockchain and generative AI, future-proofing the selected solutions is crucial. Encourage a culture of continuous improvement and upskilling within the organization to adapt to technological advancements.

6.9 References

Ali, M., Khan, S. U., & Vasilakos, A. V. (2015). Security in cloud computing: Opportunities and challenges. *Information sciences, 305*, 357-383.

Arner, D. W., Barberis, J., & Buckley, R. P. (2015). The evolution of Fintech: A new post-crisis paradigm. *Geo. J. Int'l L., 47*, 1271.

Atzori, L., Iera, A., & Morabito, G. (2010). The internet of things: A survey. *Computer networks, 54*(15), 2787-2805.

Bernstein, D. J., & Lange, T. (2017). Post-quantum cryptography. *Nature, 549*(7671), 188-194.

Böhme, R., Christin, N., Edelman, B., & Moore, T. (2015). Bitcoin: Economics, technology, and governance. *Journal of economic Perspectives, 29*(2), 213-238.

Bourne, J. (2023). AI's exponential advancement could spur $15.7T in economic growth by 2030. *Insider Intelligence.* https://www.insiderintelligence.com/content/ai-s-exponential-advancement-could-spur-15-7t-economic-growth-by-2030

Brown, T., Mann, B., Ryder, N., Subbiah, M., Kaplan, J. D., Dhariwal, P., & Amodei, D. (2020). Language models are few-shot learners. *Advances in neural information processing systems, 33*, 1877-1901.

Chishti, S., & Barberis, J. (2016). *The Fintech book: The financial technology handbook for investors, entrepreneurs and visionaries*. John Wiley & Sons.

Choudhary, V. (2007, January). *Software as a service: Implications for investment in software development*. In 2007 40th Annual Hawaii International Conference on System Sciences (HICSS'07) (pp. 209a-209a). IEEE.

Christidis, K., & Devetsikiotis, M. (2016). Blockchains and smart contracts for the internet of things. *IEEE Access, 4*, 2292-2303.

Deloitte. (2021). *Navigating the Changing Role of the CFO.* https://www2.deloitte.com/za/en/services/consulting-deloitte/navigating-the-changing-role-of-the-cfo.html

Dixon, M. F., Lukoianova, T., & McCombs, S. (2020). Machine Learning in Finance: The Case of Deep Learning for Stock Market Prediction. *Journal of Investment Management, 18*(2), 15-34.

Drucker, P. F. (1974). *Management: Tasks, responsibilities, practices*. New York : Harper & Row.

Embedded or Autonomous Finance? - Chris Skinner's blog. (2020). Chris Skinner's Blog. https://thefinanser.com/2020/11/embedded-or-autonomous-finance

Financial services unchained: The ongoing rise of open financial data. (2021). McKinsey & Company. https://www.mckinsey.com/industries/financial-services/our-insights/financial-services-unchained-the-ongoing-rise-of-open-financial-data

Gao, Q. (2010). Recent developments on applying biometrics in cryptography. *Journal of Applied Security Research, 5*(1), 107-137.

Hagerup, A. (2023, July 10). Can A Finance Organization Become Truly Autonomous? *Forbes*. https://www.forbes.com/sites/forbestechcouncil/2023/07/10/can-a-finance-organization-become-truly-autonomous/

Harvard Business Review. (2020). A Simple Tactic That Could Help Reduce Bias in AI. https://hbr.org/2020/11/a-simple-tactic-that-could-help-reduce-bias-in-ai

Harvey, C. R., Ramachandran, A., & Santoro, J. (2021). *DeFi and the future of finance*. John Wiley & Sons.

Hevner, A., & Chatterjee, S. (2010). *Design research in information systems: theory and practice (Vol. 22)*. Springer Science & Business Media.

Jain, A. K., & Ross, A. (2015). Bridging the gap: from biometrics to forensics. *Philosophical Transactions of the Royal society B: Biological sciences, 370*(1674), 20140254.

Jain, A. K., Ross, A., & Prabhakar, S. (2004). An introduction to biometric recognition. *IEEE Transactions on circuits and systems for video technology, 14*(1), 4-20.

Khalil, U., Malik, O. A., Uddin, M., & Chen, C. L. (2022). A comparative analysis on blockchain versus centralized authentication architectures for IoT-enabled smart devices in smart cities: a comprehensive review, recent advances, and future research directions. *Sensors, 22*(14), 5168.

Kourouthanassis, P. E., Boletsis, C., Bardaki, C., & Chasanidou, D. (2017). Privacy and security in the age of augmented and virtual reality: The paradox of perception. *Journal of Virtual Reality and Broadcasting, 14*(3), 201-220.

Lacity, M., & Willcocks, L. (2016). *Robotic process automation: the next transformation lever for shared services the outsourcing unit*, working research paper series paper 16/01.

Li, S. Z. (2009). *Encyclopedia of Biometrics: I - Z*. Springer Science & Business Media.

Maurer, B., Nelms, T. C., & Swartz, L. (2013). "When perhaps the real problem is money itself!": the practical materiality of Bitcoin. *Social semiotics, 23*(2), 261-277.

Mazuryk, T., & Gervautz, M. (1996). History, applications, technology and future. *Virtual Reality, 72*(4), 486-497.

Mell, P., & Grance, T. (2011). The NIST definition of cloud computing. *National Institute of Standards and Technology, 53*(6), 50.

Milgram, P., & Kishino, F. (1994). A taxonomy of mixed reality visual displays. *IEICE TRANSACTIONS on Information and Systems, 77*(12), 1321-1329.

Mougayar, W. (2016). *The business blockchain: promise, practice, and application of the next Internet technology.* John Wiley & Sons.

Nakamoto, S. (2008). *Bitcoin: A peer-to-peer electronic cash system*. Retrieved from https://bitcoin.org/bitcoin.pdf

Nielsen, M. A., & Chuang, I. L. (2010). *Quantum computation and quantum information*. Cambridge university press.

Orús, R., Mugel, S., & Lizaso, E. (2019). Quantum computing for finance: Overview and prospects. *Reviews in Physics, 4*, 100028.

Pautasso, C., Zimmermann, O., & Leymann, F. (2008, April). *Restful web services vs." big"'web services: making the right architectural decision*. In Proceedings of the 17th international conference on World Wide Web (pp. 805-814).

Publishing, N. (2019). *Intelligence is the Ability to Adapt to Change - Stephen Hawking: Daily Motivation Quotes To Do List for Work, School, and Personal Writing - 6x9 120 Pages*.

Roman, R., Zhou, J., & Lopez, J. (2013). On the features and challenges of security and privacy in distributed internet of things. *Computer networks, 57*(10), 2266-2279.

Samet, A., & Samet, A. (2023). How open banking and bank APIs are boosting fintech growth. *Insider Intelligence*. https://www.insiderintelligence.com/insights/open-banking-api-trends-explained/

Schuemie, M. J., Van Der Straaten, P., Krijn, M., & Van Der Mast, C. A. (2001). Research on presence in virtual reality: A survey. *Cyberpsychology & behavior, 4*(2), 183-201.

Stankovic, J. A. (2014). Research directions for the internet of things. *IEEE internet of things journal, 1*(1), 3-9.

Statista. (2022). Global Public IT Cloud Services Spending by Segment [Data]. https://www.statista.com/statistics/370305/global-public-it-cloud-services-spending-by-segment/

Tapscott, D., & Tapscott, A. (2016). *Blockchain revolution: how the technology behind bitcoin is changing money, business, and the world*. Penguin.

Varmantchaonala Moudina, F., Smith, J. A., & Doe, J. M. (2000). Quantum technology and its implications for finance: Modeling, risk, and security. *Journal of Financial Innovation, 15*(4), 345-367.

Vigna, P., & Casey, M. J. (2016). *The age of cryptocurrency: how bitcoin and the blockchain are challenging the global economic order*. Macmillan.

Weinhardt, C., Anandasivam, A., Blau, B., Borissov, N., Meinl, T., Michalk, W., & Stößer, J. (2009). Cloud computing–a classification, business models, and research directions. *Business & Information Systems Engineering, 1*, 391-399.

Whitmore, A., Agarwal, A., & Da Xu, L. (2015). The Internet of Things—A survey of topics and trends. *Information systems frontiers, 17*, 261-274.

Wüst, K., & Gervais, A. (2018, June). Do you need a blockchain?. *In 2018 crypto valley conference on blockchain technology (CVCBT)* (pp. 45-54). IEEE.

Zachariadis, M., & Ozcan, P. (2017). *The API economy and digital transformation in financial services: The case of open banking*.

Zetzsche, D. A., Buckley, R. P., Barberis, J. N., & Arner, D. W. (2017). Regulating a revolution: From regulatory sandboxes to smart regulation. *Fordham J. Corp. & Fin. L.*, 23, 31.

Zohar, A. (2015). Bitcoin: under the hood. *Communications of the ACM, 58*(9), 104-113.

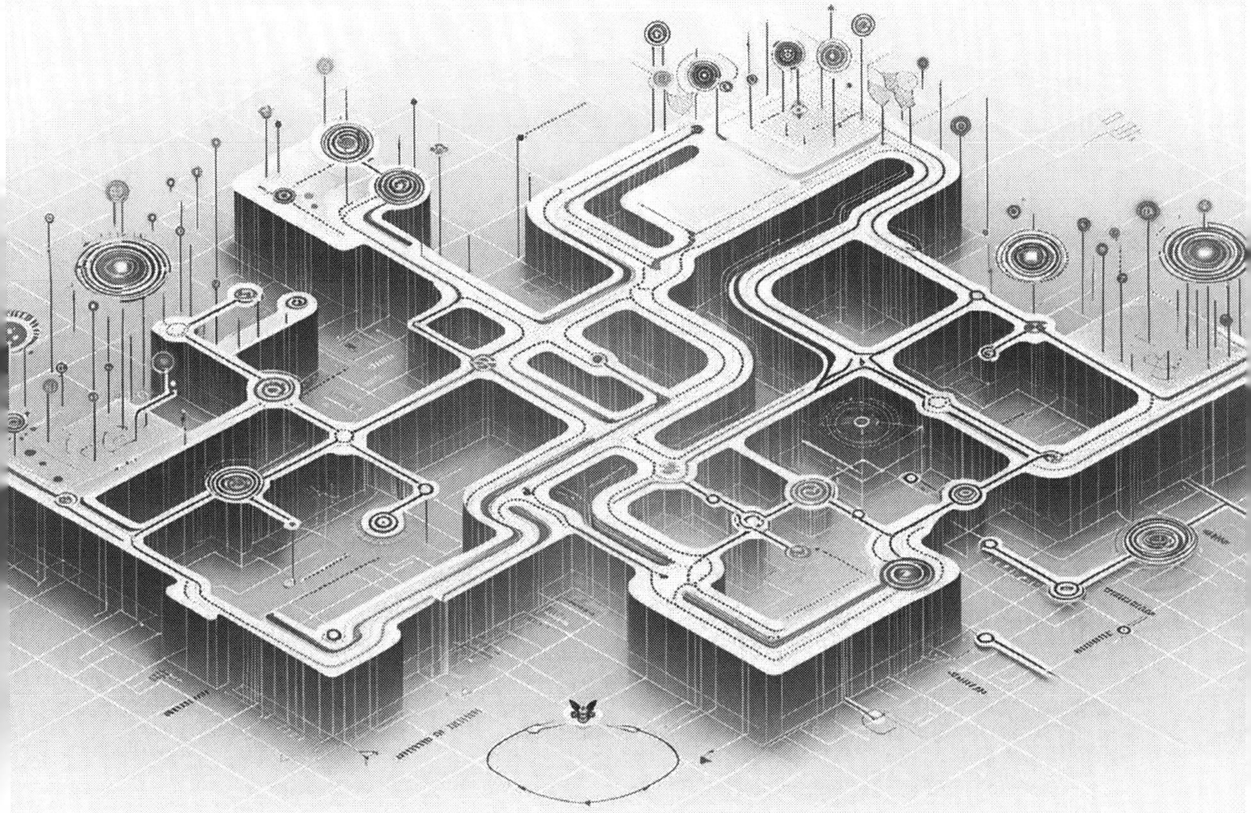

SECTION III

Blueprint for Success

Key Considerations in Autonomous Finance

Section 3 is a holistic compass for successfully adopting and integrating Autonomous Finance within organizations. It is an anthology of meticulously curated chapters, each addressing key facets, ethical considerations, and technological requisites pivotal for flourishing in the Autonomous Finance ecosystem.

Chapter 7, *"The Age of Autonomous Finance: A Strategic Transformation Guide,"* sets the stage by emphasizing the centrality of visionary leadership and well-defined objectives in accelerating the adoption of Autonomous Finance.

Chapter 8, *"Synergistic Strategies: The Role of Partnerships in Autonomous Finance,"* transitions to exploring external synergies, highlighting the catalytic impact of strategic alliances with fintech innovators, industry consortia, and regulatory entities.

Chapter 9, *"Cohesive Connectivity: Integration and Interoperability in Autonomous Finance,"* delves into the technological backbone, accentuating the indispensability of seamless interoperability among diverse systems and platforms.

Chapter 10, *"The Influence of Insight: Data and Analytics in Autonomous Finance,"* illuminates the transformative power of data analytics, its ethical considerations, and its capacity to enhance decision-making and customer engagement.

Chapter 11, *"Beyond Transactions: Crafting the Autonomous Finance Customer Experience,"* highlights the end-user, elucidating how Autonomous Finance transcends mere transactional interactions to redefine customer experiences.

Chapter 12, *"Playing by the Rules: Regulatory and Ethical Landscapes of Autonomous Finance,"* navigates the labyrinthine terrain of regulatory compliance and ethical imperatives, offering a roadmap for navigating this complex landscape.

Chapter 13, *"Unveiling the Mystique: Ethical and Explainable AI & ML in Autonomous Finance,"* demystifies the ethical nuances and the importance of explainability in AI & ML applications, stressing transparency, fairness, and accountability.

Chapter 14, *"The Digital Fortress: Cybersecurity in the Realm of Autonomous Finance,"* shifts focus to the ever-critical domain of cybersecurity, underscoring its role in fortifying the integrity and resilience of Autonomous Finance systems.

Chapter 15, *"The Symbiotic Dance: Autonomous Finance and the Future of Work,"* culminates Section 3 by examining the transformative interplay between Autonomous Finance and evolving work paradigms, offering strategic navigational insights into this intricate choreography.

Upon completion of Section 3, readers will possess a robust blueprint for piloting their journey through the multifaceted world of Autonomous Finance. This comprehensive guide sets the stage for the final voyage into the visionary terrains of Autonomous Finance, as unveiled in Section 4.

CHAPTER 7

The Age of Autonomous Finance

A Strategic Transformation Guide

"All change is hard at first,
messy in the middle,
and gorgeous at the end."

- ROBIN SHARMA (STONE, 2021)

7.0 Key Insights

- **The Autonomous Finance Maturity Model:** Navigating the transformation to Autonomous Finance requires a sophisticated framework beyond mere technological considerations. The FOCUS Framework provides a comprehensive pathway, guiding organizations through a series of progressive stages, each representing a significant milestone towards fully autonomous, intelligent finance operations.
- **Pilot Projects and Experimentation:** The journey toward Autonomous Finance is laden with uncertainties and unique challenges. Through systematic pilot projects, proofs of concept, and thoughtful experimentation, organizations can foster a culture of continuous learning, iterating and evolving their approach to meet the multifaceted demands of this transformative journey.
- **The Symphony of Technology and Strategy Insight:** Implementation, rollouts, and scaling in Autonomous Finance are not linear processes. They demand a harmonious blend of technology, human capital, operational alignment, and strategic vision. Embracing cutting-edge realms like AIOps and MLOps can play a vital role in orchestrating this complex symphony.
- **Ethical Stewardship and Societal Responsibility:** Autonomous Finance is not a soulless machine but imbued with ethical considerations, societal responsiveness, and a commitment to sustainability. Ensuring transparency, accountability, and social responsibility in this transformation shapes the organization and reflects a broader commitment to a humane, responsible future.
- **Continuous Improvement and Innovation:** The transformation to Autonomous Finance is not a destination but an ongoing journey. Continuous improvement, innovation, and adaptability are vital for sustaining success. The practices and methodologies embraced in support, maintenance, and continuous innovation breathe life into Autonomous Finance, keeping it agile, relevant, and aligned with the ever-evolving business and technology landscape.

7.1 Introduction

Emerging from the intricate technological fabric meticulously spun in preceding chapters, this chapter stands on the precipice of a monumental transformation—a transformation poised to redefine the financial function from a static operational unit into an autonomous, self-regulating entity.

A recent Accenture study succinctly captures the paradox of anticipation and readiness, revealing that while three-quarters of finance executives anticipate a revolution in Autonomous Finance within the next five years, only a minority feel adequately prepared to embrace it (Accenture, 2023). A contrasting report by McKinsey underlines the urgency of the situation, estimating that firms lagging in adopting Autonomous Finance risk falling behind by up to 50% in operational efficiency within the same timeframe (McKinsey & Company, 2022).

John Hagel III, the esteemed founder of Deloitte's Center for the Edge, once encapsulated the complexity of this transformation by stating, *"The real challenge is not technology, but rather the business models and institutions that were never designed for the exponential world"* (Hagel et al., 2014). The crisis lies in the mismatch between legacy systems and the rapidly evolving FinTech landscape. Autonomous Finance emerges as the catalyst that promises to bridge this gap, offering an innovative solution to an age-old problem.

This chapter aspires to illuminate the labyrinthine pathway to Autonomous Finance. It will serve as a comprehensive guide, delineating a maturity model that navigates the intricacies from vision to implementation and from ethical considerations to societal impact. Real-world case studies will be interspersed to lend credibility and offer practical insights into the challenges and triumphs that characterize this transformative journey.

As the narrative unfolds, ethical imperatives, societal responsiveness, and sustainability concerns will be embraced. Grounded in the principle that technology must serve humanity, this section will explore the ethical dimensions of Autonomous Finance, thereby addressing the call for responsible stewardship in FinTech.

The orchestration of this chapter aims to resonate with a diverse audience—from finance executives to technology pioneers to industry innovators. The ultimate goal is empowering stakeholders to navigate Autonomous Finance's labyrinth with vision, strategy, and ethical integrity. *"The best way to predict the future is to create it,"* as management guru Peter Drucker profoundly stated (Cohen, 2009). In that spirit, this chapter aims to predict and help shape a future where finance transcends its functional boundaries to become a dynamic catalyst that fuels business success, societal well-being, and transformative triumph.

7.2 The Roadmap to Autonomous Finance Transformation

The transition to Autonomous Finance is a multi-dimensional journey that requires strategic vision, careful planning, innovative experimentation, and continuous growth. It is a transformative process redefining the core of financial operations, leveraging cutting-edge technology to create a seamless, intelligent, and adaptive financial ecosystem. This chapter recommends a roadmap to guide organizations through the complex landscape of Autonomous Finance, providing insights, methodologies, and tools for success.

The roadmap for an Autonomous Finance Transformation initiative encompasses six critical stages, as shown in Figure 7.1, each with its unique characteristics, challenges, and opportunities:

a) **Vision, Alignment, and Commitment:** This stage sets the foundation, defining the overarching vision for Autonomous Finance within the organization. It requires alignment with the organizational strategy and unwavering commitment from leadership.

b) **Assessment using the FOCUS Framework:** A thorough assessment using the specially curated FOCUS framework helps understand the current state and readiness for transformation. It guides the organization through the multifaceted stages of maturity, aligning technology with strategic objectives.

c) **Strategic Prioritization and Planning:** Detailed strategic planning involves alignment with other business domains and governance, risk, and compliance considerations.

This phase ensures that the initiative aligns with broader organizational goals and is meticulously planned for success.

d) **Pilot Projects, Proof of Concepts (POCs), and Experiments:** This stage involves exploration and validation through pilot projects, POCs, and experimentation. Multiple initiatives may be conducted in parallel to uncover opportunities and validate solutions. This topic is covered in more detail in Section 7.6 below.

e) **Implementation, Rollout, and Scaling:** The successful pilots and experiments lead to phased implementation, rollout, and scaling across the organization. This stage can encompass various parallel paths tailored to different areas within the organization or processes.

f) **Ongoing Support, Maintenance, and Continuous Improvement:** The journey culminates in a single track for ongoing support, maintenance, and continuous innovation. It ensures long-term sustainability and the evolution of the Autonomous Finance system.

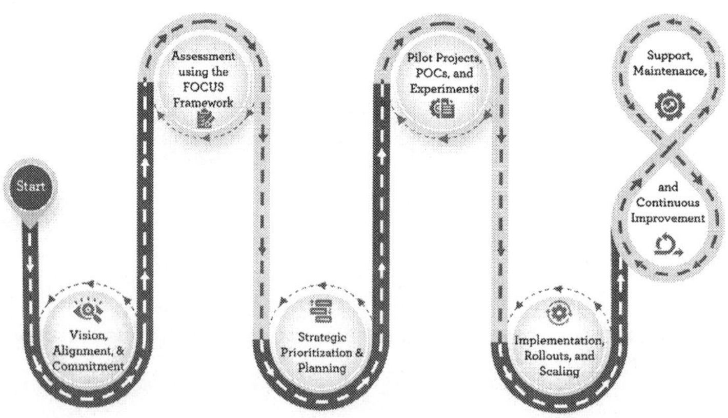

FIGURE 7.1: Roadmap for Autonomous Finance Transformation

The subsequent sections of this chapter delve into each stage of the roadmap, providing in-depth insights, methodologies, and practical guidance. From vision setting to continuous improvement, this roadmap serves as a comprehensive guide to embarking on the remarkable journey of Autonomous Finance transformation.

7.3 Vision, Alignment, & Commitment: Setting the Course

The transformation to Autonomous Finance heralds a new era in finance, where technology and human intelligence merge to create a symphony of innovation, efficiency, and value. It is a paradigm shift that promises to redefine the essence of finance, steering organizations toward a future that's both visionary and grounded in reality. This journey, however, is not a linear path; it is a complex and intricate voyage that requires a precise and adaptable roadmap.

Creating a Shared Vision

In the uncharted waters of Autonomous Finance, the strategic vision is the compass that guides the way. Developing such a vision in an organization's journey toward the Autonomous Finance paradigm is not merely about setting a direction; it is about understanding the destination, appreciating the nuances of the terrain, and crafting a vision that resonates with the very soul of the organization.

Considering the Future of Finance as envisioned in Chapter 3 earlier, such a vision must encapsulate the broader trends expected in the short-term and long-term future of Finance, such as the increasing role of AI, the prominence of real-time decision-making, and the centrality of data-driven insights (Reinventing FP&A for the Pandemic and Beyond, 2020). More than that, it must align with the organization's core values, resonate with its goals, and harmonize with the broader dynamics of the market and industry ecosystems. This vision must also be imbued with ethical considerations. As algorithms and automation assume increasingly complex roles, the commitment to transparency, accountability, and social responsibility becomes paramount (Mittelstadt et al., 2016). It is a vision that recognizes the profound implications of Autonomous Finance, not just for the organization but for society at large.

Building Alignment

Alignment with stakeholders across different departments and teams is essential to ensure that everyone understands and shares the goals of the transformation. This alignment fosters a sense of shared ownership and commitment, enabling the vision to be embraced by the entire organization. Effective communication plays a vital role in this alignment. Transparent and frequent communication of the vision and strategic goals to all levels of the organization ensures that the vision is understood and lived. It requires a tailored communication strategy that resonates with different stakeholders, fostering a culture of trust and collaboration.

Gaining Organizational Commitment

A pivotal element in this visioning process is leadership commitment. The transformation to Autonomous Finance is a complex and profound journey that requires unwavering dedication from the highest echelons of the organization. It demands that leaders endorse and actively champion the vision, nurturing an environment that thrives on innovation, collaboration, and continuous learning. It is a commitment beyond mere endorsement, encompassing resource allocation, strategic alignment, and a dedication to ethical and transparent practices (*Kotter's 8-Step Model of Change*, 1996).

Together, these elements create a strategic and visionary vision, a compass that guides the organization through the complexities of the Autonomous Finance landscape, steering toward a future that's innovative, efficient, and ethically responsible. Listed below are some examples of vision statements that encapsulate different facets of what organizations might seek to achieve from the Autonomous Finance Transformational journey:

a) **Empower the Future of Finance:** Leverage cutting-edge technologies and innovative practices to transform finance into a predictive, adaptive, and strategic partner, driving unprecedented efficiency and value across the entire organization and ecosystem.

b) **Redefine Financial Leadership:** Pioneer a new era of Autonomous Finance through intelligent automation, seamless collaboration, and ethical stewardship, fostering a culture of continuous learning and innovation that positions the organization at the forefront of financial excellence.

c) **Unlock Financial Potential:** Embrace the possibilities of Autonomous Finance to unlock unparalleled insights, agility, and resilience, creating a finance function that supports and leads the organization's growth and success in a rapidly evolving global landscape.

The examples above reflect a forward-thinking approach, embracing technological innovation, ethical leadership, collaboration, agility, and a focus on driving organizational success. However, the vision for Autonomous Finance must also be adaptable and ready to respond to the ever-changing Finance landscape. It requires a strong commitment from leadership and alignment with stakeholders at all levels, fostered through transparent communication and collaboration. As organizations consider their path toward Autonomous Finance, they must ask themselves: Are we ready to redefine the future of finance within our organization? The journey begins with a clear, shared vision that inspires innovation, commitment, and a relentless pursuit of excellence.

7.4 Conduct an Assessment using the FOCUS Framework

The journey toward Autonomous Finance is not a mere technological advancement but a transformative process redefining the core finance function in an organization. It is an evolution that takes an organization from the traditional manual methodologies to a futuristic world where finance becomes self-driving, self-learning, and self-healing. The pathway to this visionary transformation necessitates a well-defined framework to assess, guide, and nurture growth in this direction. The need for a framework uniquely tailored to the complex landscape of Autonomous Finance gives rise to the FOCUS Framework.

Overview of the FOCUS Framework

This framework represents the logical progression of a finance team through five stages: Formalizing, Organizing, Converging, Unifying, and Synchronizing. Each stage encapsulates specific dimensions and considerations that reflect a comprehensive view of what makes Autonomous Finance genuinely transformative. Unlike generic maturity models, the **FOCUS Framework** is created to address the specific nuances and challenges of Autonomous Finance, aligning technological innovations with strategic objectives. Figure 7.2 elucidates the acronym representing the five stages of maturity, shedding light on the comprehensive nature of Autonomous Finance's transformative potential.:

a) **F: Formalizing:** This stage represents the building block of the Autonomous Finance transformation journey where the organization acknowledges the need for Autonomous Finance. At this stage, the emphasis is on creating awareness and building an initial strategy. For example, an organization may begin by recognizing the inefficiencies in manual reconciliation processes and implementing essential automation tools to reduce the effort and time needed for reconciliation.

b) **O: Organizing:** This stage involves integrating existing systems and data to establish a cohesive structure for Autonomous Finance. The organization starts implementing the strategy, experimenting with AI models, and addressing integration challenges. An example could be a multinational organization aligning its regional financial systems to ensure uniform data flow, enabling more accurate global financial forecasting.

c) **C: Converging:** The Converging stage marks the transition from inside-out to outside-in focus. The transformation to the Autonomous Finance paradigm is not just about transforming the finance function but forming alliances with other functions within the organization like Supply Chain, Manufacturing, HR, etc. This stage is where the Autonomous Finance initiative begins to integrate and work in harmony with other autonomous systems supporting other functions within the organization. For instance, integrating finance with supply chain systems to enable real-time tracking of goods, linking purchase orders to invoices, and thus streamlining the entire procurement process.

d) **U: Unifying:** The Unifying stage is about achieving a seamless alignment of strategy, technology, and operations. It is about creating a unified platform where systems make financial decisions autonomously. This stage might involve the creation of a centralized AI-driven dashboard that provides an organization-wide view of financial health, risk factors, and potential growth areas. It is where the boundaries between different financial functions blur, creating a harmonious flow of data and decision-making.

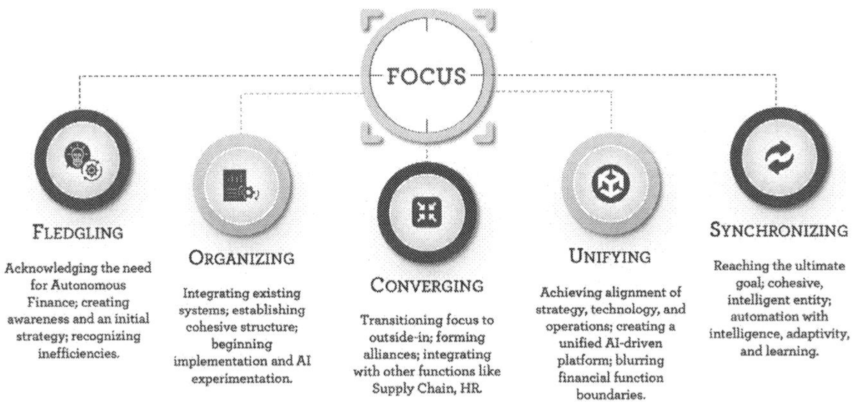

FIGURE 7.2: Maturity Levels in the FOCUS Framework

e) **S: Synchronizing:** The pinnacle of the journey, the Synchronizing stage, represents the ultimate goal of Autonomous Finance. It is where all elements come together to work as a cohesive and intelligent entity. At this stage, the finance functions are not just automated but are intelligent, adaptive, and capable of learning and evolving. A bank, for example, might employ advanced AI algorithms that detect fraudulent transactions in real-time, predict potential fraud patterns, and adapt security measures accordingly.

These stages are not isolated but form a logical progression where each stage builds upon the previous one, leading the organization toward the ultimate goal of fully synchronized, intelligent finance operations.

The FOCUS Framework is not a static model but a dynamic guide that recognizes the evolving nature of technology and each organization's unique needs. The transformation journey through the FOCUS Framework is guided by six critical dimensions that represent the core aspects of Autonomous Finance. These dimensions, as seen in Figure 7.3, were chosen to cover the entire transformation spectrum, from strategy to execution, reflecting a comprehensive view of what makes Autonomous Finance truly transformative:

a) **Strategic Alignment** refers to integrating the Autonomous Finance vision with the overall business strategy and objectives. This dimension ensures that the journey toward Autonomous Finance is not a siloed effort but is aligned with the broader organizational goals. It encompasses setting a clear vision, defining specific objectives, and ensuring that all efforts are directed toward fulfilling these strategic aims. For example, a financial institution aiming to enhance customer satisfaction might align its Autonomous Finance strategy with customer-centric goals, leveraging AI and analytics to deliver personalized services and insights.

b) **Data Intelligence**: The Data Intelligence dimension emphasizes the intelligent use of data to drive financial decisions and automate processes. This goes beyond data collection to include data quality, integration, analysis, and actionable insights. In the context of Autonomous Finance, Data Intelligence involves using data analytics and ML to identify patterns, predict trends, and provide decision support. A retail company, for instance, may use data intelligence to predict cash flow trends, enabling proactive financial planning and optimization.

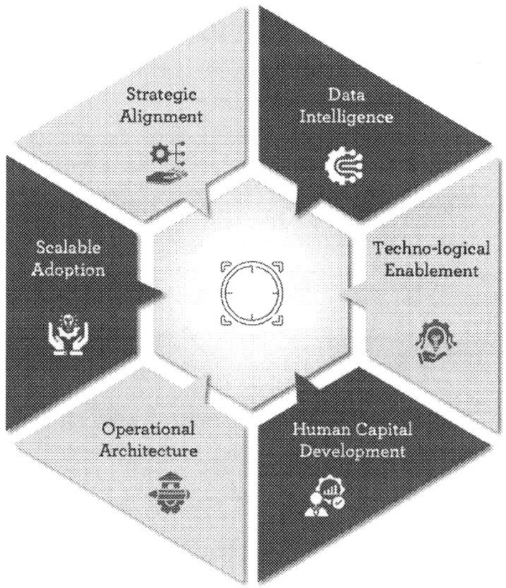

FIGURE 7.3: Dimensions of FOCUS Framework

c) **Technological Enablement:** Technological Enablement signifies the adoption and integration of technology that facilitates the autonomous operation of financial functions. It is not just about having the latest tools but about selecting, implementing, and optimizing the right technologies that enable automation, analysis, and intelligence. For a manufacturing firm, this might include implementing RPA to handle invoice processing or using AI-driven forecasting tools for better budgeting.

d) **Human Capital Development:** Human Capital Development focuses on the people aspect of the transformation, recognizing that skills, talents, and cultural readiness are vital for Autonomous Finance success. This involves training and development, change management, and fostering a culture that embraces innovation and agility. A global corporation, for instance, might invest in continuous learning programs to upskill its finance team in data analytics and ML, ensuring that human expertise complements technological advancements.

e) **Operational Architecture:** The Operational Architecture dimension involves the design and structure that facilitates autonomous operation within the finance function. This includes the processes, workflows, controls, and governance structures that enable efficient and effective automation. For a healthcare provider, this might entail reengineering payment processes to incorporate AI-driven fraud detection, ensuring that payments are processed swiftly yet securely.

f) **Scalable Adoption:** Scalable Adoption emphasizes the strategic expansion of Autonomous Finance throughout an organization, focusing on scaling success and learning from failures to ensure a sustainable transformation.

	Level 1 Formalizing	Level 2 Organizing	Level 3 Connecting	Level 4 Unifying	Level 5 Synchronizing
Strategic Alignment	No clear alignment between Autonomous Finance and organizational strategy	Initial efforts to align autonomous finance with business goals; lacks formalization.	Clear alignment with some business objectives; ongoing integration.	Strong alignment with most business objectives; strategic collaboration across functions.	Full alignment with organizational strategy, autonomous finance is a core business driven.
Data Intelligence	Limited data collection; no systematic analysis of insights.	Basic data analytics; some insights for decision-making.	Integrated data analytics, regular insights driving some financial decisions	Advanced analytics and AI-driven insights; data-driven culture.	Fully autonomous data intelligence; continuous insights and predictions
Technology Enablement	Limited or outdated technology; no automation	Introduction of some automation and digital tools; lacks integration.	Integrated technology solutions, some AI and automation in place.	Advanced technology stack; AI-driven automation across most functions.	Fully autonomous technology ecosystem; seamless integration and innovation.
Human Capital Development	No focus on skill development or change management	Initial training in technology and change management.	Ongoing skill development , cultural readiness for change.	Continuous learning culture; strong alignment with autonomous finance needs.	Full integration of human and machine capabilities; agile and innovative culture.
Operational Architecture	Manual processes; no autonomous capabilities	Some process automation; initial design for autonomy.	Integrated processes; some autonomous workflows.	Advanced autonomous operations, strong governance and controls.	Fully autonomous operations; optimal efficiency and effectiveness.
Scalable Adoption	No scalability; isolated efforts.	Initial scalability in limited areas; lacks systematic approach.	Scalable across some functions, learning from successes and failures.	Systematic scalability across most functions; continuous improvement.	Fully scalable adoption; autonomous finance across the entire organization

FIGURE 7.4: FOCUS Framework For Journey to Excellence in Autonomous Finance

The six dimensions of the FOCUS Framework offer a holistic approach to adopting Autonomous Finance, blending technological advancements with organizational strategies for a complete transformation. This framework aids in navigating the complexities of Autonomous Finance, providing a roadmap through various maturity levels, highlighting key milestones, and addressing potential challenges. Refer to Appendix B for an in-depth look at each dimension's capabilities, implications, benefits, challenges, and recommendations at each maturity level.

Using the FOCUS Framework, organizations can segment their transformation into smaller, manageable initiatives. Each can be assessed on its own merits, ensuring alignment with broader organizational objectives. This tailored approach facilitates a detailed understanding of each dimension's impact, guiding strategic decisions across the maturity spectrum.

Conduct Maturity Assessments

Using the FOCUS Framework, organizations can break down the transformational program into meaningful smaller initiatives. Each initiative can be evaluated separately, considering its unique attributes and alignment with the organization's goals.

a) **Identify the Functional Areas to be evaluated:** The first task in breaking down the Autonomous Finance Transformation Program's scope is to identify the Functional Areas within the organization that are ripe for transformation. This list will be a subset of the Functional Areas shown in Figure 5.1 of this book. Note that the selected areas should align with the organization's strategic objectives and offer substantial potential benefits through Autonomous Finance.

b) **Identify the Initiatives to be evaluated:** For the initiatives to be evaluated, define each project's objectives, expected results, and KPIs, considering their relevance across geographies, business units, specific functions, or technology platforms, ensuring they contribute to the Functional Area's overarching strategy.

c) **Consider Integration with Initiatives in Other Business Domains:** Integrating Autonomous Finance with various business domains is not a mere alignment of functions but a strategic fusion that creates a cohesive, intelligent, and agile organizational ecosystem. Organizations can break down silos, foster collaboration, and enable more informed, data-driven decisions by forming synergies across different domains. The result is a holistic approach to business operations where Autonomous Finance acts as a catalyst, driving efficiency, innovation, and success across the entire organization. Whether aligning with the supply chain for streamlined procurement, synchronizing with HR for intelligent workforce planning, or collaborating with IT for technological innovation, integrating Autonomous Finance is critical in achieving true organizational excellence. Figure 7.5 epitomizes these synergistic intersections of Autonomous Finance with other business domains.

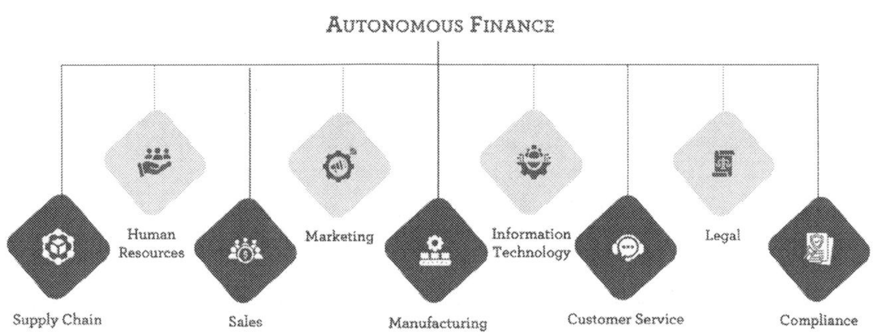

FIGURE 7.5: Synergy between Autonomous Finance and Other Business Domains

d) **Conduct Assessments at the Initiative Level:** The success of the transformation towards Autonomous Finance relies heavily on the precise and thoughtful execution of each initiative. Conducting detailed assessments at the initiative level is crucial for understanding the complexities, challenges, and opportunities of every aspect of the transformation. The FOCUS Framework is a systematic guide for conducting these assessments, providing a structured approach that aligns with the organization's strategic vision. Refer to Appendix B for an extensive list of key questions to assess the maturity level by that dimension.

7.5 Strategic Prioritization & Planning

The Strategic Prioritization and Planning stage lays the groundwork for the Autonomous Finance Transformation initiative, encompassing key initiatives, alignment with organizational goals, and meticulous orchestration of the transformation journey. This comprehensive planning involves several critical aspects:

Strategic Prioritization: Data-Driven Comparative Analysis of Initiatives

Strategic Prioritization is pivotal in evaluating potential transformation initiatives for Autonomous Finance. This process should align with an organization's broader vision, considering essential factors such as expected benefits, costs, risks, resource constraints, and organizational goals (OnStrategy, 2021). By assessing these aspects, organizations can systematically identify and prioritize the most valuable initiatives that align with their long-term strategies.

One essential aspect of this process should be the computation of a Business Value Score (BVS). This score should represent a numerical evaluation of each initiative's potential worth. It should be calculated using a formula that combines the Expected Business Value per Initiative with the Total Estimated Cost and adjusted for the Risk Factor. The expected business value should reflect the potential benefits of the initiative, while the total estimated cost should include all projected expenses. On the other hand, the risk factor should quantify the potential risks associated with implementing the initiative, giving a nuanced insight into its feasibility.

The next stage should involve a Comparative Analysis, where different initiatives should be examined side by side using the BVS and other relevant criteria. This objective and data-driven analysis should be visually represented through matrices or charts, enabling clear comparisons. This should facilitate informed decision-making, ensuring that concrete data back the initiatives chosen and align with the organization's overarching goals.

Planning: Orchestrating the Transformation Journey

Once key initiatives for the Autonomous Finance transformation are identified and prioritized, a meticulous plan should be crafted to guide their implementation. This process should begin with Roadmap Development, where a detailed outline should be created, encompassing the sequence, timeline, resources, dependencies, and milestones for each initiative. This roadmap should serve as the compass for the transformation journey, harmonizing the various initiatives to ensure they are orchestrated in sync and aligned with the organization's strategic goals.

The next essential component is Resource Allocation. This step recognizes that every initiative requires proper support and infrastructure to thrive. Whether it is human capital, technology, budget, or time, carefully allocating these resources is vital to each initiative's successful execution. It is about laying the groundwork, ensuring all the necessary elements are in place to translate vision into reality.

Monitoring and Evaluation form the pulse of the transformation journey (Mehta, 2020). With continuous monitoring and periodic evaluations, progress should be tracked and measured against defined KPIs and other metrics. This vigilant oversight allows for necessary adjustments, ensuring the initiatives stay on track and align with the core objectives.

Strategic Prioritization and Planning are the backbone of the Autonomous Finance transformation journey. They provide the structure, direction, and tools to navigate the transformation's complex terrain and align the organization's vision with grounded and practical approaches. By channeling resources and efforts towards the most promising initiatives and striking a balance between ambition and pragmatism, these components set the stage for a transformation that is both visionary and achievable. The art of the possible is harmonized with the pragmatism of the practical, creating a pathway that leads to transformative success.

7.6 Pilot Projects, POCs, and Experimentation

The transformational journey toward Autonomous Finance is like venturing into uncharted waters, filled with possibilities and uncertainties. It is a voyage that requires careful navigation guided by exploration, experimentation, and validation. This is the realm of pilot projects, POCs, and experimentation in Autonomous Finance.

With the initiatives and related projects selected, the next step is to define success criteria. What does success look like? How will it be measured? These questions guide establishing clear and measurable success criteria, encompassing financial, operational, and strategic metrics. This definition of success serves as the North Star, guiding the project's direction and evaluation.

Designing and planning the pilot or POC is akin to crafting a well-structured blueprint. It involves technology selection, process mapping, and integration planning. Every detail matters, from the algorithms that will drive decision-making to the ethical considerations surrounding data handling.

Engaging key stakeholders from the beginning ensures alignment and support. It builds a coalition of champions who understand the vision, share the excitement, and are committed to making it a reality. This engagement is not a one-time event but a continuous dialogue, fostering collaboration and shared ownership.

The development phase embodies an agile and iterative approach, allowing for rapid testing, learning, and refinement. It is a dynamic dance between technology and human insight, where continuous monitoring and regular evaluations provide insights into performance and potential improvements.

Measuring success is the culmination of this journey, assessing performance against defined KPIs, gathering feedback, evaluating alignment with strategic goals, and documenting lessons learned. It is a reflective moment, reflecting on the path traversed, celebrating successes, acknowledging failures, and gleaning insights for future endeavors.

In Autonomous Finance, pilot projects, POCs, and experimentation are more than trial runs or isolated initiatives. They are the building blocks of transformation, the laboratories of innovation, and the validation of vision. By embracing a structured, thoughtful, and measured approach, organizations can lay a solid foundation for broader implementation and success, setting the stage for a future where finance is not just a function but a proactive, intelligent, and adaptive entity that drives business success.

7.7 Implementation, Rollouts, and Scaling

The success of a pilot project or POC is but a prelude to the broader journey of implementing Autonomous Finance. It is a glimpse of potential, a spark that must now be fanned into a roaring flame that lights up the entire organization. The transition from a successful pilot to full-scale implementation is complex, fraught with challenges, but filled with immense possibilities. It requires a strategic approach, well-planned execution, and an understanding of the nuances of scaling.

Implementation Strategy

The implementation of Autonomous Finance is not a linear progression but a carefully orchestrated symphony of interconnected elements. It begins with formulating a clear and coherent strategy that aligns with the organizational vision and goals. This strategy is a roadmap guiding every implementation aspect, from technology selection to change management.

One of the fundamental decisions in crafting this strategy is choosing the right technology stack. It is about selecting tools and platforms that meet the immediate needs and have the flexibility and scalability to evolve with the organization. This selection process is an intricate evaluation of functionality, interoperability, cost-effectiveness, and long-term viability.

The alignment with other business domains is a pivotal aspect of the implementation strategy. Autonomous Finance does not exist in isolation; it interfaces with various functions such as Supply Chain, HR, and Sales. This alignment ensures seamless integration, enabling a holistic and interconnected view of the organization's operations.

Change management is the heartbeat of successful implementation. It is about preparing the organization for a new way of working, fostering a culture of acceptance, and building the skills needed to thrive in an autonomous environment. It is a people-centric approach that recognizes the human element in the transformation, addressing fears, building confidence, and nurturing a culture of innovation and agility.

Rollouts and Scaling

Rollouts and scaling are about translating strategy into action. It is the execution phase where plans are tested, theories meet reality, and the rubber meets the road.

The rollout process is often staged, starting with a controlled deployment in specific areas, learning from the experience, and gradually expanding across the organization. This phased approach allows for continuous monitoring, evaluation, and refinement, identifying and addressing potential challenges early.

Scaling is not merely about replicating success; it is about understanding the unique dynamics of different areas within the organization, adapting the approach, and tailoring solutions. It is a nuanced and iterative process that recognizes the diversity of needs, constraints, and opportunities across the organization.

Challenges and Best Practices

The road to implementation and scaling has bumps and curves. Challenges are inevitable, from technological hurdles to resistance to change. Navigating these challenges requires foresight, resilience, and adaptability.

Some common challenges include integration complexities, data quality issues, lack of skilled resources, and organizational resistance. These challenges are manageable but require thoughtful planning, proactive management, and a commitment to continuous improvement.

Best practices in navigating these challenges include regular communication, stakeholder engagement, robust governance structures, ongoing training, and alignment with industry standards and regulations. It is about creating an ecosystem that supports innovation, values collaboration, and is committed to excellence.

The journey of implementing and scaling Autonomous Finance is a tapestry of strategy, execution, challenges, and triumphs. It is a journey that transforms the fabric of the finance function, weaving technology, people, and processes into a cohesive and dynamic entity that drives organizational success. It is a journey that redefines what finance can be, unlocking new potentials, creating new opportunities, and setting the stage for a truly autonomous future.

7.8 Support, Maintenance, and Continuous Improvement

Embarking on the journey of Autonomous Finance is an ongoing commitment that extends beyond the initial implementation and scaling. It is a continuous support, maintenance, and innovation process that ensures the organization's long-term success and evolution of the Autonomous Finance paradigm. This section delves into the critical aspects of ongoing support, maintenance, and the pursuit of continuous improvement and innovation.

Ongoing Support and Maintenance

The transformation to Autonomous Finance is not a one-time event but an evolving ecosystem that requires constant nurturing, monitoring, and support. This ongoing support encompasses various dimensions as elucidated in Figure 7.6:

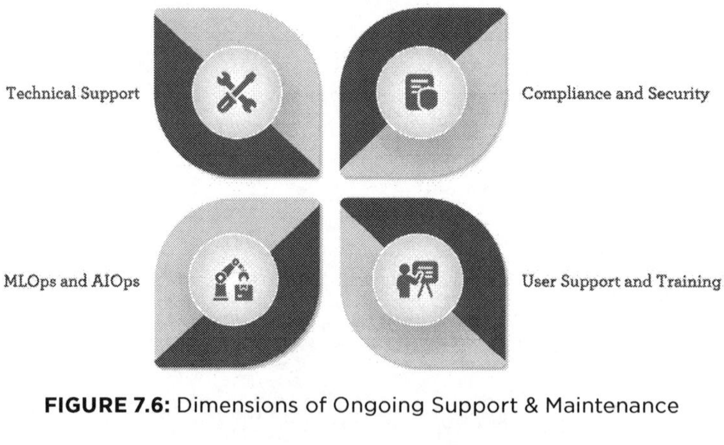

FIGURE 7.6: Dimensions of Ongoing Support & Maintenance

a) **Technical Support** involves regular monitoring, troubleshooting, and optimization of the autonomous systems. Ensuring that the systems operate smoothly, updating them to adapt to new technological advancements, and promptly addressing any issues is vital.

b) **Compliance and Security:** Autonomous Finance must comply with various regulations and security protocols. Regular updates, audits, and adherence to compliance standards safeguard the system's integrity and protect sensitive information.

c) **User Support and Training:** As the Autonomous Finance system evolves, continuous training and support are essential to ensure the team can adapt to new features, functionalities, and working methods. The continuous training and support ensure that the human aspect of the system remains in sync with technological advancements.

d) **MLOps and AIOps:** Implementing MLOps (Machine Learning Operations) practices ensures a seamless transition from development to deployment and monitoring of ML models. AIOps takes this further, integrating various AI-based systems within the organization and ensuring their seamless interplay. These practices contribute to the Autonomous Finance system's scalability, robustness, and agility.

Ongoing support and maintenance are the bedrock of sustainable Autonomous Finance. It ensures the system remains robust, compliant, efficient, and aligned with the organization's changing needs.

Continuous Improvement and Innovation

Autonomous Finance is dynamic, fueled by continuous technological advancements, regulation shifts, and ever-evolving organizational needs. A commitment to continuous improvement and innovation is essential to stay at the forefront of this dynamic landscape. The multifaceted dynamics of Autonomous Finance are captured and simplified in Figure 7.7.

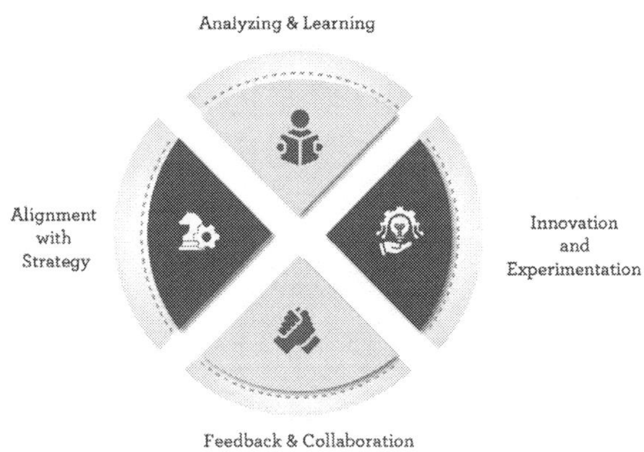

FIGURE 7.7: The Continuous Improvement and Innovation Cycle

a) **Analyzing and Learning:** Regularly reviewing the performance of the Autonomous Finance system, analyzing trends, and learning from successes and failures is crucial. This analysis fosters a culture of continuous learning and improvement.

b) **Innovation and Experimentation:** Encouraging innovation and experimentation keeps the Autonomous Finance system vibrant and responsive. Whether exploring new algorithms, integrating emerging technologies, or redesigning workflows, a spirit of innovation ensures that the system continues to evolve and adapt.

c) **Feedback and Collaboration:** Engaging stakeholders, collecting feedback, and fostering collaboration across various domains within the organization strengthens the alignment and synergy of Autonomous Finance with the broader organizational goals.

d) **Alignment with Strategy:** As organizational strategies evolve, so must the Autonomous Finance system. Regular alignment checks and strategic reviews ensure that the system supports and drives the organization's goals.

Pursuing continuous improvement and innovation is not just about maintaining the status quo. It is about pushing boundaries, exploring new horizons, and ensuring that Autonomous Finance remains a strategic asset contributing to the organization's success. It is about constantly looking for ways to improve processes, products, and services in the finance domain. It is also about being open to new ideas and willing to take risks, about constantly learning and growing, and about being the best that we can be.

7.9 Pragmatic Steps and Best Practices

Adopting and implementing a framework such as the FOCUS framework is not merely an administrative exercise but a transformative journey that shapes the future of finance within an organization. This journey, while beneficial, is laden with complex decisions, investment considerations, and change management efforts. This section summarizes the pragmatic steps and best practices discussed in previous sections of this chapter to help organizations navigate this path and successfully transform from traditional to Autonomous Finance.

a) **Maturity Assessment:** Embarking on this journey begins with an honest assessment of the current state across the eight dimensions of the maturity model. Such an assessment entails understanding existing capabilities, shortcomings, and the level of maturity within each dimension. By doing so, organizations identify the gap between their current and desired state and gain a deeper understanding of the interconnectedness and dependencies across different dimensions.

b) **Goal Setting:** The next critical step involves goal setting. Goals should align with the organization's broader strategic vision and business objectives. It is vital to establish a clear, shared understanding of these goals across all levels of the organization, from executives to frontline employees. Establishing realistic timelines, milestones, and success metrics for these goals promotes transparency and accountability.

c) **Charting the Roadmap:** An organization can chart its roadmap once goals are set. This roadmap should outline the sequence of actions, investments, and initiatives. Given the interdependencies across the dimensions, the sequence matters. For instance, an organization may prioritize Process Automation and Data Analytics before focusing on Decision-Making Autonomy. This roadmap is dynamic and should be periodically revisited and revised to accommodate changing business contexts, technology advancements, and learning from implementation.

d) **Implementation Plan & Execution**: Implementing the maturity model necessitates investment decisions, both financial and non-financial. Beyond financial investments in technology and infrastructure, organizations must invest in talent development, change management, and cultivating a culture that supports Autonomous Finance. It is also crucial to ensure the availability of necessary resources and commitment from key stakeholders throughout this journey.

e) **Continuous Improvement**: Lastly, remember that the journey to Autonomous Finance maturity is a continuous improvement rather than a one-time effort. It requires ongoing reassessment and recalibration. Monitoring progress against the set metrics, learning from successes and setbacks, and iterating on the strategy are all integral to this journey.

In conclusion, embarking on the journey to implement Autonomous Finance is a digital transformation initiative that allows organizations to revolutionize their Finance function and reap substantial benefits. While the journey may be challenging, with a clear vision, strategic planning, and a steadfast commitment, organizations can successfully navigate this path and position themselves at the forefront of the Autonomous Finance revolution.

7.10 Conclusion

As elaborated in this chapter, the journey toward Autonomous Finance is neither linear nor monolithic; it is a complex, multidimensional transformation characterized by a series of interconnected milestones. The exploration of the FOCUS Framework, the alignment of organizational vision, and the intricate strategies for pilot projects and rollouts substantiates the thesis proffered at the chapter's outset. Such a robust validation is further corroborated by data: a recent Accenture study revealed that companies implementing comprehensive frameworks saw a 20% improvement in operational efficiency within the first year (Accenture, 2023).

As finance continues its evolutionary ascent, a vista of transformative possibilities unfolds. Questions, profound in their implications, arise: How will AI-imbued finance functions achieve self-repairing capabilities? In what ways will the nascent field of AIOps catalyze operational paradigms? These questions, supported by studies like those from Gartner, which predict that AI in the finance market will reach $11 billion by 2025 (Gartner, 2021), beckon a future where finance transcends mere functionality to become an intelligent, adaptive catalyst of enterprise-wide success.

However, in an era of interconnectedness, the triumph of Autonomous Finance is irrevocably tied to synergistic strategies and partnerships, the focus of the ensuing chapter. The narrative will elucidate how these collaborations form the sinews that connect technology, strategy, and execution.

In closing, the odyssey toward Autonomous Finance is not a journey to a distant future but a transformative reality manifesting in the present. It represents technological advancements and a seminal paradigm shift—a reconceptualization of finance into an intelligent, adaptive catalyst for enterprise-wide success.

7.11 Key Recommendations

a) **Develop a Comprehensive Maturity Model:** Embrace a maturity model like the FOCUS Framework, tailored to the specific needs and challenges of Autonomous Finance. Such a model should guide the organization through progressive stages, integrating strategic alignment, data intelligence, technology enablement, human capital development, operational architecture, and scalable adoption.

b) **Invest in Pilot Projects and Experimentation:** Encourage a culture of innovation by investing in pilot projects, proofs of concept, and thoughtful experimentation. This approach allows for careful risk assessment, iterative learning, and alignment with organizational needs, offering a safe space to explore and adapt to the unique challenges of Autonomous Finance.

c) **Implement with Vision and Strategy:** Approach implementation and scaling with a clear vision and a robust strategy. Incorporate cutting-edge practices such as AIOps and MLOps to ensure seamless technology and human capital integration. Understand that the implementation process requires a multifaceted approach, considering technology, organizational alignment, and ethical considerations.

d) **Commit to Ethical Practices and Social Responsibility:** Ensure the transformation to Autonomous Finance adheres to the highest ethical standards, which include a commitment to transparency, accountability, social responsibility, and sustainable practices. Ethical stewardship should be at the heart of the journey, reflecting a responsible and humane approach to finance.

e) **Foster a Culture of Continuous Improvement and Innovation:** Recognize that the journey to Autonomous Finance is an ongoing process that demands continuous improvement, innovation, and adaptability. Implement methodologies and practices for ongoing support, maintenance, and continuous innovation. Keep abreast of emerging trends, technologies, and market dynamics to ensure that Autonomous Finance remains agile, relevant, and responsive to the evolving landscape.

f) **Invest in AIOps and MLOps Capabilities:** As Autonomous Finance grows more complex, traditional operational paradigms will become obsolete. Investment in AIOps and MLOps is a technological enhancement and a strategic imperative for sustainable scalability.

7.12 References

Accenture. (2023). *Accenture Life Trends 2023 (formerly Fjord Trends).* https://www.accenture.com/us-en/insights/song/accenture-life-trends#

Cohen, W. A. (2009). *Drucker on leadership: New Lessons from the Father of Modern Management.* John Wiley & Sons.

Dhar, V. (2013). Data science and prediction. *Communications of the ACM, 56*(12), 64-73. https://doi.org/10.1145/2500499.

Gartner. (2021). Title of the Document. https://www.gartner.com/en/documents/4009799.

Hagel, J., Brown, J. S., & Kulasooriya, D. (2014). *Institutional Innovation: How to Help Your Organization Learn Faster and Thrive.*

Kotter's 8-step Model of Change. (1996). https://www.managementstudyguide.com/kotters-8-step-model-of-change.htm.

Mehta, N. (2020). Your digital transformation journey is incomplete without Digital Experience Monitoring! *www.linkedin.com*. https://www.linkedin.com/pulse/your-digital-transformation-journey-incomplete-without-mehta/.

Mittelstadt, B., Allo, P., Taddeo, M., Wachter, S., & Floridi, L. (2016b). The ethics of algorithms: Mapping the debate. *Big Data & Society, 3*(2), 205395171667967. https://doi.org/10.1177/2053951716679679.

OnStrategy. (2021). The Strategic Planning Process in 4 Steps | OnStrategy. *OnStrategy*. https://onstrategyhq.com/resources/strategic-planning-process-basics/.

Provost, F., & Fawcett, T. (2013). *Data science for business: What you need to know about data mining and data-analytic thinking*. O'Reilly Media, Inc., https://dl.acm.org/citation.cfm?id=2564781.

Reinventing FP&A for the pandemic and beyond. (2020). Deloitte United States. https://www2.deloitte.com/us/en/pages/finance/articles/cfo-insights-reinventing-fpa-for-the-pandemic-and-beyond.html.

Stone, L. (2021, May 19). *"Change is hard at first, messy in the middle, and gorgeous at the end" - Robin Sharma • She leads change*. She Leads Change. https://sheleadschange.org/change-is-hard-at-first/.

The promise and challenge of the age of artificial intelligence. (2018). McKinsey & Company. https://www.mckinsey.com/featured-insights/artificial-intelligence/the-promise-and-challenge-of-the-age-of-artificial-intelligence.

7.13 Appendix B: Dimensions of the FOCUS Framework

The FOCUS Framework is a systematic approach to guiding organizations through the complex journey of Autonomous Finance transformation. It emphasizes six critical dimensions, each representing a core aspect of the transformation. Together, they create a comprehensive roadmap that aligns with the organization's unique goals, capabilities, and challenges. The following sections delve into these dimensions, providing an in-depth look at their definitions, detailed assessment criteria, and key questions to guide the assessment.

Dimension #1: Strategic Alignment

Strategic Alignment refers to the conscious integration of the Autonomous Finance vision with the overall business strategy and objectives. This dimension ensures that the journey toward Autonomous Finance is harmonized with broader organizational goals, values, and mission. It encompasses setting a clear vision, defining specific milestones, and creating a roadmap that unites all efforts toward fulfilling these strategic objectives. For example, an energy company striving for sustainability might align its Autonomous Finance strategy with green initiatives, utilizing AI and analytics to optimize energy consumption, reduce costs, and support environmental stewardship. Through Strategic Alignment, the organization ensures that Autonomous Finance becomes a strategic enabler, driving alignment, innovation, and value creation across the enterprise.

Table 7.1 elucidates the capabilities, benefits, challenges, and recommendations pertinent to five key stages—Formalizing, Organizing, Connecting, Unifying, and Synchronizing in the Strategic Alignment dimension of the FOCUS framework.

TABLE 7.1: Detailed Assessment of Strategic Alignment Across Maturity Levels

Strategic Alignment	F - Formalizing	O - Organizing	C - Connecting	U - Unifying	S - Synchronizing
Capabilities	Identification of Autonomous Finance-related strategic goals; Establishment of initial communication channels.	Development of alignment pathways across business units; Definition of key alignment metrics.	Embedding of aligned strategies within departmental processes; Establishment of alignment review cycles.	Full integration of Autonomous Finance strategy with business strategy; Alignment with external stakeholders.	Achieving strategic harmony across organizations, including suppliers and partners.
Benefits	Foundation laid for alignment; Initial connection between finance and organizational vision.	Enhanced collaboration and synergy between departments; Improved alignment tracking.	Streamlined alignment across organizational processes; Regular alignment checks.	Holistic strategic alignment: Alignment with industry and market dynamics.	Seamless alignment with all stakeholders; Continuous alignment adaptation.
Challenges	Conflicting priorities; Lack of clarity in strategic direction.	Misalignment with business units; Resistance to change.	Potential misalignment due to conflicting departmental goals.	Balancing alignment with evolving business needs and external pressures.	Constant realignment needs due to changing business dynamics and regulations.
Recommend-ations	Develop clear communication strategies between strategic planning and finance teams; Conduct alignment workshops.	Facilitate cross-departmental collaboration; Regularly review alignment metrics.	Implement regular alignment reviews; Update strategies as needed.	Maintain alignment with both internal and external factors; Implement continuous alignment monitoring.	Implement alignment feedback loops; Continuously assess and realign as needed.

Key Questions to Assess the Maturity Level in Dimension #1:

- How are finance strategic goals identified and mapped to organizational objectives?
- What channels are in place to communicate alignment strategies?
- How are alignment pathways developed across different business units?
- What metrics track strategic alignment, and how are they reviewed?
- How are conflicting priorities and resistances to alignment addressed?
- How is alignment embedded within departmental processes?
- What cycles are in place for regular alignment reviews?
- How is alignment maintained with external stakeholders, such as suppliers and partners?
- How is continuous alignment monitoring conducted?
- What are the mechanisms for continuous alignment adaptation?

- How is alignment balanced with evolving business needs and external pressures?
- How are alignment feedback loops implemented and managed?
- How does the organization assess and realign strategic alignment as needed?
- What are the tools and technologies used for alignment tracking and management?
- How are alignment success stories celebrated and shared within the organization?
- What training and development programs are in place to support alignment strategies?
- How are alignment challenges anticipated, and what proactive measures are taken?
- What is the role of leadership in fostering and maintaining strategic alignment?
- How does the organizational culture support or hinder strategic alignment?
- What strategies are implemented to achieve holistic alignment with internal and external factors?

Dimension #2: Data Intelligence

Data Intelligence emphasizes the intelligent use of data within financial functions, encompassing more than mere data collection. It comprehensively handles data quality, integration, analysis, and actionable insights. Within the Autonomous Finance context, Data Intelligence translates into using data analytics, ML, and AI to identify patterns, predict trends, and offer decision support. For example, an e-commerce company might use Data Intelligence to analyze customer purchasing behavior, predict inventory needs, and optimize supply chain management.

Table 7.2 portrays the evolutionary path from the preliminary 'Formalizing' stage to the advanced 'Synchronizing' stage in the Data Intelligence dimension of the FOCUS framework, shedding light on the shifting landscape of capabilities, benefits, challenges, and crucially, the recommendations for each.

TABLE 7.2: Detailed Assessment of Data Intelligence Across Maturity Levels

Data Intelligence	F - Formalizing	O - Organizing	C - Connecting	U - Unifying	S - Synchronizing
Capabilities	Basic data collection; no systematic analysis or insights.	Introduction of data analytics tools; some insights for decision-making.	Integration of data analytics across functions; regular insights driving decisions.	Advanced analytics and AI-driven insights; data-driven culture.	Fully autonomous data intelligence; continuous insights and predictions.
Benefits:	Limited insights; potential underutilization of data.	Improved decision-making; better use of existing data.	Enhanced data-driven decisions; improved alignment across functions.	Strategic use of data; significant efficiency and effectiveness gains.	Full leverage of data for strategic advantage; continuous innovation.

Data Intelligence	F - Formalizing	O - Organizing	C - Connecting	U - Unifying	S - Synchronizing
Challenges	Data quality issues; lack of analytical skills.	Integration challenges; need for data governance.	Complexity in handling large data sets; alignment with business objectives.	Ensuring data security; managing AI and advanced analytics.	Maintaining up-to-date technology and continuous alignment with business needs.
Recommend-ations	Invest in data quality; initiate analytics training.	Establish data governance; integrate analytics tools.	Align data strategies with business goals; invest in technology.	Strengthen security measures; foster a data-driven culture.	Continuously monitor and update technology; align with strategic objectives.

Key Questions to Assess the Maturity Level in Dimension #2:

- How is data collected, stored, and managed within the organization?
- What tools and technologies are in place for data analytics, and how are they integrated?
- How is data quality ensured, and what mechanisms are used for governance?
- To what extent are data analytics and ML utilized to identify patterns and predict trends?
- How are data insights translated into actionable strategies and decisions within the finance function?
- What challenges are faced in handling and analyzing large data sets, and how are they addressed?
- How does the organization ensure data security, especially when using advanced analytics and AI?
- How are data strategies aligned with overall business goals and objectives?
- What is the level of investment in data analytics technology, and how is it justified?
- How does the organization foster a data-driven culture, and what training programs are in place?
- How are data insights shared across different functions, and how do they contribute to collaboration?
- What are the main data sources, and how is external data integrated with internal data?
- How is the effectiveness of data analytics measured, and what KPIs are used?
- What plans are in place for continuous development and updating of data analytics technology?
- How does the organization balance using human expertise and machine-driven insights in decision-making?
- What role do data analytics play in risk management and compliance within the finance function?

- How are customers and other stakeholders engaged in the use of data, and how is their privacy ensured?
- How does the organization ensure ethical AI and data analytics practices?
- What successes and failures have been experienced using data intelligence, and how have they informed future strategies?
- How does the organization envision the future of data intelligence in driving Autonomous Finance, and what plans are in place to achieve this vision?

Dimension #3: Technology Enablement

Technology Enablement signifies the deliberate integration of technology that facilitates autonomous operation within the financial function. It is not solely about acquiring the latest tools but strategically selecting, implementing, and optimizing the technologies that empower automation, analysis, and intelligence. In the healthcare sector, Technology Enablement might involve implementing RPA for claims processing and AI-driven tools for predictive health analytics.

Table 7.3 elucidates the capabilities, benefits, challenges, and recommendations pertinent to five key stages—Formalizing, Organizing, Connecting, Unifying, and Synchronizing in the Technology Enablement dimension of the FOCUS framework.

TABLE 7.3: Detailed Assessment of Technology Enablement Across Maturity Levels

Technology Enablement	F - Formalizing	O - Organizing	C - Connecting	U - Unifying	S - Synchronizing
Capabilities	Limited technology infrastructure; no automation.	Introduction of some automation tools; lacks full integration.	Integrated technology solutions; some AI and automation in place.	Advanced technology infrastructure; AI-driven automation across functions.	Fully autonomous technology ecosystem; seamless integration and continuous innovation.
Benefits:	Limited efficiency; manual processes dominate.	Improved efficiency through automation; more accurate processing.	Enhanced collaboration and decision-making; reduced manual interventions.	Strategic technology alignment; significant gains in efficiency and accuracy.	Full technology leverage for strategic advantage, continuous innovation, and growth.
Challenges	Outdated technology; resistance to change.	Integration challenges; managing change and adoption.	Balancing technology and human intervention; alignment with business objectives.	Ensuring technology security and managing complex AI and automation processes.	Maintaining up-to-date technology; ensuring continuous alignment with business needs.
Recommend-ations	Invest in technology upgrades; initiate change management.	Integrate automation tools; foster a technology-friendly culture.	Align technology strategies with business goals; invest in continuous development.	Strengthen security measures; foster an innovative technology culture.	Continuously update technology; align with strategic objectives and market trends.

Key Questions to Assess the Maturity Level in Dimension #3:

- What technologies are currently in place to support financial functions, and how are they integrated?
- To what extent is automation applied in financial processes, and how effectively is it managed?
- How aligned are the current technology solutions with the organization's business objectives?
- What are the strategies for adopting emerging technologies such as AI, machine learning, and blockchain in finance?
- How is technology innovation fostered within the finance department?
- How are technology investments prioritized, and what criteria are used for decision-making?
- What measures are in place to ensure the security and compliance of financial technologies?
- How is the balance between human intervention and technology automation maintained in financial operations?
- What is the level of collaboration between the finance department and IT in technology decision-making?
- How are technology-related changes communicated and managed across the organization?
- What training and development programs are in place to ensure the finance team's proficiency in new technologies?
- How is the success and effectiveness of technology implementations measured and evaluated?
- What strategies are in place to ensure the scalability of technology solutions as the organization grows?
- How does the organization stay abreast of technological advancements relevant to finance?
- What challenges have been encountered in implementing and integrating new technologies, and how were they addressed?
- How does the organization ensure technology implementations align with ethical considerations and industry regulations?
- What role does technology play in enhancing the customer experience in financial services?
- How does the organization plan for technology obsolescence and ensure continuous technology refreshment?
- What provisions are in place to manage technology-related risks, including cybersecurity?
- How does the organization envision the future of technology enablement in finance, and what plans are in place to achieve this vision?

Dimension #4: Human Capital Development

The dimension of Human Capital Development casts a spotlight on the essential human facets of transformation within the context of Autonomous Finance. It acknowledges the indispensable contributions of individual skills, inherent talents, and the readiness of organizational culture to realize the full potential of this new financial paradigm. At the core of this focus lies the continuous nurturing of abilities, the strategic management of change, and the cultivation of an organizational environment conducive to innovation, nimbleness, and synergistic collaboration. Consider the scenario of a global corporation intricately crafting learning journeys for its workforce. These pathways are designed to equip employees with competencies in emerging fields such as data analytics, AI, and strategic discernment, thereby aligning the human element with the continually shifting requisites of Autonomous Finance.

Table 7.4 portrays the evolutionary path from the preliminary 'Formalizing' stage to the advanced 'Synchronizing' stage in the Human Capital Development dimension of the FOCUS framework, shedding light on the shifting landscape of capabilities, benefits, challenges, and crucially, the recommendations for each.

TABLE 7.4: Detailed Assessment of Human Capital Development Across Maturity Levels

Human Capital Development	F - Formalizing	O - Organizing	C - Connecting	U - Unifying	S - Synchronizing
Capabilities	No focused skill development or change management.	Initial training in technology; introduction of change management.	Ongoing skill development; fostering a culture ready for change.	Continuous learning culture; strong alignment with Autonomous Finance needs.	Full integration of human and machine capabilities; agile and innovative culture.
Benefits	Limited human capital readiness; resistance to change.	Improved skills and readiness; better alignment with technology.	Enhanced collaboration and innovation; agility in adapting to change.	Strategic human capital alignment; significant gains in efficiency and innovation.	Full human capital leverage for strategic advantage; continuous growth and development.
Challenges	Lack of skills in emerging technologies; cultural resistance.	Balancing technology and human intervention; managing change.	Aligning human capital strategies with business objectives; continuous skill development.	Ensuring continuous alignment with technology and managing complex change processes.	Maintaining up-to-date skills, ensuring continuous alignment with business needs and technology trends.
Recommendations	Invest in process redesign; initiate change management.	Integrate automation tools; foster a process-friendly culture.	Align operational strategies with business goals; invest in continuous development.	Strengthen security measures; foster an innovative operational culture.	Continuously update processes; align with strategic objectives and market trends.

Key Questions to Assess the Maturity Level in Dimension #4:

- What strategies are in place to align human capital with the Autonomous Finance vision?
- How is the organization fostering a culture of continuous learning and innovation?
- What training programs are available to upskill employees in emerging technologies?
- How is change management integrated into the human capital development strategy?
- How are employees' roles and responsibilities evolving with Autonomous Finance initiatives?
- What mechanisms exist to identify and fill skill gaps related to Autonomous Finance?
- How is the organization balancing human intervention with technological advancements?
- What initiatives are in place to promote collaboration between finance, technology, and other departments?
- How is the organization measuring the effectiveness of its human capital development strategies?
- How are employees engaged and involved in the Autonomous Finance transformation journey?
- What are the main challenges and obstacles in developing human capital for Autonomous Finance?
- How does the organization ensure that human expertise complements technological advancements?
- What strategies are in place to retain key talents and skills relevant to Autonomous Finance?
- How does the organization promote agility and adaptability in its workforce?
- How are diversity and inclusion integrated into the human capital development strategy?
- How is the organization preparing its leaders to drive and support the Autonomous Finance transformation?
- What are the key success factors in the organization's human capital development approach?
- How is the organization leveraging external partnerships, networks, and resources to enhance human capital development?
- How are ethical considerations and values embedded in human capital strategies?
- What are the long-term goals and visions for human capital development in Autonomous Finance?

Dimension #5: Operational Architecture

Operational Architecture is concerned with the design, structure, and governance that enable the autonomous operation of the financial function. This dimension includes the planning and execution of processes, workflows, controls, and governance structures that support efficient and effective automation. In the manufacturing sector, Operational Architecture might mean reengineering procurement processes to include AI-driven supplier risk assessments, ensuring that procurement decisions are swift and informed.

Table 7.5 elucidates the capabilities, benefits, challenges, and recommendations pertinent to five key stages—Formalizing, Organizing, Connecting, Unifying, and Synchronizing in the Operational Architecture dimension of the FOCUS framework.

TABLE 7.5: Detailed Assessment of Operational Architecture Across Maturity Levels

Operational Architecture	F - Formalizing	O - Organizing	C - Connecting	U - Unifying	S - Synchronizing
Capabilities	Manual processes; no design for autonomy.	Some process automation; initial design for autonomy.	Integrated processes with some autonomous workflows.	Advanced autonomous operations with strong governance and controls.	Fully autonomous operations with optimal efficiency and effectiveness.
Benefits:	Limited efficiency; potential for errors.	Improved process efficiency; reduced manual interventions.	Enhanced collaboration and decision-making; increased accuracy.	Strategic, operational alignment; significant gains in efficiency and accuracy.	Full leverage of operations for strategic advantage, continuous innovation, and growth.
Challenges	Inefficient processes; resistance to change.	Integration challenges; managing change in operations.	Balancing autonomy and control; alignment with business objectives.	Ensuring operational security and managing complex autonomous processes.	Maintaining up-to-date processes, ensuring continuous alignment with business needs.
Recommend-ations	Invest in process redesign; initiate change management.	Integrate automation tools; foster a process-friendly culture.	Align operational strategies with business goals; invest in continuous development.	Strengthen security measures; foster an innovative operational culture.	Continuously update processes; align with strategic objectives and market trends.

Key Questions to Assess the Maturity Level in Dimension #5:

- What processes are currently in place for the financial functions, and how are they structured?
- How much automation is integrated into the existing operational architecture?
- What governance structures are in place to oversee autonomous operations?
- How are workflows designed to facilitate efficient and effective automation?
- What challenges have been encountered in implementing autonomous operations?
- How are security measures implemented within the operational architecture?
- What mechanisms are in place to ensure data integrity and maintain system security within autonomous operations?

- How are changes to operational processes managed, and who oversees these changes?
- How are operational efficiencies measured, and what metrics are used?
- What strategies are in place for scaling autonomous operations across different functions?
- How are failures or errors in autonomous operations handled, and what remediation processes are in place?
- How is the alignment between operational architecture and broader business objectives ensured?
- What training and development are provided to staff to manage and optimize autonomous operations?
- How are emerging technologies and innovations integrated into the operational architecture?
- How is the balance between human intervention and automation managed within operations?
- What collaboration exists between different functions to ensure cohesive autonomous operations?
- How is continuous improvement in operational efficiency encouraged and tracked?
- What role does leadership play in fostering a culture that supports autonomous operations?
- How does the organization plan for future developments in technology and process design that might affect operational architecture?
- How are customer feedback and external stakeholder input considered in the design and execution of operational processes?

Dimension #6: Scalable Adoption

Scalable Adoption underscores the necessity to grow and expand Autonomous Finance capabilities across the organization systematically. This dimension focuses on scaling successes, learning from failures, and building a sustainable and adaptable transformation. For a technology startup, Scalable Adoption might involve the gradual expansion of autonomous capabilities, starting with automating A/P and then systematically extending these capabilities across other financial functions, learning and adapting along the way.

Table 7.6 portrays the evolutionary path from the preliminary 'Formalizing' stage to the advanced 'Synchronizing' stage in the Scalable Adoption dimension of the FOCUS framework, shedding light on the shifting landscape of capabilities, benefits, challenges, and crucially, the recommendations for each.

TABLE 7.6: Detailed Assessment of Scalable Adoption Across Maturity Levels

Scalable Adoption	F - Formalizing	O - Organizing	C - Connecting	U - Unifying	S - Synchronizing
Capabilities	No scalability; isolated efforts.	Initial scalability in limited areas; lacks systematic approach.	Scalable across some functions, learning from successes and failures.	Systematic scalability across most functions; continuous improvement.	Fully scalable adoption; Autonomous Finance across the entire organization.
Benefits:	Limited scalability; potential underutilization of successes.	Improved scalability; better use of existing successes.	Enhanced scalability; improved alignment across functions.	Strategic scalability alignment; significant efficiency and effectiveness gains.	Full Leverage of scalability for strategic advantage, continuous innovation, and growth.
Challenges	Lack of scalable models; resistance to expansion.	Integration challenges; need for scalable governance.	Complexity in scaling across functions; alignment with business objectives.	Ensuring scalability security; managing complex scaling processes.	Maintaining up-to-date scalability and continuous alignment with business needs.
Recommend-ations	Invest in scalable models; initiate scalability management.	Establish scalable governance; integrate scalability tools.	Align scalability strategies with business goals; invest in technology and process design.	Strengthen security measures; foster a scalability-friendly culture.	Continuously monitor and update scalability; align with strategic objectives and market trends.

Key Questions to Assess the Maturity Level in Dimension #6:

- How is scalability currently defined and measured within the organization?
- What mechanisms are in place to ensure that successful initiatives can be replicated and scaled?
- How does the organization identify and prioritize areas for scalable adoption?
- What challenges have been encountered in previous attempts to scale initiatives, and how were they addressed?
- How is the scalability of Finance aligned with the overall business strategy?
- What processes and frameworks are in place to manage the scalability of Finance across different functions?
- How does the organization ensure that scalability is sustainable and adaptable to changing business needs?
- What technologies and tools facilitate scalable adoption, and how are they integrated?
- How does the organization foster a culture that supports and encourages scalability?
- What metrics and KPIs are used to monitor and evaluate the success of scalable adoption?

- How are employees trained and developed to support scalable adoption across various functions?
- How does the organization maintain security and compliance during the scaling process?
- What collaboration and communication channels exist to facilitate scalable adoption across different departments?
- How does the organization leverage external partnerships and collaborations to enhance scalability?
- What budgeting and resource allocation strategies are used to support scalable adoption?
- How are failures and setbacks in scalable adoption analyzed and learned from?
- How does the organization meet customer needs and expectations are met as initiatives are scaled?
- What role do leadership and governance play in guiding and overseeing scalable adoption?
- How are innovations and new technologies evaluated for their scalability potential?
- How does the organization plan for the future scalability of Autonomous Finance, considering emerging trends and market dynamics?

In summation, the tables encompassed within this Appendix serve as an exhaustive analytical compendium, delineating the nuanced capabilities, far-reaching implications, multifaceted challenges, and prescriptive recommendations germane to each distinct dimension across the maturity continuum within the FOCUS Framework. In the aggregate, these tables transcend mere informative utility; they coalesce to form a strategic cartography, meticulously crafted to shepherd organizations through the labyrinthine corridors of transformation towards an Autonomous Finance frontier.

CHAPTER 8

Synergistic Strategies

The Role of Partnerships in Autonomous Finance

"Coming together is a beginning;
keeping together is progress;
working together is a success."

- HENRY FORD (FORD & CROWTHER, 1922)

8.0 Key Insights

- **Unlocking Growth and Innovation through Strategic Partnerships:** The Autonomous Finance era emphasizes the importance of strategic partnerships for growth, innovation, and success. Collaborative efforts are vital in driving improvements and opening new value streams.
- **The Benefits of Strategic Partnerships:** Strategic partnerships offer three key value propositions: economic, strategic, and operational. Understanding these values can guide organizations in selecting and prioritizing partnerships.
- **Evaluating, Integrating, and Cultivating Strategic Partnerships:** Building a partner ecosystem requires careful evaluation of potential partners' technology capabilities, financial health, cultural fit, and strategic alignment. The integration of partners into an ecosystem should be designed to enhance collective capabilities.
- **Effective Communication and Coordination in Partnership Relationships:** Maintaining and optimizing partnership relationships requires clear communication, effective coordination, and regular performance monitoring.
- **Proactively Managing Risks in Strategic Partnerships:** Risk management and mitigation in strategic partnerships are crucial, involving systematic identification, assessment, and prioritization of risks. A robust risk management plan should include contingency plans, regular audits and assessments, and proactive communication strategies.
- **Building and Managing Strategic Partnerships for Optimal Outcomes:** A roadmap for building and managing strategic partnerships can guide organizations through the process, ensuring the best possible outcomes. This roadmap should include identifying needs, conducting market research, evaluating potential partners, forging, and integrating the partnership, maintaining regular communication, tracking performance, and periodic review and adjustment.

8.1 Introduction

In a world where the landscape of Autonomous Finance is perpetually evolving, strategic partnerships ascend from being mere ancillary constructs to vital cogs in the machinery of financial transformation. This chapter unfurls its narrative upon the tapestry woven in Chapter 7, where the transformation to Autonomous Finance was elucidated in detail. As Henry Ford aptly stated, *"Coming together is a beginning; keeping together is progress; working together is a success"* (Ford & Crowther, 1922)—a principle that serves as the lodestar for the exploration herein.

The acceleration of fintech investments to a staggering $105.3 billion in the Americas alone in 2021, a growth rate of 15% year-over-year (Ruddenklau & Caplain, 2022), magnifies the imperative of crafting strategic alliances. This chapter, therefore, embarks on an intellectual odyssey, dissecting the variegated layers of partnerships in Autonomous Finance. From fintech collaborations and technology alliances to cross-sector partnerships and social impact alliances, each strategic relationship bears its unique imprint on the financial landscape.

However, this chapter transcends mere description. It seeks to codify a definitive roadmap for the conceptualization, execution, and optimization of partnerships. This exploration will delve into the tripartite framework encompassing economic, strategic, and operational value propositions, offering tactical insights applicable to enterprises across the spectrum. However, the course is fraught with hazards. Thus, a segment of this chapter will be dedicated to the art and science of risk mitigation in strategic partnerships.

In an era marked by exponential technological progress and complex market dynamics, mastering strategic partnerships in Autonomous Finance is no longer optional—it emerges as a quintessential catalyst for transformative success. Through the lens of this chapter, organizations will find the navigational compass and the strategic arsenal required not merely to traverse but conquer the labyrinthine realm of financial partnerships.

8.2 Dimensions of Strategic Partnerships

Embracing the changes brought on by the Autonomous Finance revolution can be a complex endeavor that calls for a diverse network of skilled allies. In the context of Autonomous Finance, strategic partnerships are pivotal in enhancing capabilities, optimizing operations, and unlocking new opportunities. However, the breadth and depth of these partnerships can span a broad spectrum, each with its unique offerings and implications.

The following subsections offer a nuanced exploration of these dimensions, providing a comprehensive map to navigate partnerships in Autonomous Finance. Also, Figure 8.1 provides an executive summary of the strategic partnerships that coalesce to form a multifaceted strategy for excelling in Autonomous Finance.

a) **Fintech Partners:** Fintech companies have been at the forefront of the financial services revolution, introducing disruptive technologies and innovative solutions that have reshaped the industry landscape. Collaborations between traditional finance teams and fintech firms can create a symbiotic relationship that delivers enhanced customer experiences and improved operational efficiency. For example, Santander's partnership with Kabbage, a leading fintech platform, has enabled the bank to provide small businesses with quick and seamless access to working capital loans (*Online Working Capital Solution to Give UK SMEs Access to Same Day Funding | Santander UK*, 2016). Selecting the right fintech partners is critical for organizations seeking to drive growth and innovation. Factors to consider include the partner's technological capabilities, alignment with business objectives, and cultural fit.

b) **Technology Alliances:** Advanced technologies such as AI, ML, RPA, Big Data, Advanced Analytics, Cloud Computing, SaaS, and APIs are reshaping the finance landscape. Organizations can access cutting-edge tools and platforms that enhance decision-making, streamline operations, and unlock new growth opportunities by forging strategic alliances with technology providers. AI-driven solutions are increasingly important in financial services, enabling organizations to improve risk management, automate routine tasks, and deliver personalized customer experiences. For instance, JPMorgan Chase's collaboration with IBM has allowed the bank to leverage IBM's Watson AI technology to enhance its wealth management services and deliver tailored investment recommendations to clients (Lamonte & Lamonte, 2020). Establishing strategic technology alliances can help organizations stay ahead of the curve and capitalize on emerging innovations in the Autonomous Finance era. To ensure long-term success, organizations should prioritize partners that share a common vision and can demonstrate a proven track record of delivering results.

c) **Industry Networks and Cross-Sector Partnerships:** Industry networks, such as consortia and trade associations, can serve as valuable platforms for collaboration and knowledge sharing, enabling organizations to stay informed about the latest trends, regulatory developments, and best practices. Additionally, these networks can facilitate partnerships with non-financial players with complementary capabilities, opening new avenues for growth and innovation. Cross-sector partnerships can help organizations unlock new value streams and drive innovation by combining the strengths of various industries. For example, the collaboration between Mastercard and R3, a leading blockchain software provider, aims to develop a new blockchain-enabled cross-border payment solution, leveraging the power of distributed ledger technology to streamline payment processes and reduce transaction costs (Eisen, 2019). Successful cross-industry collaborations can serve as a blueprint for organizations seeking to explore new opportunities in the Autonomous Finance era. One such example is the partnership between Barclays and Safaricom, a Kenyan mobile network operator, which led to the creation of M-Shwari. This mobile banking platform offers microloans and savings accounts to millions of unbanked customers.

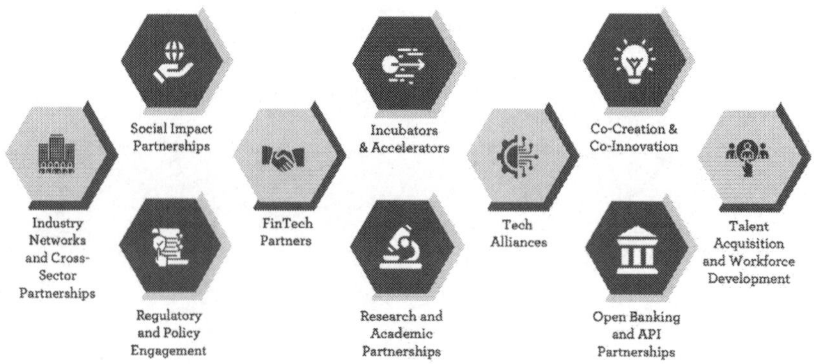

FIGURE 8.1: Partnerships in Autonomous Finance

d) **Regulatory and Policy Engagement:** As the financial services industry undergoes rapid transformation, the regulatory landscape is evolving to address emerging risks and promote stability. By proactively engaging with regulators and policymakers, organizations can better anticipate and adapt to changing regulations, ensuring compliance, and fostering stakeholder trust. Transparency and compliance with regulations are essential to building trust with customers, investors, and regulators. Collaborating with regulators to develop future-ready solutions can help organizations demonstrate their commitment to ethical practices and responsible innovation. Organizations can work closely with regulators to influence policy and develop solutions that address emerging challenges in the Autonomous Finance era. For example, the Global Financial Innovation Network (GFIN), a group of over 70 organizations, aims to promote cross-border collaboration and knowledge sharing, supporting the development of innovative financial services that align with regulatory objectives (*GFIN*, 2021).

e) **Research and Academic Partnerships:** Collaborating with research institutions and academia can provide organizations with access to the latest research, enabling them to stay at the forefront of innovation and develop solutions that address the unique challenges of the Autonomous Finance era. Academic partnerships can help organizations foster a culture of innovation and continuous learning, encouraging employees to think creatively and embrace new ideas. For instance, BBVA's partnership with MIT aims to drive innovation in finance through joint research initiatives, knowledge sharing, and talent development programs (Communications, 2023). Academic partnerships can also support talent development by creating a pipeline of skilled professionals equipped to navigate the challenges of the Autonomous Finance era. Collaborations with universities and other educational institutions can facilitate internships, training programs, and employee exchange initiatives that enhance workforce skills.

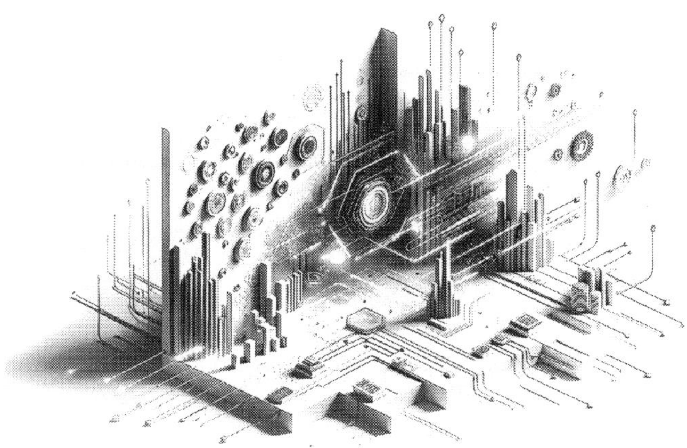

f) **Open Banking and API Integration:** Open banking, enabled by the widespread adoption of APIs, has emerged as a powerful driver of collaboration in the financial services industry. Banks can foster a more diverse and competitive ecosystem by providing third-party developers access to their data and services, unlocking new business models and revenue streams. Integrating API ecosystems can enable organizations to become more agile and responsive to market needs, allowing them to develop and deploy new products and services quickly. For example, Capital One's Developer Portal provides third-party developers access to various APIs, enabling them to create innovative financial applications that enhance customer experiences (Taft, 2021). Open banking initiatives, such as the European Union's (EU's) Revised Payment Services Directive (PSD2), have spurred the development of innovative financial services that deliver greater value to customers. For instance, the UK-based fintech company Yolt has leveraged open banking to create a personal finance management app that allows users to view and manage their accounts from multiple banks in a single interface, simplifying financial management and promoting better financial decision-making.

g) **Talent Acquisition and Workforce Development Partnerships:** The shift toward Autonomous Finance demands a workforce equipped with the skills to harness advanced technologies and drive innovation. By partnering with specialized talent acquisition firms and workforce development organizations, businesses can source the right talent to navigate the challenges of this new era. Collaborative training initiatives and employee exchange programs can help organizations develop their talent pool and foster a culture of continuous learning. For example, Coursera and Goldman Sachs have partnered to provide a digitized curriculum and interactive training at no charge to women business owners and employees in developing markets (*Goldman Sachs 10,000 Women*, n.d.). Encouraging a learning mindset and promoting skill development across the organization is critical to staying competitive in the Autonomous Finance era. By partnering with external organizations and leveraging internal resources, businesses can create a culture of continuous improvement and adaptability.

h) **Co-creation and Co-innovation Partnerships:** Co-creation and co-innovation partnerships bring together organizations with complementary capabilities and expertise to develop new products, services, and business models. By fostering a collaborative environment and leveraging the strengths of each partner, these partnerships can drive innovation and create significant competitive advantages. The strategic alliance between Citigroup and Feedzai exemplifies the potent efficacy of co-creation in today's Autonomous Finance landscape. Integrating Feedzai's expertise in real-time risk management through AI with Citigroup's expansive financial network, the partnership has crafted a state-of-the-art fraud detection system. Employing ML algorithms, this collaborative venture significantly enhances security protocols, fortifies regulatory compliance, and amplifies customer trust. It stands as a beacon of how converging competencies can engender transformative solutions in the ever-evolving ecosystem of Autonomous Finance (Feedzai, 2023). To succeed in the Autonomous Finance era, organizations must cultivate a culture of collaborative innovation, encouraging employees to engage with external partners and share ideas openly. By embracing a co-creation mindset, businesses can unlock new opportunities and drive continuous improvement.

i) **Incubators and Accelerators:** Incubators and accelerators can play a critical role in fostering Autonomous Finance innovation. By providing startup mentorship, funding, and resources, these programs can help drive the development of disruptive solutions that challenge the status quo and reshape the financial services landscape. Barclays' Rise accelerator program is a prime example of a successful initiative that supports fintech startups and drives innovation in finance (*Barclays Accelerator Programme | Barclays*, n.d.). The program has nurtured over 200 fintech companies, many of which have gone on to secure significant investments and forge strategic partnerships with established industry players. Engaging with incubators and accelerators can help organizations tap into a pool of innovative startups and forge strategic partnerships that drive growth and innovation. By closely monitoring these programs and actively participating in their activities, businesses can identify potential partners and stay abreast of emerging trends and technologies.

j) **Social Impact Partnerships:** Organizations must consider their action's broader social and environmental implications. By forging social impact partnerships, businesses can develop solutions that address pressing societal challenges while driving growth and creating shared value. One notable example of a social impact partnership is the collaboration between Mastercard and the World Bank, which aims to promote financial inclusion and digital financial services in developing countries (*UNDP/Mastercard Lighthouse MASSIV Partnership Programme | United Nations Development Programme*, n.d.). Through this partnership, both organizations develop and implement innovative solutions that expand access to formal financial services and empower underserved communities. Social impact partnerships can help organizations build a more sustainable and inclusive financial ecosystem, fostering long-term growth and resilience. Organizations can create a positive brand image, attract conscious investors, and drive customer loyalty by prioritizing social impact goals and aligning them with business objectives.

It is crucial for organizations to understand the different types of partnerships they can form and to choose the ones that align with their strategic objectives.

8.3 The Value Proposition of Strategic Partnerships

What are the quantifiable benefits of strategic partnerships for organizations on the journey toward Autonomous Finance? Turning attention to Figure 8.2, it is evident that the value triad - **economic, strategic, and operational** - forms the essence of these partnerships.

a) **Economic Value:** A partner's offerings can help organizations reduce costs or increase revenue. For instance, by partnering with a technology provider, an organization can leverage the partner's expertise and solutions instead of investing heavily in developing proprietary technology.

b) **Strategic Value:** Partnerships can offer significant value, helping organizations navigate new markets, access unique capabilities, or strengthen their competitive position. A case in point is the partnership between banking entities and fintech companies. The fintech companies bring innovation and agility, while banks offer a broad customer base and regulatory experience (Schueffel, 2016).

FIGURE 8.2: Value Proposition of Strategic Partnerships in Autonomous Finance

c) **Operational Value:** On a functional level, partnerships can improve the efficiency of various processes, enhancing the organization's operational agility and responsiveness. Partnerships with data providers or AI technology firms can enhance the organization's ability to process large volumes of financial data more efficiently and accurately.

Understanding the potential value derived from a strategic partnership can guide the selection and prioritization of partners, just as a lighthouse guides ships safely to their destination.

8.4 Ecosystem Dynamics: Partner Selection & Integration

The process of evaluating potential partners requires a keen understanding of the environment and foresight into the future. The critical considerations during this process include:

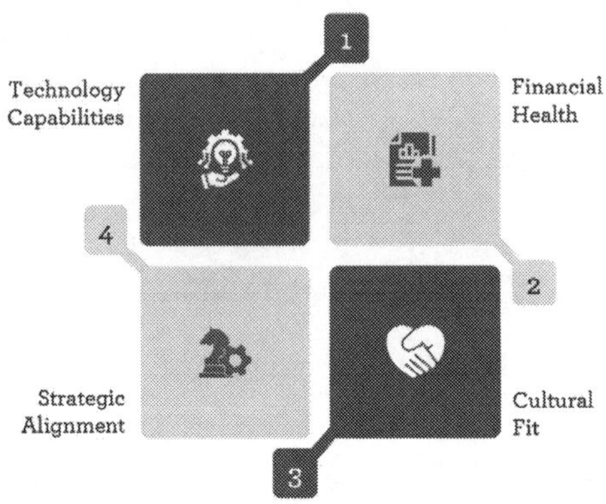

FIGURE 8.3: Primary Criteria for Partner Selection

a) **Technology Capabilities:** A partner's technology capabilities can determine how effectively they can support an organization's Autonomous Finance objectives. For instance, a partner specializing in AI and ML can offer invaluable expertise for developing and implementing Autonomous Finance solutions.

b) **Financial Health:** The financial stability of a potential partner is crucial, as it impacts their ability to meet their obligations and contribute to the long-term partnership.

c) **Cultural Fit:** Like a well-coordinated ship crew, partners must work cohesively toward a common goal. A good cultural fit ensures the partners can collaborate effectively, sharing a similar vision and values.

d) **Strategic Alignment:** This involves ensuring the partner's strategic objectives align with the organization's objectives. A lack of alignment can lead to conflicts and inefficiencies, hindering the journey toward Autonomous Finance.

Once potential partners have been evaluated and selected, they should be integrated into the organization's ecosystem. Each partner should complement the organization and other partners, enhancing the collective capabilities of the ecosystem.

For example, a partnership with an AI technology provider can complement another partnership with a data provider, enabling the organization to leverage high-quality data for AI-driven financial processes. Similarly, a partnership with a regulatory consulting firm can be invaluable in navigating the complex regulatory landscape of Autonomous Finance, complementing technology-driven partnerships.

8.5 Maintaining & Optimizing Partner Relationships

Maintaining and optimizing the relationships between strategic partners is an ongoing process necessary to weather storms and reach the intended destination. To maintain a strong relationship with strategic partners, regular communication is essential. Transparent and honest dialogue can help build trust and address potential issues proactively. Additionally, setting clear expectations and responsibilities can prevent misunderstandings and foster a sense of shared ownership.

Performance tracking is another important aspect of relationship management. Regular reviews of KPIs can help ensure that the partnership is achieving its intended objectives and contributing positively to the Autonomous Finance journey.

As the landscape of Autonomous Finance evolves, strategic partnerships should be flexible enough to adapt and grow. Regularly reassessing and realigning the partnership strategy can help organizations stay relevant and competitive in the dynamic world of Autonomous Finance.

8.6 Risk Management in Strategic Partnerships

Just as a seasoned sailor anticipates and mitigates risks on the sea, so should organizations manage potential challenges in their strategic partnerships. These partnerships, though providing immense value, are not devoid of risks and complications.

The potential risks and pitfalls in strategic partnerships can range from conflicts of interest and misalignment of objectives to technology incompatibility and regulatory non-compliance. Understanding these risks early on aids in charting a secure course.

A robust risk management plan involves systematic identification, assessment, and prioritization of risks. This involves open and honest conversations with potential partners about areas of concern and areas for improvement. Once the risks are identified, developing effective risk mitigation strategies is next. These include contingency plans, regular audits and assessments, and proactive communication strategies.

A well-crafted partnership contract, including clauses that protect the interests of both parties, is a crucial component of risk mitigation. It can include terms for dispute resolution, intellectual property rights, data privacy, and regulatory compliance. Regular audits and assessments help in the proactive identification and mitigation of risks. These can include technology audits, compliance checks, and financial assessments.

8.7 Building & Managing Strategic Partnerships

Navigating the complex ocean of strategic partnerships demands a clearly defined roadmap. This will guide organizations through identifying, selecting, and managing their strategic partners, ensuring the best possible outcomes. Figure 8.3 provides a visual representation of the key steps to establishing and maintaining strategic partnerships in the age of Autonomous Finance:

a) **Identifying Needs:** The first step is to identify the organization's strategic needs that could be addressed through a partnership.

b) **Market Research:** Once the needs are identified, it is crucial to conduct market research to identify potential partners to meet them.

c) **Evaluating Potential Partners:** Next, potential partners must be evaluated on various parameters, including technology capabilities, financial health, cultural fit, and strategic alignment.

d) **Forging the Partnership:** The partnership terms must be negotiated and formalized once a suitable partner is found.

e) **Integration:** Post formalization, integration of systems, processes, and teams takes place.

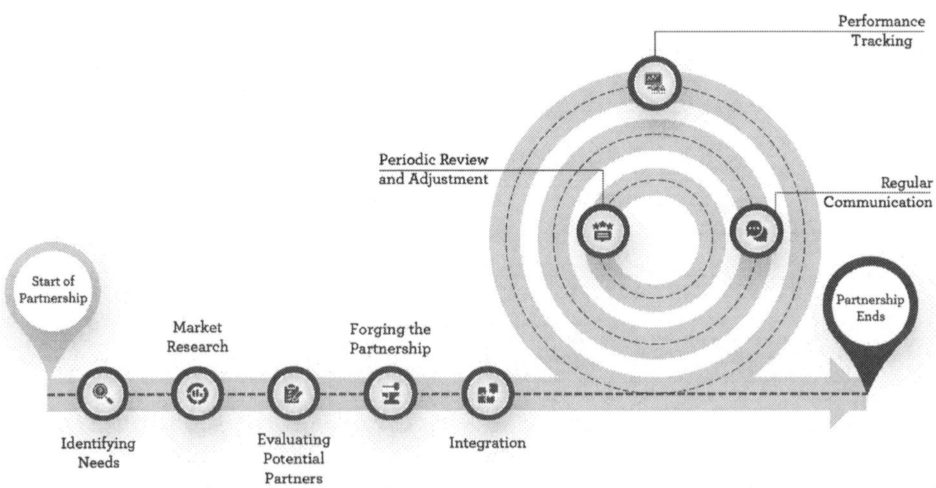

FIGURE 8.4: Roadmap for Maintaining and Optimizing Partnership Relationships

Maintaining a successful partnership requires continuous effort and ideally should include:

a) **Regular Communication:** Regular and open communication is key to addressing issues as they arise and ensuring that both parties are aligned regarding objectives.

b) **Performance Tracking:** Regularly tracking the partnership's performance against agreed KPIs will help identify improvement areas and take corrective action if required.

c) **Periodic Review and Adjustment:** The partnership agreement should be periodically reviewed and adjusted to account for changes in the business environment or strategic objectives.

8.8 Conclusion

The exploration undertaken in this chapter illuminates the indispensability of strategic partnerships in the ecosystem of Autonomous Finance. Far from being auxiliary constructs, these alliances emerge as the bedrock upon which organizations anchor their ambitions for uncharted growth, innovation, and transformative success. The tripartite structure—encompassing economic, strategic, and operational facets—is an unequivocal testament to the multifarious value inherent in such collaborations.

To navigate this complex terrain, an emphasis was laid on meticulous partner selection, integration strategies, and relationship management paradigms. The role of risk mitigation and continuous optimization in fortifying these partnerships against unforeseen exigencies was also elucidated. Consequently, strategic partnerships have ascended to a pivotal role, indubitably shaping the future of Autonomous Finance.

As the gaze shifts toward an ever-expanding horizon, limitless opportunities beckon. The synergy born from the amalgamation of disparate competencies and resources has the potential to drive unprecedented growth, catalyze societal transformation, and seed sustainable practices within the financial realm.

The forthcoming chapter heralds the advent of Interoperability—a lynchpin that synergizes with strategic partnerships to create a seamless, efficient, and resilient Autonomous Finance ecosystem. A venture into this critical realm will offer insights into how strategic alliances facilitate harmonizing disparate data sets, technologies, and services, forming the bedrock of future success in Autonomous Finance.

Echoing the profound wisdom encapsulated in Marie Curie's words, *"Nothing in life is to be feared, it is only to be understood"* (Steele, 2006), the quest for a deeper understanding of Interoperability promises to arm organizations and individuals with the tools and knowledge to face an uncertain future with unwavering confidence and courage.

8.9 Key Recommendations

a) **Promote Collaboration and Social Impact:** Develop a collaborative culture that aligns business objectives with social impact goals within your organization. This approach can enhance innovation, attract conscious investors, and foster customer loyalty.

b) **Engage with External Innovation Platforms:** Actively participate in activities hosted by incubators, accelerators, or other innovation platforms to identify potential partners that can contribute to your growth and innovation.

c) **Careful Partner Selection and Integration:** Prioritize potential partners based on their technology capabilities, financial health, cultural fit, and strategic alignment with your organization. Following selection, integrate them into your ecosystem to enhance collective capabilities.

d) **Optimize Partnership Relationships and Manage Risks:** Maintain strong partnerships through clear communication and performance tracking. Develop comprehensive risk management strategies to identify, assess, and manage partnership-related risks, including contingency plans and regular audits.

e) **Ensure Flexibility and Adaptability:** As the Autonomous Finance landscape evolves, your partnership strategy should be flexible enough to adapt and grow with these changes. Regularly reassess and realign the strategy to stay relevant and competitive.

8.10 References

Barclays Accelerator Programme | Barclays. (n.d.). https://home.barclays/who-we-are/innovation/barclays-accelerator/

Communications. (2023). BBVA teams up with MIT to improve card fraud detection. NEWS BBVA. https://www.bbva.com/en/innovation/bbva-teams-up-with-mit-to-improve-card-fraud-detection/

Eisen, S. (2019). Mastercard and R3 partner to develop new Blockchain-Powered Cross-Border payments solution. *Mastercard Newsroom.* https://www.mastercard.com/news/press/2019/september/mastercard-and-r3-partner-to-develop-new-blockchain-powered-cross-border-payments-solution/

Feedzai. (2023, July 20). Citi, FeedZai Partner on Machine Learning Payment solutions, *Feedzai.* https://feedzai.com/pressrelease/citi-partners-with-feedzai-to-provide-machine-learning-payment-solutions/

Ford, H., & Crowther, S. (1922). *My life and work.* Doubleday, Page & Company.

GFIN. (2021). *GFIN.* https://www.thegfin.com/

Goldman Sachs 10,000 women. (n.d.). Coursera. https://www.coursera.org/collections/goldman-sachs-10000-women

Lamonte, T., & Lamonte, T. (2020). When Morpheus met Watson: JPMC teams with IBM Data Science and AI Elite. *IBM Blog.* https://www.ibm.com/blog/when-morpheus-met-watson-jpmc-teams-with-ibm-data-science-and-ai-elite/

Online working capital solution to give UK SMEs access to same-day funding. Santander, UK. (2016). https://www.santander.co.uk/about-santander/media-centre/press-releases/online-working-capital-solution-to-give-uk-smes-access

Ruddenklau, A., & Caplain, J. (2022, February 7). Pulse of Fintech H2 2021 – *Global. KPMG.* https://kpmg.com/xx/en/home/insights/2022/01/pulse-of-fintech-h2-2021-global.html

Schueffel, P. (2016). Taming the Beast: A scientific definition of fintech. *Journal of Innovation Management, 4*(4), 32-54.

Steele, P. (2006). Marie Curie: *"Nothing in Life is to be Feared. It is Only to be Understood."*

Taft, D. K. (2021, August 24). Capital One launches DevExchange Developer Portal. *eWEEK.* https://www.eweek.com/development/capital-one-launches-devexchange-developer-portal/

UNDP/Mastercard Lighthouse MASSIV Partnership Programme | United Nations Development Programme. (n.d.). UNDP. https://www.undp.org/sdg-accelerator/undp/mastercard-lighthouse-massiv-partnership-programme

CHAPTER 9

Cohesive Connectivity

Integration & Interoperability in Autonomous Finance

"Interoperability is key to the paradigm shift
from a world where we are intelligent
to a world where our environment is intelligent."

- DAVE RAGGETT (SETHI & SARANGI, 2017)

9.0 Key Insights

- **Integration and Interoperability:** The true prowess of Autonomous Finance is manifested through the integration and interoperability of diverse financial systems, platforms, and applications. Integration enables a seamless flow of data and transactions, while interoperability unifies these systems despite inherent differences. Together, they form a harmonious nexus, turning Autonomous Finance into a revolutionary conduit for financial excellence.
- **Technology Enablers:** APIs and Blockchain: APIs and blockchain technology are pivotal to integration and interoperability in Autonomous Finance. APIs facilitate interaction between different software applications, and blockchain creates a unified source of truth across diverse systems, reducing errors and eliminating the need for data reconciliation.
- **Levels of Integration in Autonomous Finance:** Integration takes place at various levels: UI, application, process, and data. This multilevel integration enhances different aspects of Autonomous Finance, including user experience, functional capabilities, workflow streamlining, and information flow.
- **Security, Compliance, and Standardization:** Ensuring robust security measures, compliance with financial regulations, and standardization in integration mechanisms is essential for building a reliable Autonomous Finance ecosystem. These factors foster user trust, enable seamless data exchange, and maintain data integrity.
- **Decentralization, Scalability, and User Experience:** Decentralization empowers individuals to control their financial transactions while scalability accommodates increasing transaction volumes. Simultaneously, integration and interoperability pave the way for enhanced user experiences, enabling seamless transactions and providing hyper-personalization and transparency.

9.1 Introduction

In the preceding chapter, the exploration ventured into the labyrinthine intricacies of strategic partnerships in Autonomous Finance—a landscape where alliances serve as the compass guiding organizations through a kaleidoscope of challenges and opportunities. This chapter steers the discourse toward another essential paradigm: Integration and Interoperability in Autonomous Finance.

The financial realm is undergoing a seismic shift catalyzed by various technological marvels—AI, ML, RPA, Big Data, Advanced Analytics, Cloud Computing, SaaS, and APIs. The global market size for Big Data and business analytics is projected to grow to $274.3 billion by 2022 (*Worldwide Big Data and Analytics Spending Guide*, n.d.), accentuating the scale and impact of these technologies. However, the symbiosis of these technologies through integration and interoperability unleashes their true transformative power.

In this intricate landscape, what do the terms integration and interoperability signify? Integration encompasses the seamless connectivity of various financial platforms, applications, and systems, obviating manual intervention. Interoperability is the glue, melding these heterogeneous systems into a cohesive whole. For instance, a consumer's financial life is often dispersed across a spectrum of platforms—banking, investment, lending, and digital wallets. The magic of integration and interoperability coalesces these disparate elements, offering a unified financial panorama.

The ensuing narrative will journey through the complexities of integration and interoperability, revealing their indelible impact on Autonomous Finance. What differentiates Autonomous Finance from traditional financial systems? How do integration and interoperability fuel a revolution redefining how finance operates? These compelling queries will be meticulously addressed, offering a comprehensive guide to understanding these critical dimensions.

So, as the narrative embarks on this transformative voyage, each turn promises to unveil novel insights, groundbreaking mechanics, and endless possibilities. The stage is set for an enthralling journey through integration and interoperability in Autonomous Finance—a landscape teeming with opportunity and innovation.

9.2 Interplay in Autonomy: Integration & Interoperability

FIGURE 9.1: Integration & Interoperability in Autonomous Finance

The intricacies of integration and interoperability may initially appear as complex technological jargon. However, a closer examination reveals their immense significance within Autonomous Finance. As the journey through this chapter unfolds, a thorough comprehension of these concepts and their vital role in shaping the future of Autonomous Finance becomes indispensable.

Integration, in the context of Autonomous Finance, refers to the process by which disparate financial systems, platforms, and applications are connected. This metaphorical bridge allows distinct platforms to communicate with each other, facilitating seamless data exchange and transactions without manual intervention (*III. the Future Monetary System*, 2022). Integration forms the bedrock on which the potential of Autonomous Finance is constructed. The absence of it leaves the disparate elements of financial lives in isolation, curbing the power of autonomous decision-making and automated management.

Contrastingly, interoperability ensures these integrated systems harmonize in their operations. Discrepancies in design, functioning, and data formats are natural among different platforms. Interoperability is the mechanism that allows these systems to function cohesively despite these disparities, facilitating different systems to use and interpret shared data, thus fostering collaboration and cooperation within a diverse ecosystem (Ide & Pustejovsky, 2010).

Table 9.1 provides a critical comparative analysis that disentangles the intricate yet distinct constructs of Integration and Interoperability within Autonomous Finance ecosystems.

TABLE 9.1: Integration vs. Interoperability

NO.	PARAMETER	INTEGRATION	INTEROPERABILITY
1	Definition	Combining multiple systems, applications, or services into a unified whole, where they can work together seamlessly.	The ability of different systems, applications, or services to communicate and exchange information while remaining distinct entities.
2	Example	For example, a bank may integrate its banking application with a financial management tool, allowing customers to view their account details and manage their budget within the same interface.	For example, a robo-advisor platform may have interoperability with various investment brokerages. This allows the robot advisor to place trades on different brokerage platforms on behalf of the user, but the user still needs to maintain separate accounts on each platform.
3	Requirements	Integration often requires custom programming and configuration to create a smooth user experience and ensure all components work together without glitches.	Interoperability requires standardization of data formats and protocols but does not necessitate a high customization or configuration-like integration.
4	Benefits	It provides a seamless user experience. It can enhance the functionality of an application, allowing it to provide more comprehensive services. It can also improve efficiency and accuracy by eliminating manual data transfer between systems.	It allows for flexibility and choices. It allows users to use different systems and services in tandem, choosing the best from each. It fosters competition and innovation by preventing vendor lock-in.

Integration and interoperability in Autonomous Finance can be likened to a symphony orchestra. An orchestra encompasses multiple instruments, each with its unique sound and role. Integration represents the sheet music coordinating these instruments, providing guidance on when and how each should play. Interoperability, within this context, symbolizes the ability of the instruments to produce harmonious music together despite their inherent differences, responding to each other's cues and complementing their sounds.

Promising a financial world as harmonious as a symphony, integration and interoperability pave the way for the free flow of data, intelligent financial decisions, and seamless customer experience across platforms. As this chapter progresses, these concepts are further explored in greater detail, revealing different aspects of integration and interoperability, their significance, and their potential to revolutionize the finance domain.

9.3 Mastering the Art of Integration

As the world of finance becomes increasingly automated and interconnected, integration emerges as the master key for unlocking unprecedented opportunities. This section explores the possibilities available when leveraging integration at various levels in Autonomous Finance, namely, UI, application, process, and data levels, powered by intelligent automation.

Figure 9.2 captures the essence of this spectrum across four crucial axes: Automation Sophistication, Degree of Automation, Degree of Risk, and Business Impact.

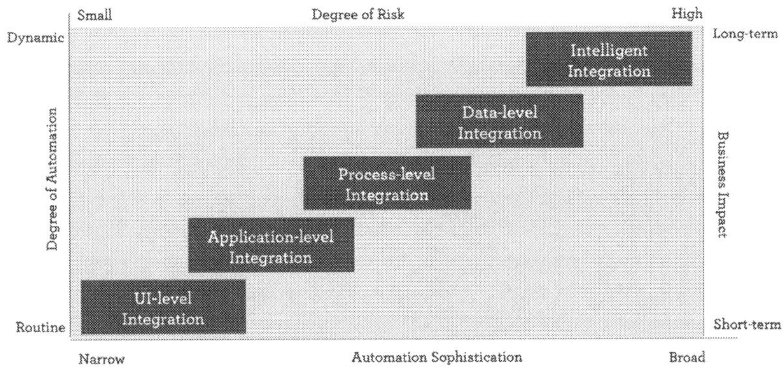

FIGURE 9.2: The Art of Integration in Autonomous Finance

a) **UI-Level Integration: Unifying Customer Experience:** UI-level integration creates a consistent, streamlined user interface (UI) across various financial applications. Here, it is less about integrating back-end systems and more about the seamless user experience on the front end. Users should be able to interact with different applications as if they were one unified platform. For instance, Open Banking, a revolution in banking services, is thriving mainly due to effective UI-level integration, providing a seamless experience across various financial services. This implies that customers can manage their finances from multiple providers through a single digital app and avoid using several platforms.

b) **Application-Level Integration: Bridging Functional Capabilities:** Application-level integration is about connecting the functional capabilities of different software applications. For instance, linking a loan management system with a credit scoring application automates the approval process. This enables businesses to enhance their functional capabilities and create innovative solutions that otherwise would be complex and time-consuming to build from scratch.

c) **Process-Level Integration: Streamlining Workflows:** At this level, integration is focused on aligning and automating various business processes across different systems. This typically involves using RPA tools to replicate human actions in completing tasks. For instance, in the context of loan processing, process-level integration could involve automating data gathering from different sources, evaluating creditworthiness, sending loan approval or rejection notices, and creating an end-to-end automated workflow.

d) **Data-Level Integration: Fostering Information Flow:** Data-level integration involves ensuring that data flows seamlessly across various systems, enhancing the decision-making capabilities of Autonomous Finance platforms. With data-level integration, finance platforms can pull in data from different sources, analyze it in real-time, and use it to provide personalized financial advice, detect fraudulent activity, or predict market trends.

e) **The Power of Intelligent Integration:** Intelligent Integration plays a pivotal role across these integration levels. By leveraging AI and ML, intelligent integration automates tasks and learns, adapts, and improves over time. This leads to improved efficiency, reduced errors, and better decision-making in Autonomous Finance scenarios.

In summary, the framework portrayed in Figure 9.2 elucidates a shift from compartmentalized technicalities to an interconnected, strategic continuum. This multidimensional framework beckons industry leaders to view integration not merely as a series of tasks but as a pivotal strategy that foments organizational agility, enriches customer engagement, and pioneers innovation in finance. With integration and intelligent automation working hand in hand, Autonomous Finance platforms can now deliver superior value, heralding a new era in finance where systems work in unison, enhancing customer experience and business value.

9.4 Tech Giants on the Horizon

Stepping into the technological underpinnings of integration and interoperability in Autonomous Finance, a couple of tech titans stand out – APIs and blockchain, as seen in Figure 9.3 below. These technologies act as gateways, helping Autonomous Finance platforms intermingle and converse, delivering more value.

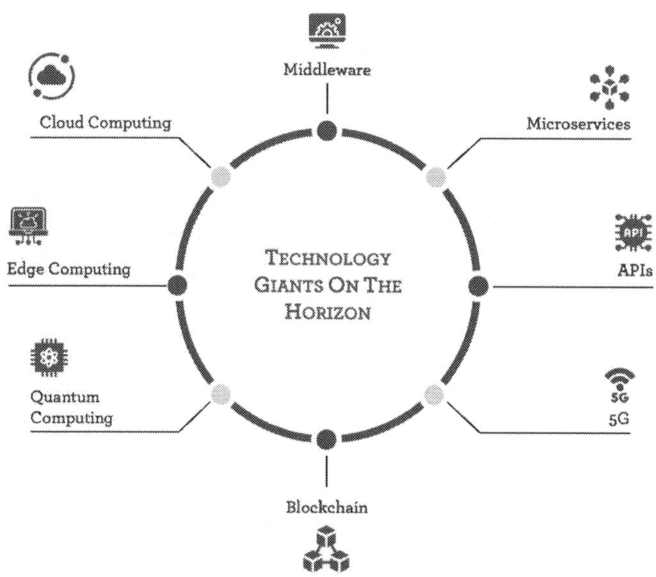

FIGURE 9.3: Technology Giants on the Horizon in Autonomous Finance

a) **APIs - The Connective Tissue in Fintech:** APIs can be aptly described as the 'connective tissue' in the body of fintech. These interfaces allow different software applications to interact, exchange data, and leverage the functional capabilities of each other. APIs have played a pivotal role in the emergence of the Open Banking paradigm, where banking data can be securely shared between different entities, promoting innovation and competition. APIs in Autonomous Finance can enable various scenarios like fetching real-time stock market data, integrating with payment gateways, conducting identity verification, and much more. They are a fundamental building block enabling the seamless operation of various fintech applications and services.

b) **Blockchain - A Paradigm Shift in Interoperability:** Blockchain technology is known for its potential to disrupt various industry verticals, and finance is no exception. In the context of integration and interoperability, blockchain can create a single source of truth across various systems and networks, eliminating the need for data reconciliation and reducing errors (Mougayar, 2016). DeFi, powered by blockchain, is creating a paradigm shift in finance. It is bringing about a new level of interoperability where financial assets and services are not confined within the boundaries of specific institutions but can freely move across different platforms and services. Moreover, smart contracts, a feature of blockchain technology, automate contractual obligations, creating more efficient and transparent financial processes. This has significant implications for integration, as smart contracts can be programmed to trigger actions across various systems based on predefined conditions (Tapscott & Tapscott, 2016).

c) **Exploring Beyond - Future Tech in Integration:** As the fintech landscape continues to evolve, newer technologies like 5G, Quantum Computing, and Edge Computing are set to redefine the possibilities of integration and interoperability. With greater computational power and faster data transmission, the boundaries of what is possible in Autonomous Finance are expected to expand significantly.

Through these advanced technologies, Autonomous Finance is moving toward a seamless, interconnected ecosystem where the sum of integrations surpasses the individual parts. It is a world where Autonomous Finance platforms are not merely transactional entities but a vibrant, interconnected ecosystem enhancing the overall user experience and delivering superior value.

9.5 Security, Compliance, & Standardization

The robustness of integration and interoperability mechanisms in Autonomous Finance is substantially determined by the trifecta of security, compliance, and standardization. These factors are crucial for fostering trust among users, facilitating seamless data exchange, and ensuring the integrity and confidentiality of data.

a) **Security - The Cornerstone of Integration:** Security is a pivotal concern in financial integrations. As financial applications are often a prime target for cyber threats, ensuring robust security in integration mechanisms becomes critical. Data exchange through APIs or other integration methods must be securely encrypted to avoid data breaches. Authentication mechanisms, like OAuth, can provide an additional layer of security in API-based integrations, allowing only authorized systems to access data. Furthermore, implementing blockchain technology in financial integrations can heighten security by eliminating the risk of a single point of failure and providing an immutable record of transactions (Mougayar, 2016).

b) **Compliance - Navigating Regulatory Landscapes:** Compliance with local and international financial regulations is another significant aspect of financial integrations.

Regulations like the GDPR in Europe and the California Consumer Privacy Act (CCPA) in the U.S. provide guidelines for securing customer data. In the banking sector, compliance with Open Banking regulations and the PSD2 ensures that data sharing and integration APIs adhere to defined standards and follow secure protocols.

c) **Standardization - A Universal Language for Integration:** Standardization in integration mechanisms promotes interoperability by ensuring that different systems and platforms communicate effectively. Standards such as the ISO 20022 for financial messaging provide a universal framework for developing financial applications and facilitating integration (*ISO 20022 | ISO20022*, n.d.). In the blockchain realm, Ethereum's ERC standards have helped to standardize smart contracts, thus enabling interoperability across different blockchain applications (Buterin, 2014). Similarly, using standardized APIs in Open Banking ensures that all banks and fintech applications 'speak' the same language, enhancing integration efficiency.

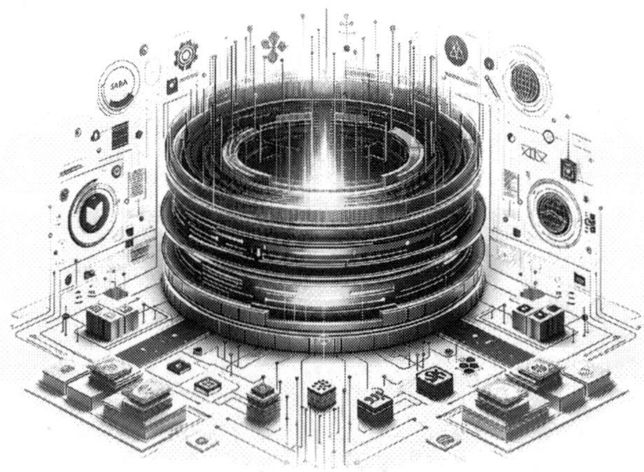

Thus, managing security, compliance, and standardization while facilitating integration is like guarding the fort in Autonomous Finance. These factors are not merely regulatory hurdles or technological challenges but vital components for building a resilient, reliable, and user-friendly Autonomous Finance ecosystem.

9.6 Decentralization & Scalability

The convergence of decentralization and scalability paints an exciting picture for the future of Autonomous Finance. This section elucidates these concepts' significance and impact on the fintech ecosystem.

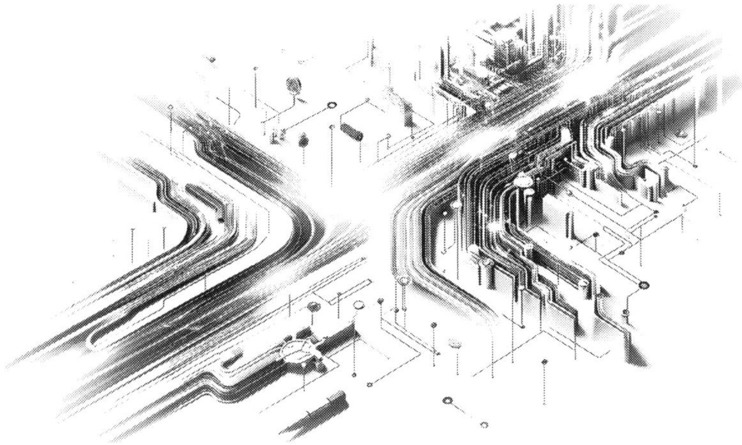

Decentralization: A Paradigm Shift in Finance

In a conventional financial system, transactions occur through centralized authorities, such as banks and finance teams. Decentralization, however, redefines this norm by eliminating intermediaries and empowering individuals to take control of their financial transactions (Zetzsche et al., 2020).

The use of blockchain technology in Autonomous Finance epitomizes this shift. Blockchain provides a decentralized ledger system, enhancing security and trust through transparency and immutability. For example, DeFi applications leverage blockchain to provide services such as lending and borrowing without an intermediary.

Decentralization also propels the concept of 'financial sovereignty,' where individuals have complete control over their assets, thereby fostering financial inclusion (Tapscott & Tapscott, 2016).

Scalability: Fueling Growth in Autonomous Finance

The ability to accommodate increasing transaction volumes without sacrificing speed, efficiency, or scalability is critical to Autonomous Finance.

Scalability is critical in the context of blockchain, where a surge in transactions can slow down the network. Innovations such as Layer 2 solutions, like the Lightning Network for Bitcoin and Plasma for Ethereum, are emerging to address these challenges (Bashir, 2022).

Within the API ecosystem, the challenge of scalability lies in effectively handling a surge in API requests without compromising system performance. To achieve this goal, techniques such as implementing a microservices architecture and using API gateways are leveraged (Newman, 2021).

Both decentralization and scalability are redefining the contours of the Autonomous Finance landscape. As these technologies mature, they open up infinite horizons for creating more robust, secure, and inclusive financial systems.

9.7 Crafting Superior User Experiences

In Autonomous Finance, the harmony achieved through seamless integration and interoperability transcends functional benefits and impacts user experience.

a) **Effortless Transactions, Empowered Users:** As digital finance platforms evolve, consumers are empowered to complete financial tasks with increased ease, precision, and speed. Integration plays a pivotal role in enabling this empowerment. For instance, integrating multiple bank accounts and credit cards on a single platform simplifies financial management, giving users a comprehensive overview of their financial status at a glance (OECD, 2021). Moreover, the interoperability between different financial service providers enables users to perform transactions across platforms without switching between apps or websites. This reduces friction and makes the user experience smoother and more convenient (Boar et al., 2021).

b) **Hyper-Personalization-The New Standard:** Autonomous Finance, powered by AI, ML, and RPA, offers a high degree of personalization to its users. For example, integrated data from various sources can provide customized financial advice, product recommendations, and proactive alerts tailored to each individual's unique financial situation and goals (*Building the AI Bank of the Future*, 2021).

c) **Confidence through Transparency:** Integrating blockchain technology in Autonomous Finance gives users greater transparency in their financial transactions. Each transaction is recorded and cannot be altered, providing users with a transparent and trustworthy history (Habib et al., 2022).

The integration and interoperability in Autonomous Finance have ushered in an era where technology is not just a functional tool but a strategic enabler of enhanced user experience. As finance functions evolve in the Autonomous landscape, the focus on user experience will remain paramount. For more details, please refer to *"Chapter 11: Beyond Transactions: Crafting the Autonomous Finance Customer Experience."*

9.8 The Future of Integration and Interoperability

The marriage of finance and technology has witnessed unprecedented growth in recent years. However, with the advent of integration and interoperability, the future of Autonomous Finance promises a new era. Figure 9.4 distills four monumental trends that shall serve as compass points in our exploratory sojourn into what lies ahead:

d) **The Rise of the API economy:** APIs have long been the linchpin of seamless integrations and will continue to drive the future of interoperable financial ecosystems. As per a report by Deloitte, the use of APIs in finance is expected to grow by 22% annually, reaching $3 billion in market size by 2025 (Deloitte, 2023).

e) **The Age of DeFi:** Blockchain technology, with its promise of secure and decentralized transactions, will pave the way for DeFi. In such a system, financial products will be available on a public decentralized blockchain network. This can provide secure and transparent financial services to those currently underserved by traditional banking systems (Sun et al., 2023).

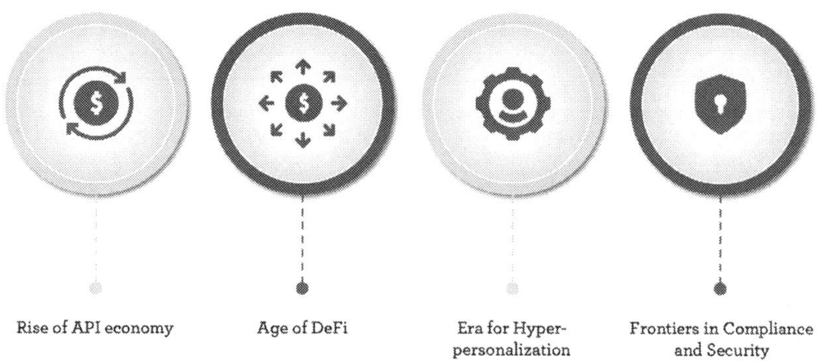

FIGURE 9.4: The Future of Integration & Interoperability

f) **The Era of Hyper-Personalization:** With an increased emphasis on personalized user experiences, the fintech industry is expected to leverage the power of AI and ML more than ever. Personalized product recommendations, investment advice, and risk assessments will become the norm rather than the exception (Sirisena, n.d.).

g) **New Frontiers in Compliance and Security:** With the expansion of the digital finance landscape, compliance and security will remain paramount. As the complexity of integration and interoperability increases, innovative solutions will address these challenges to ensure data privacy, transaction security, and compliance with global financial regulations (Hussain et al., 2021).

With the harmonious blend of integration and interoperability, the future of Autonomous Finance promises to be exciting, dynamic, and full of opportunities. In this ever-evolving landscape, a single constant will be the focus on creating superior user experiences through seamless integrations and interoperability.

9.9 Conclusion

The exploration within this chapter has illuminated integration and interoperability as cornerstones in the architecture of Autonomous Finance. These elements are not mere technological imperatives but the arteries that link a complex financial ecosystem, enabling it to function as a unified whole. As Steve Jobs insightfully stated (Mejia, 2018), "*True innovation lies in discernment—saying NO to all but the most crucial features.*" This ethos finds its embodiment in the principles of integration and interoperability, allowing Autonomous Finance to focus on key elements that amplify user value and experience.

Drawing from the in-depth examination of various integration levels—UI, application, process, and data—critical insights into user experience, workflow efficiency, and intelligent decision-making have been gleaned. However, these technologies do not operate in a vacuum. They are bolstered by robust security protocols, compliance frameworks, and standardization, thereby enhancing trust and reliability within the Autonomous Finance ecosystem.

Looking beyond the present, the chapter has shed light on the imminent future of Autonomous Finance. Here, the proliferation of the API economy, the disruptive force of DeFi, and the advent of AI-powered hyper-personalization stand poised to redefine the financial landscape. As cited by Chaitman in 2018 (Chaitman et al., 2018)), Albert Einstein once remarked, *"The significant problems we face cannot be solved at the same level of thinking we were when we created them."* Integration and interoperability invite a paradigmatic shift in thinking, a departure from traditional financial boundaries. They beckon a future where finance evolves beyond transactional functionality to become an empowered, interconnected ecosystem.

As the narrative unfolds, the ensuing chapters will delve deeper into the transformative facets of Autonomous Finance. The expedition continues, each chapter promising to unravel new layers of insights and innovations, forever changing the understanding of what finance can be.

9.10 Key Recommendations

a) **Implement Robust Data Protection Measures:** The prioritization of data privacy and security is paramount, necessitating the deployment of robust protective measures to shield delicate information from unauthorized access and cyber incursions. The sanctity of data can be maintained and safeguarded through frequent audits, the application of data encryption, and the establishment of stringent access controls.

b) **Invest in Advanced Analytics:** Harness the power of advanced analytics, including AI and ML, to extract insights from the large amounts of data generated in the financial world. These technologies can significantly enhance decision-making capabilities, improve efficiency, and deliver highly personalized customer services.

c) **Adopt a Data-Driven Approach:** Embrace a data-driven culture that prioritizes data in decision-making processes. Leverage analytics to generate insights and make informed decisions. This approach will improve efficiency and foster a culture of transparency and accountability.

d) **Establish Data Management Best Practices:** Implement best practices to ensure data quality, consistency, and reliability. This includes maintaining data integrity, standardizing data collection and storage processes, and keeping data up-to-date.

e) **Stay Ahead of Regulatory Changes:** Regularly monitor changes in data protection regulations and ensure compliance. Given the sensitive nature of financial data, staying compliant with evolving regulatory requirements is vital for avoiding legal issues and maintaining customer trust.

9.11 References

III. The future monetary system. (2022). https://www.bis.org/publ/arpdf/ar2022e3.htm

Bashir, I. (2022). Blockchain. In *Blockchain Consensus: An Introduction to Classical, Blockchain, and Quantum Consensus Protocols* (pp. 167-205). Berkeley, CA: Apress.

Boar, C., Claessens, S., Kosse, A., Leckow, R., & Rice, T. (2021). *BIS Bulletin No 49 Interoperability between payment systems across borders*. https://www.bis.org/publ/bisbull49.pdf

Building the AI bank of the future. (2021, May 18). McKinsey & Company. https://www.mckinsey.com/industries/financial-services/our-insights/building-the-ai-bank-of-the-future

Buterin, V. (2014). *Ethereum: A Next-Generation Smart Contract and Decentralized Application Platform. By Vitalik Buterin (2014)*. https://ethereum.org/669c9e2e2027310b6b3cdce6e1c52962/Ethereum_Whitepaper_-_Buterin_2014.pdf

Chaitman, B. R., Brooks, M. M., Fox, K., & Lüscher, T. F. (2018). ORBITA revisited: what it really means and what it does not? *European Heart Journal, 39*(11), 963–965. https://doi.org/10.1093/eurheartj/ehx796

Chen, Y., Lu, Y., Bulysheva, L. *et al.* Applications of Blockchain in Industry 4.0: a Review. *Inf Syst Front* (2022). https://doi.org/10.1007/s10796-022-10248-7

Deloitte. (2023). Future of Fintech: The rise of the API economy. *Fintech Reports, 19*(2), 10-14.

Habib, G., Sharma, S., Ibrahim, S., Ahmed, I., Qureshi, S., & Ishfaq, M. M. (2022). Blockchain Technology: Benefits, Challenges, Applications, and Integration of Blockchain Technology with Cloud Computing. *Future Internet, 14*(11), 341. https://doi.org/10.3390/fi14110341

Hussain, M., Nadeem, M. W., Iqbal, S., Mehrban, S., Fatima, S. N., Hakeem, O., & Mustafa, G. (2021). Security and privacy in FinTech: a policy enforcement framework. In *Research anthology on concepts, applications, and challenges of FinTech* (pp. 372-384). IGI Global.

Ide, N., & Pustejovsky, J. (2010, January 18-20). What does interoperability mean, anyway? Toward an operational definition of interoperability for language technology. In *Proceedings of the Second International Conference on Global Interoperability for Language Resources*. Hong Kong, China.

ISO 20022 | ISO20022. (n.d.). ISO20022. https://www.iso20022.org/

Mejia, Z. (2018, October 2). Steve Jobs: Here's what most people get wrong about focus. *CNBC*. https://www.cnbc.com/2018/10/02/steve-jobs-heres-what-most-people-get-wrong-about-focus.html

Mougayar, W. (2016). *The business blockchain: promise, practice, and application of the next Internet technology*. John Wiley & Sons.

Newman, S. (2021). *Building microservices*. O'Reilly Media, Inc.

OECD. (2021). *Artificial Intelligence, Machine Learning and Big Data in Finance - OECD*. www.oecd.org. https://www.oecd.org/finance/artificial-intelligence-machine-learning-big-data-in-finance.htm

Penguin Cao, L., Yang, Q., & Yu, P. S. (2021). Data science and AI in FinTech: An overview. *International Journal of Data Science and Analytics, 12*, 81-99.

Sethi, P., & Sarangi, S. R. (2017). Internet of Things: architectures, protocols, and applications. *Journal of Electrical and Computer Engineering, 2017*, 1–25. https://doi.org/10.1155/2017/9324035

Sirisena, P. (n.d.). *Revolutionizing Fintech Product Development: Enhancing Decision-Making and Risk Management with AI*. Classic Informatics: Top Web Development Company in India. https://www.classicinformatics.com/blog/revolutionizing-fintech-product-development-enhancing-decision-making-and-risk-management-with-ai

Sun, Y., Li, S., & Wang, R. (2023). Fintech: From budding to explosion - an overview of the current state of research. *Review of Managerial Science, 17*(3), 715-755.

Tapscott, D., & Tapscott, A. (2016). *Blockchain revolution: how the technology behind bitcoin is changing money, business, and the world*. Penguin.

Worldwide Big Data and Analytics Spending Guide. (n.d.). IDC: The Premier Global Market Intelligence Company. https://www.idc.com/getdoc.jsp?containerId=IDC_P33195

Zetzsche, D. A., Arner, D. W., & Buckley, R. P. (2020). Decentralized finance (defi). *Journal of Financial Regulation, 6*, 172-203.

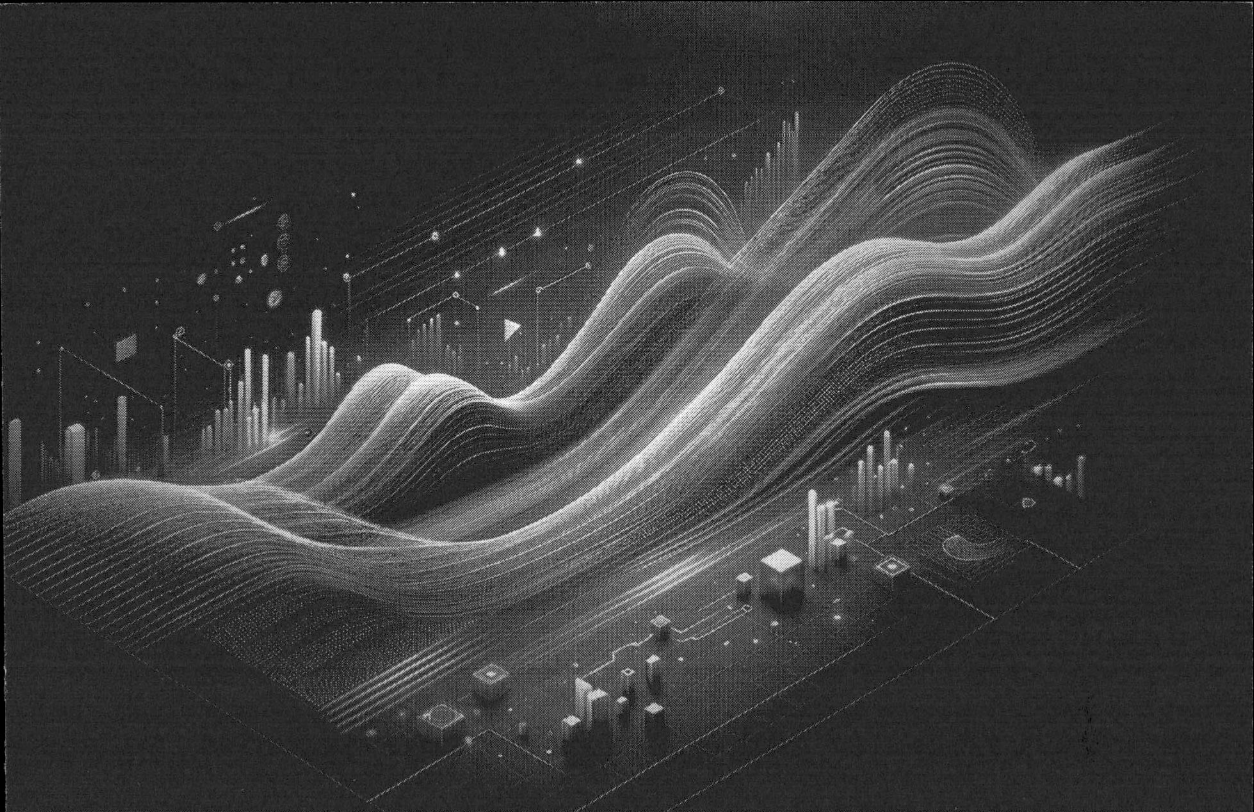

CHAPTER 10

The Influence of Insight

Data & Analytics in Autonomous Finance

"Without data,
you are just another person
with an opinion."

- W. EDWARDS DEMING (DEMING, 1982)

10.0 Key Insights

- **Insights Driving Financial Decisions: Data Integration in Autonomous Finance:** A comprehensive view of an organization's financial landscape is achieved by effectively integrating complex and heterogeneous financial data.
- **Guiding the Way: AI and ML in Autonomous Finance:** With their capabilities to predict market trends, optimize portfolio management, and offer personalized financial advice, AI and ML technologies are leading the way in Autonomous Finance.
- **Illuminating the Path: Data Visualization in Autonomous Finance:** Finance teams use data visualization to decipher complex datasets, allowing them to identify patterns and trends effectively for strategic decision-making.
- **Fortifying the Treasury: Data Security and Protection in Autonomous Finance:** Ensuring continuous protection against potential data breaches is crucial in Autonomous Finance.
- **Seizing Opportunities: The Potential of Big Data and Predictive Analytics in Finance:** Despite the challenges associated with data quality, privacy, and volume, the potential of Big Data and predictive analytics offers vast opportunities and clear direction in the financial landscape.

10.1 Introduction

In the realm of Autonomous Finance, data and analytics serve as the foundational pillars, comparable to the hull of a formidable ship navigating uncharted seas. This chapter builds upon the exploration of integration and interoperability undertaken in the previous chapter, focusing on the transformative power of data and analytics as the linchpins of Autonomous Finance. The germane words of Steve Jobs (2005), who once said, *"You cannot connect the dots looking forward; you can only connect them looking backward,"* provide a poignant backdrop for this discourse. In 2022 alone, an estimated 2.5 quintillion bytes of data were generated every day, and by 2025, the worldwide data sphere is expected to grow to 175 zettabytes, a staggering increase from 33 zettabytes in 2018 *(Worldwide Big Data and Analytics Spending Guide, 2019)*.

Data is the lifeblood that fuels the potent engines of AI and ML in Autonomous Finance. It is an invaluable resource, teeming with the volume, velocity, and variety required to enable intricate algorithms to learn, adapt, predict, and navigate. However, a ship is only as good as its navigational tools. In this analogy, Analytics is the compass and astrolabe, skillfully guiding organizations through the data maze to extract actionable insights, forecast trends, and enable informed decisions.

However, this expedition is not without its challenges. For instance, the looming specters of data privacy and governance have become increasingly pronounced, especially in light of legislative frameworks like GDPR and CCPA. This chapter will meticulously examine these issues and the strategies to navigate them.

The primary thesis of this chapter is to deliver an incisive exploration into the indispensable roles of data and analytics in Autonomous Finance. This journey will not merely demystify the mechanics but will provide a comprehensive guide for leveraging these elements to achieve a seamless, effective, and sustainable financial ecosystem.

10.2 Data Management

As a fishing ship sets sail, its net sinks into the sea to haul in the treasure beneath the waves. In Autonomous Finance, our treasure is data; the ocean represents a complex, diverse, and vast universe of information. Like the ocean's inhabitants, data comes in myriad shapes, sizes, and forms, each harboring unique characteristics and potential for use. The net symbolizes the systems responsible for acquiring and managing this invaluable asset (Katal et al., 2013).

FIGURE 10.1: Data Management in Autonomous Finance

a) **Data Acquisition:** The process of data acquisition in Autonomous Finance refers to the collection, filtering, and cleansing of data from various sources, such as transaction records, customer profiles, market trends, social media interactions, and regulatory compliance. Like a skilled fisherman, efficient data acquisition involves meticulous selection and sieving through diverse data to retain what is most valuable and relevant to the organization's goals (Chen et al., 2012).

b) **Data Quality Management:** Data quality management is the vigilant gatekeeper in the vast ocean of information, ensuring that only the most pristine and reliable data are allowed passage. This process ensures the information is trustworthy, prepared, and refined for further processing by meticulously validating, enriching, and deduplicating data. It is a critical task that mirrors the careful examination of a catch for quality, a task that guarantees the integrity of the very foundation of Autonomous Finance.

c) **Data Integration:** The interconnection of various data sources is vital in creating a comprehensive view. Data integration in Autonomous Finance involves harmonizing disparate data sets, allowing for a unified and coherent understanding of financial landscapes.

d) **Data Storage and Management:** The catch must be appropriately stored once hauled onboard. In the data context, this involves organizing, maintaining, and ensuring accessibility for future use. Good data storage and management ensure that information is classified securely and retrieved efficiently (Redman, 1997).

e) **Data Security and Privacy:** Protecting the catch from external threats is paramount, as is the security and privacy of data. Implementing robust safeguards against unauthorized access and complying with privacy regulations are crucial to maintaining integrity (Davenport, 2014).

f) **Data Governance:** Guides the entire process and sets the policies, standards, and procedures that ensure high-quality data management. The captain steers the ship, ensuring adherence to best practices and compliance (Weber et al., 2009).

g) **Data Analytics and Metadata Management:** Navigating through the vast ocean of data requires precise tools. Data analytics and metadata management transform raw numbers into actionable insights, like the compass and maps guiding the ship's course (Wang & Strong, 1996).

h) **Master Data Management:** Coordinating various data sources into a single master file, master data management acts as the ship's log, providing a consistent and unified view of data across the organization.

Within Autonomous Finance, these elements form the bedrock upon which sophisticated analytics can be performed. Figure 10.1 illustrates how comprehensive data allows finance teams to assess risks better, develop targeted financial products, and improve decision-making processes.

However, like the perilous sea, the task is fraught with challenges such as data silos, quality issues, privacy concerns, and ethical dilemmas. Each aspect demands attention, as a lapse in any can compromise the entire data venture, leading to a potential shipwreck in this ocean of opportunity.

In the words of Andre Gide, *"Man cannot discover new oceans unless he dares to lose sight of the shore"* (Gide 1925). Pursuing data mastery in Autonomous Finance requires courage, vision, and skill. As we navigate this complex world, we must remain vigilant, innovative, and committed to leveraging data as a transformative agent in the financial landscape.

10.3 Data Quality and Integrity

Just as the keel of a ship provides stability, strength, and shape against tumultuous waves, data quality and integrity serve as the structural foundation of Autonomous Finance. They are essential for maintaining the robustness and reliability of all data-driven operations, as shown in Figure 10.2 below.

a) **Data Quality Management:** Data quality refers to the degree to which data accurately represents the real-world entities or events it is supposed to depict (Wang & Strong, 1996). It involves several key aspects:

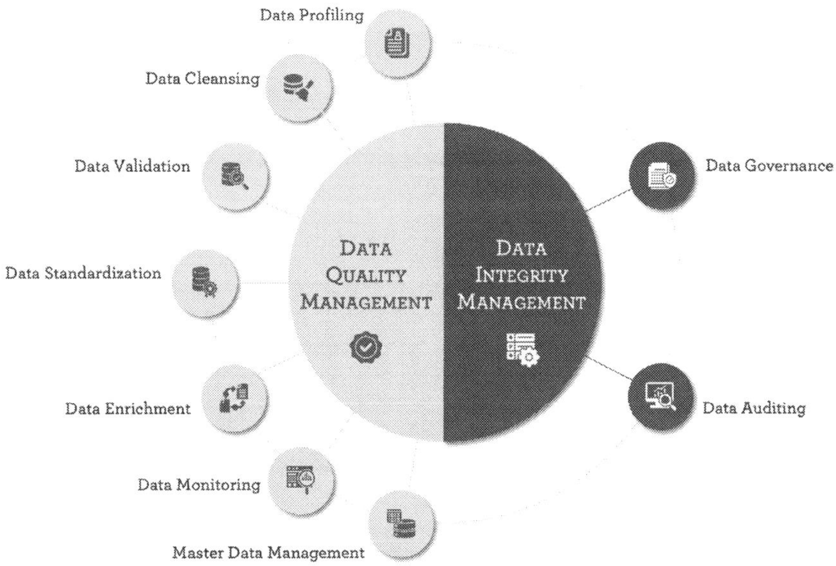

FIGURE 10.2: Data Quality and Integrity in Autonomous Finance

- **Data Profiling:** Examining the data to understand its structure, relationships, and patterns.
- **Data Cleansing:** Detecting, correcting, or removing corrupt or inaccurate records (Rahm & Do, 2000).
- **Data Validation:** Ensuring that only valid data enters the system, adhering to predefined formats and rules.
- **Data Standardization:** Aligning data to common standards for consistency across various platforms.
- **Data Enrichment:** Enhancing data with additional information to increase its value for analytical purposes.
- **Data Monitoring:** Continuously observing data to detect anomalies and maintain quality.
- **Master Data Management:** Coordinating and managing the primary data within an organization to ensure a consistent and unified view.

These processes ensure accuracy, completeness, timeliness, consistency, and relevancy. Data quality directly influences the credibility of insights and predictions derived from it, and poor data quality can lead to misinformed decisions, veering an organization off course.

b) **Data Integrity Management:** Data integrity ensures that data remains intact, consistent, and unchanged during storage, retrieval, and processing (Romdhane et al., 2019). In the unpredictable ocean of Autonomous Finance, data integrity is akin to ensuring that the keel remains unbreached, holding the ship together amidst the buffeting waves. Key aspects include:

- **Data Governance:** Establishing policies and procedures to manage data quality and integrity.
- **Data Auditing:** Reviewing the data management process to identify and rectify inconsistencies, ensuring high data integrity.

In the context of Autonomous Finance, maintaining data quality and integrity is paramount. Every function depends on data reliability, from credit scoring and fraud detection to portfolio management and regulatory reporting. A minor error in transaction data can lead to significant financial misstatements, just as a small crack in the keel can sink a ship (Chen et al., 2012).

The mastery of data quality and integrity management is essential for organizations navigating the intricate world of Autonomous Finance. As the ship's keel, they form the very basis upon which the structure stands, and their continuous maintenance ensures that the journey through the financial seas remains steadfast and true.

The Data Quality Dilemma

Embarking upon the journey to Autonomous Finance represents a seismic shift—a departure from traditional financial practices towards a future imbued with AI and ML. This is not merely a technological shift but an ontological metamorphosis in enterprise management. However, as with any significant change, questions arise. One such question of pivotal importance is whether to cleanse data before integrating it into an Autonomous Finance platform. Figure 10.3 epitomizes one such question of pivotal importance - "To Cleanse or Not to Cleanse" before integrating it into an Autonomous Finance platform.

GIGO, "Garbage In, Garbage Out" (Provost & Fawcett, 2013) is a common aphorism in data science and AI. The dictum argues for the imperative of impeccable data as the foundation for any analytics or ML initiative. Within the context of Autonomous Finance, this principle is especially poignant. Cleansed data, particularly critical data elements, ensures the integrity of the intelligence generated and fortifies the automation strategies implemented. However, an astute observation highlights that the quest for data perfection need not be absolutist. The incremental gain in moving from high to very high data quality may not significantly enhance outcomes but could consume disproportionate resources.

Indeed, while the initial training data sets the course, it is not the be-all and end-all for Autonomous Finance systems. These AI models are not static; they are dynamic and adaptive, learning continually from data ingested post-go-live (Dhar, 2013). Over time, the influence of the training data wanes, overtaken by real-time data that continually fine-tune the model. Therefore, while pristine initial data is desirable, it does not monopolize the system's long-term efficacy. The architecture is designed to evolve, adapt, and refine its algorithms based on the fresh data it continually ingests.

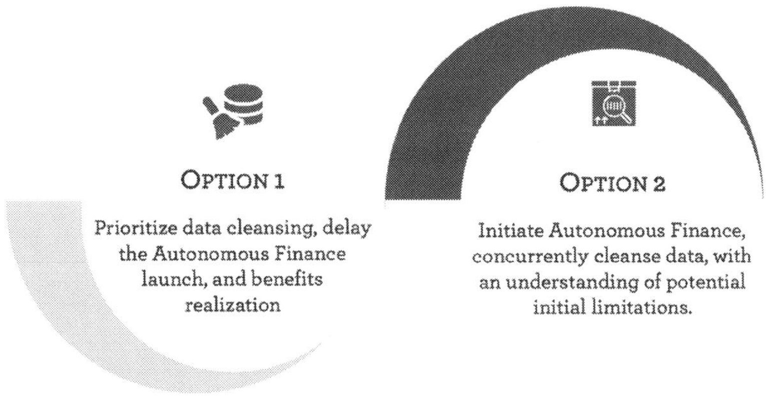

FIGURE 10.3: The Dilemma: To Cleanse or Not To Cleanse?

To synthesize, the decision to cleanse or not to cleanse data precludes a monolithic answer. It demands a nuanced approach sensitive to the organization's existing data maturity, the capabilities of the Autonomous Finance system, and the broader strategic goals. Given these complexities, a judicious path may involve a hybrid strategy that prioritizes cleansing for critical data elements without striving for unrealistic perfection. Therein lies the optimal balance—a blend of precision and adaptability—that serves as the linchpin for successfully harnessing the unprecedented promise of Autonomous Finance.

10.4 Data Governance

Guiding the journey through the intricate landscape of Autonomous Finance, data governance acts as the navigational compass, setting direction and ensuring alignment as illustrated in Figure 10.4. It provides the framework, policies, and standards that steer how data is managed and used within an organization, maintaining the course in the ever-changing seas of data complexity (Khatri & Brown, 2010).

a) **Data Quality Management:** Data quality management, committed to the precision, uniformity, and dependability of information, constructs the bedrock upon which reliable analytics and discerning decision-making processes are built (Wang & Strong, 1996).

b) **Data Privacy and Security:** Protecting sensitive information and ensuring compliance with privacy laws, data privacy, and security measures safeguard the integrity and confidentiality of financial data (Loshin, 2010).

c) **Data Architecture:** Designing the underlying structure and relationships between various data entities, data architecture enables a cohesive and efficient data ecosystem.

d) **Data Integration and Interoperability:** Facilitating seamless communication and collaboration between disparate data sources, integration, and interoperability ensure a unified and consistent information view across platforms.

e) **Master Data Management:** Master data management orchestrates and oversees the principal data within an organization, guaranteeing a harmonized and uniform perspective across diverse functional areas.

f) **Metadata Management:** Managing the data that describes other data, metadata management supports better understanding, usage, and control of information within the system.

g) **Data Lifecycle Management:** Guiding the flow and management of data throughout its lifecycle, from creation to disposal, data lifecycle management ensures optimal utilization and compliance.

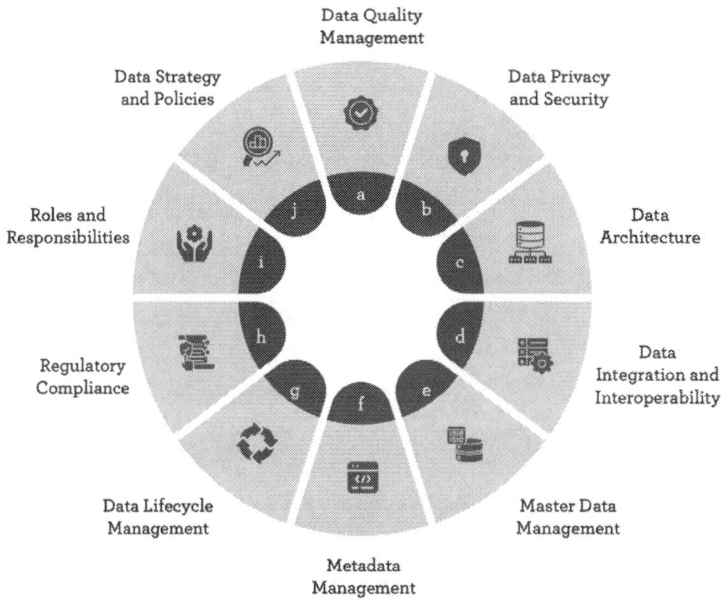

FIGURE 10.4: Data Governance in Autonomous Finance

h) **Regulatory Compliance:** Adhering to legal and industry standards, regulatory compliance ensures that data management practices align with external regulations and requirements (Otto, 2011).

i) **Roles and Responsibilities:** Defining the roles, responsibilities, and access levels of individuals involved in data management ensures clear accountability and control.

j) **Data Strategy and Policies:** Crafting the overarching strategy and policies that guide data usage, management, and governance, this function aligns data practices with organizational goals and objectives (Ladley, 2019).

In the context of Autonomous Finance, data governance plays a pivotal role in directing the correct, efficient, and compliant use of data. From credit scoring to regulatory reporting, governance establishes the protocols, access controls, privacy guidelines, and compliance measures that underpin all data-driven functions.

However, navigating the turbulent seas of Autonomous Finance is not without challenges. Clear communication, alignment between business and IT, cultural shifts, and adapting to rapidly changing global standards demand careful attention and expertise (Ladley, 2019).

Like the steadfast ship's wheel, effective data governance guides the organization's journey through Big Data's vast and intricate landscape. Its mastery ensures that the voyage remains on course, steering clear of pitfalls and propelling the organization toward success in the rapidly evolving world of Autonomous Finance.

10.5 Unification: Integration and Interoperability

The seamless unification of various systems and data sources is a cornerstone of Autonomous Finance. Integration and Interoperability enable disparate systems to work together, allowing for smooth data flow and coordination across the financial landscape.

a) **Integration:** The process of combining data from different sources to provide users with a unified view (Batini & Scannapieco, 2016).

b) **Interoperability:** The ability to connect different information systems, devices, or applications to access, exchange, and cooperatively use data (Benson & Grieve, 2020).

These concepts are vital in Autonomous Finance, facilitating improved decision-making and personalized financial services. However, they also present challenges due to varying data standards, privacy rules, and technical specifications.

For a comprehensive exploration of integration and interoperability, including detailed insights, challenges, and practical applications in Autonomous Finance, please refer to the previous chapter, Chapter 9, titled *"Cohesive Connectivity: Integration and Interoperability in Autonomous Finance."*

10.6 The Beacon of AI and ML

In the vast seas of financial data, AI and ML technologies are essential to chart our course. These powerful tools transform raw, unprocessed data into detailed maps, enabling better decision-making and performance across various industry sectors (Hale & Singh, 2020).

As illustrated in Figure 10.4, the predictive abilities of AI and ML extend to areas such as market trend analysis, portfolio optimization, personalized financial advice, and risk mitigation (Leo et al., 2019). However, they are not without challenges, such as overfitting, requiring careful calibration and attention to maintain effectiveness (Hale & Singh, 2020).

This section offers a glimpse of AI and ML's vital role in navigating Autonomous Finance's complexities. For a comprehensive exploration of AI technologies, including ML, NLP, computer vision, ethical considerations, and more specific applications, please refer to Chapter 6, Section 6.2, titled *"Role of Artificial Intelligence & Machine Learning."*

10.7 The Periscope of Data Visualization

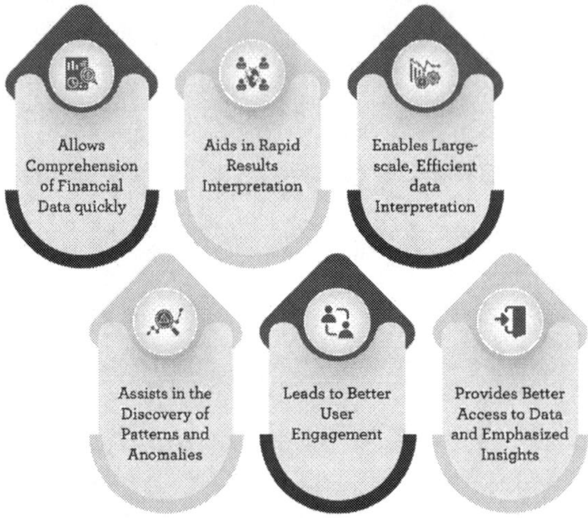

FIGURE 10.5: Data Visualization in Autonomous Finance

Much like a periscope enabling a submarine's crew to observe objects above the sea's surface, data visualization facilitates understanding complex data sets by presenting information in graphical or pictorial formats.

Data visualization, an essential aspect of data analytics, offers an interface for observing and interpreting patterns, trends, and insights from the complex depths of data. This tool translates intricate numeric and text-based data into a visual story, rendering it more intuitive and comprehensible (Few, 2009).

In Autonomous Finance, data visualization plays a vital role in deciphering financial patterns and trends. Consider the stock market, where finance teams employ charts and graphs to represent stock performance and market trends. Such graphical depictions enable traders to comprehend vast data volumes and make informed trading decisions (Kirk, 2019).

Moreover, the emergence of personal finance management applications that leverage data visualization provides users with transparent insights into spending habits. Through vivid graphs and charts, users readily discern financial allocations, enhancing budgeting and saving strategies (Shalimov, 2023).

As shown in Figure 10.5, In the turbulent waters of Autonomous Finance, data visualization functions as a periscope, furnishing essential insights for navigation. It unveils the broader picture, anticipates forthcoming scenarios, and supports necessary course adjustments. As vital as the compass for contemporary financial navigation, data visualization is indispensable for financial professionals and users.

10.8 Data Security and Privacy

In the world of Finance, where the pulse of transactions beats in real-time, the emergence of Autonomous Finance marks a transformative journey. Enabled by cutting-edge technologies such as AI, ML, RPA, Big Data, Analytics, Cloud Computing, SaaS, and APIs, Autonomous Finance promises unparalleled efficiency, personalization, and strategic decision-making. However, with great power comes great responsibility—and complexity.

FIGURE 10.6: Data Security & Privacy in Autonomous Finance

The technologies that drive the capabilities of Autonomous Finance also present intricate challenges in data security and privacy. The expansive volumes of data that fuel predictive AI and ML models, the synchronization of structured and unstructured information, and the real-time nature of financial solutions intertwine in a multifaceted web. This complexity is visually captured in Figure 10.6 below. More data enables better predictions but simultaneously becomes more cumbersome and difficult to guard.

Data security and privacy are no longer isolated technical concerns; they form a nexus that intersects technology, regulation, ethics, and human behavior. The fortress protecting sensitive financial data must be robust, adaptive, and multifaceted. It is not merely about building walls but orchestrating a symphony of protection measures, compliance strategies, ethical considerations, and human-centered practices.

The following sections delve into ten vital areas that constitute the fortress of data security and privacy in Autonomous Finance. From data protection and privacy by design to third-party risk management and employee training, each area is explored in depth, reflecting the multifaceted nature of the challenge and the complexity of the solutions. The interplay between these elements offers a comprehensive view of what it takes to navigate Autonomous Finance's intricate landscape securely.

This exploration provides a snapshot of the current state and a roadmap for the future, guiding practitioners, regulators, technologists, and decision-makers in their quest to harness the transformative power of Autonomous Finance without compromising the integrity, privacy, and security of the data that fuels it.

Data Protection

In the ever-evolving world of Autonomous Finance, data protection is a multifaceted endeavor that must be constantly reinforced and updated. Utilizing encryption protocols for data at rest and in transit is just the tip of the iceberg.

Real-time monitoring of transactions has become a vital aspect of data protection. Through advanced AI algorithms, transactions can be scanned for suspicious activities, enabling immediate detection and mitigation. These real-time systems require constant fine-tuning and validation, and the complexity is further amplified by the need to handle both structured and unstructured data.

Moreover, blockchain technology has been introduced as an innovative measure to ensure data integrity, offering a decentralized and tamper-proof ledger. However, integrating blockchain within existing systems adds a new layer of complexity and requires specialized expertise.

Another aspect of data protection is the secure handling of AI and ML models. As these models are often trained on sensitive financial data, ensuring the training and validation sets are securely handled without compromising their quality is a significant challenge.

The need for vast data sets to improve the predictive accuracy of AI models further complicates data protection. The more data, the higher the risk; therefore, the more complex the protection mechanisms must be. Security becomes a balancing act between facilitating innovation and safeguarding information.

Compliance with Data Privacy Regulations

Autonomous Finance operates within a regulatory landscape that is both complex and constantly evolving. Compliance with data privacy regulations like GDPR, CCPA, and HIPAA requires a deep understanding of legal obligations and implementing processes to fulfill them.

The need to synchronize structured and unstructured data in real-time further complicates regulatory compliance. Different jurisdictions may have varying data handling, storage, and usage requirements. Navigating this maze of regulations requires robust legal expertise and technological solutions tailored to specific regulatory environments.

Data mapping and systematic tracking of consents become essential components of compliance. This includes a clear understanding of where data is sourced, how it is processed, and where it resides. With the global nature of Autonomous Finance, this often involves cross-border considerations, adding another layer of complexity.

Automation in compliance monitoring and reporting is an emerging trend, offering efficiency and accuracy. However, developing and validating these automated systems requires extensive collaboration between legal, IT, and business departments.

Privacy by Design

Privacy by Design is a proactive approach to data privacy that involves integrating privacy measures from the inception of product or system development.

In Autonomous Finance, this principle translates into developing algorithms and systems that inherently respect user anonymity and data minimization. For example, techniques like Differential Privacy ensure that individual data cannot be reverse engineered from aggregated information.

However, implementing Privacy by Design in the complex ecosystem of Autonomous Finance is not without its challenges. It requires a profound understanding of both technological capabilities and legal requirements. Collaboration between engineers, legal experts, and business stakeholders is essential.

Furthermore, maintaining privacy without sacrificing the functionality and efficiency of AI and ML models is a complex task. Techniques like homomorphic encryption, which allows computation on encrypted data, are being explored but are still in their infancy.

Privacy by Design also involves ongoing monitoring and validation to ensure that privacy measures remain effective as technologies and regulations evolve. This continuous monitoring adds complexity but is essential to maintain user trust and legal compliance.

Data Breach Response

Data breaches are an ever-present risk in the financial industry, and the complexity of Autonomous Finance only amplifies this threat. A robust and systematic response plan must be in place to address potential breaches.

The initial containment phase involves immediate action to isolate the affected systems and prevent further damage. This requires a detailed understanding of the system architecture and the ability to react quickly without disrupting other functionalities.

Investigation follows, often requiring specialized forensic expertise to determine the breach's nature, extent, and origin. This investigation can be particularly challenging in Autonomous Finance, where AI and ML models can be intricate and opaque.

Notification of affected parties must be handled carefully, balancing the need for transparency with legal obligations and reputational considerations. In some jurisdictions, timely reporting to regulatory bodies is mandatory, adding another layer of complexity.

The recovery phase involves restoring systems, reinforcing security measures, and implementing lessons learned from the breach. This requires a multifaceted approach encompassing technological solutions, legal compliance, and communication strategies.

The use of AI in breach response is an emerging trend, with algorithms designed to detect anomalies and respond in real-time. However, these systems must be protected, creating a complex recursive security challenge.

Data Minimization

In Autonomous Finance, the paradox of data minimization is particularly pronounced. While vast data sets improve AI and ML model accuracy, collecting only essential data reduces overexposure and potential misuse risks.

Data minimization is not merely a technological challenge but a strategic one. It requires a deep understanding of what data is necessary for the desired functionality and what might be superfluous. Implementing data minimization principles often involves complex negotiations with business stakeholders, who may see more data as always better. Bridging this gap requires clear communication of the risks and benefits and possibly a cultural shift towards privacy.

Technologically, data minimization may require the development of more sophisticated models capable of extracting value from smaller data sets. This can be a significant research and development challenge. The real-time nature of many Autonomous Finance systems adds another layer of complexity. Minimizing data in real-time may require new algorithms and techniques to maintain functionality while adhering to privacy principles.

Data Disposal

This is a critical but often overlooked aspect of data security and privacy. Ensuring that data is securely disposed of when no longer needed minimizes unauthorized access risk and contributes to system integrity.

In the context of Autonomous Finance, this means not only deleting data but ensuring that it cannot be recovered. This might involve specialized technological solutions and adherence to specific regulatory requirements for data destruction. The challenge is further complicated by the need for historical data to train AI and ML models. Deciding when data is no longer needed and safely disposing of it without compromising future model development is complex.

Data disposal must also be aligned with data retention policies, which might require holding onto certain data for legal or business purposes. Navigating these conflicting requirements requires careful planning and execution. Automation in data disposal is a promising solution, but it must be handled carefully. Incorrectly configured automated disposal could lead to unintended data loss, with significant legal and business implications.

Audits and Reviews

Audits and reviews form an essential feedback loop in the security and privacy process, continuously aligning standards and practices with evolving threats and technologies.

In Autonomous Finance, audits and reviews must be multifaceted, encompassing the complex interplay between AI, Big Data, cloud computing, and more. This requires specialized expertise and a deep understanding of technology and regulatory landscapes. Third-party audits can objectively assess, but selecting the right auditors is crucial. They must understand the unique challenges of Autonomous Finance and be able to provide actionable insights and recommendations.

Internal reviews offer an ongoing opportunity to assess and refine security and privacy practices. In a rapidly evolving field, these reviews must be frequent and comprehensive, addressing not just technology but also processes and human factors.

Automation can play a role in continuous monitoring, but it can never fully replace human judgment. Striking the right balance between automated and human-driven audits is a complex but necessary task. Transparency in audits and reviews fosters trust among stakeholders, including customers. Sharing the process and results while maintaining necessary confidentiality can enhance reputational benefits.

Transparency

Transparency is a multifaceted challenge involving clear communication with users, control over their data, and adherence to guidelines such as the Fair Information Practices.

In Autonomous Finance, transparency must extend to algorithms, a significant challenge given the complexity and often opaque nature of AI and ML models. Users must understand how their data is being used, a task made more difficult by the intricate computations involved. Providing users with control over their data is not just an ethical obligation but often a legal one. This can be technologically challenging, requiring sophisticated UI and backend systems that can respond to user preferences in real-time.

Transparency is not just about technology but communication. Clear, accessible language must be used to explain complex processes, requiring collaboration between technical experts, legal teams, and communication professionals.

Third-Party Risk Management

Third-party risk management is a broad and complex field, reflecting the interconnected nature of modern financial systems.

In Autonomous Finance, third-party risks can include everything from cloud service providers to data sources for AI models. Each presents unique challenges and requires a tailored risk management approach. Due diligence is crucial but must be ongoing. A vendor that passes an initial assessment may later change practices or suffer a breach. Continuous monitoring and contractual safeguards must be in place to address these risks.

Synchronizing structured and unstructured data across multiple platforms adds another layer of complexity. Ensuring that all parties adhere to the same security and privacy standards requires careful coordination and oversight.

Employee Training

The human factor is often the weakest link in security, but it can also be a strength with the right training and culture. In Autonomous Finance, employees must understand general security principles and the specific challenges and risks associated with AI, ML, Big Data, and more. This requires ongoing, specialized training.

Training must also address the ethical dimensions of data handling, particularly in a field that relies heavily on personal and sensitive information. Employees must understand not just what to do but why it matters. Creating a culture prioritizing security and privacy can be as important as any specific training. Leadership must demonstrate commitment, and rewards and incentives must align with security and privacy goals.

Navigating the complex waters of data security and privacy in Autonomous Finance is a multifaceted challenge, intertwining technological solutions, regulatory compliance, organizational policies, and human factors. The interconnected nature of the technologies that enable Autonomous Finance, such as AI, ML, RPA, Big Data, Analytics, APIs, Cloud Computing, and SaaS, adds layers of complexity but also offers innovative solutions. By addressing the challenges head-on, with careful planning, vigilance, and adherence to best practices, the fortress of protection can be robustly maintained, ensuring the transformative journey of Autonomous Finance continues securely.

10.9 Predictive Analytics and Forecasting

In the vast ocean of financial data, predictive analytics and forecasting act as navigational tools, enabling the journey through the intricacies of Autonomous Finance. The insights provided by Figure 10.7 utilize statistical algorithms, ML techniques, and historical data to project future events and trends, offering precision and discernment in financial decision-making (Siegel, 2013).

a) **Predictive Analytics:** This is applied across various scenarios in finance, including but not limited to:

- **Credit Scoring:** By analyzing historical credit data, predictive models can assess the creditworthiness of individuals or businesses, guiding lending decisions.
- **Market Risk Management:** Predictive analytics identifies potential market risks and opportunities, enabling proactive risk mitigation strategies.
- **Portfolio Optimization:** Predictive models can optimize investment portfolios through pattern recognition and simulation, balancing risk and reward.
- **Algorithmic Trading:** Sophisticated algorithms analyze market trends to make real-time trading decisions, enhancing profitability (Mehdi Khosrow-Pour, 2019).

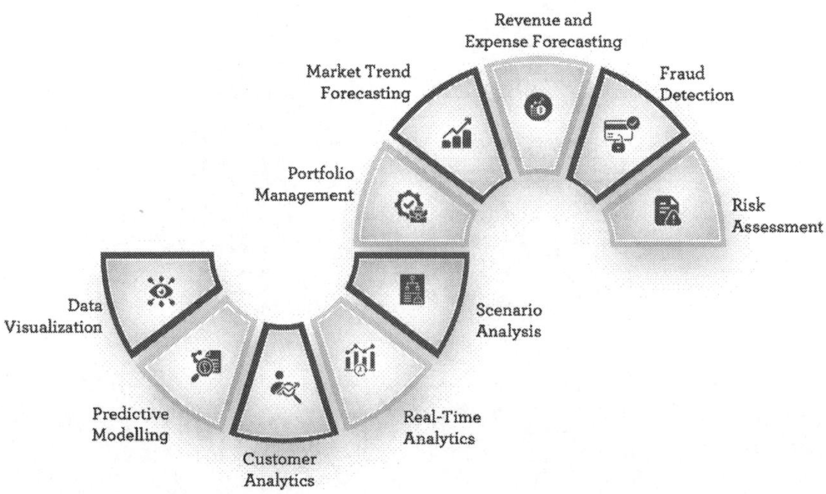

FIGURE 10.7: Predictive Analytics and Forecasting in Autonomous Finance

An illustrative example of predictive analytics in action is the rise of robo-advisors. These automated platforms utilize predictive algorithms to anticipate market trends, make data-driven investment recommendations, and manage portfolios, democratizing access to personalized investment advice (Mehdi Khosrow-Pour, 2019).

b) **Forecasting:** While predictive analytics identifies patterns and likelihoods, forecasting extends this analysis to predict specific future outcomes. In Autonomous Finance, forecasting is akin to predicting the weather for a sailor navigating uncharted waters. The applications include:

- **Market Movement Forecasting:** Utilizing historical market data, forecasting models can predict market fluctuations and guide investment strategies.
- **Customer Behavior Prediction:** Forecasting can tailor products and services to individual preferences and needs by analyzing customer behavior patterns.
- **Cash Flow Forecasting:** A critical application in personal and business finance, cash flow forecasting uses historical financial data to predict future financial positions (Samonas, 2015). Such insights empower business owners and individuals to plan investments, manage resources, and mitigate financial risks.

Predictive analytics and forecasting are not merely tools but a synthesis of mathematics, technology, and financial acumen. They enable financial practitioners to read the signs within the data, guiding decisions with foresight and intelligence. In Autonomous Finance, where complex algorithms power real-time decisions, the role of predictive analytics and forecasting is paramount. They illuminate the path, reduce uncertainty, and offer a strategic edge, turning the vast and chaotic sea of data into a navigable realm of insights and opportunities.

10.10 Surfing the Data Wave: The Significance of Big Data

In Autonomous Finance, Big Data represents both a mighty wave of opportunities and a tumultuous sea filled with challenges. It encompasses structured, semi-structured, and unstructured data from diverse sources, including financial transactions, social media, IoT devices, and beyond. This expansive ocean of data presents an unprecedented opportunity for organizations to extract insights, drive decision-making, and gain a competitive edge (Chen et al., 2012). Figure 10.8 outlines the dualities of challenges and opportunities organizations steering through the seas of Big Data face on the journey to Autonomous Finance.

a) **Opportunities:** Big Data provides the following opportunities for Autonomous Finance:

- **Financial Forecasting:** Advanced analytics and ML algorithms sift through large datasets to forecast market trends, enabling more accurate investment decisions.
- **Risk Assessment Enhancement:** Autonomous Finance systems can offer a nuanced understanding of financial risks by analyzing diverse data sources, aiding in robust risk management strategies.
- **Personalization:** Big Data allows for personalized financial advice and services tailored to individual preferences and needs.

- **Process Automation:** The ability to process and analyze vast data volumes in real-time paves the way for automation in various financial processes, enhancing efficiency.

b) **Challenges:** Despite its potential, Big Data presents formidable challenges:

- **Data Quality:** Ensuring the accuracy and reliability of vast data sets is a significant hurdle.
- **Privacy and Security:** Safeguarding the privacy and security of information within this expansive data landscape requires vigilant measures.
- **The 3Vs of Big Data:** The sheer volume, velocity, and variety of Big Data present complex challenges for Autonomous Finance systems. Managing this trifecta requires sophisticated tools and methodologies (Katal et al.,2013).
- **Deriving Value:** The most intricate challenge is transforming raw data into actionable insights. This transformation demands advanced data analytics skills and technologies that may not be readily available to all organizations.

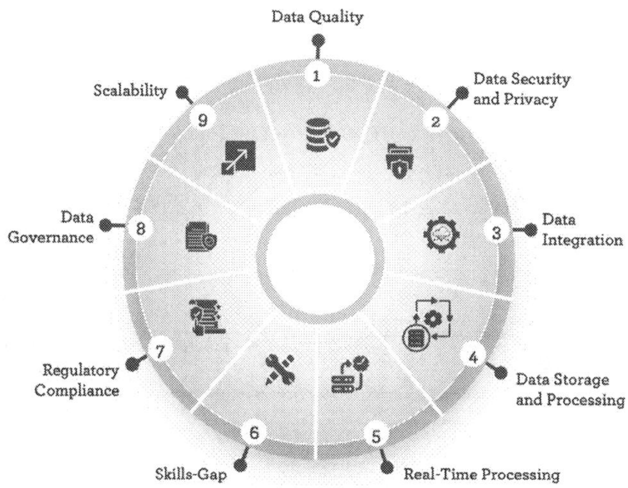

FIGURE 10.8: Navigating the Big Data Challenges in Autonomous Finance

The significance of Big Data in Autonomous Finance transcends mere data accumulation. It is about surfing the wave of opportunities while skillfully navigating the stormy seas of challenges. The integration of AI and ML emerges as a powerful compass, guiding the journey through this expansive ocean, adeptly overcoming obstacles, and unveiling the latent potential hidden within.

In essence, Big Data is not merely a phenomenon in Autonomous Finance; it is a dynamic force that propels innovation, fosters growth, and demands strategic acumen and technological prowess. Mastering Big Data is akin to charting a confident course across uncharted waters, turning challenges into opportunities, and transforming the financial landscape.

10.11 Conclusion

In the intricate tapestry of Autonomous Finance, data and analytics emerge not merely as threads but as the weft and warp that give the fabric its integrity and design. This chapter has meticulously explored these core elements, casting them as the foundational underpinnings of this transformative journey. To borrow from Jeff Weiner (2012), former CEO of LinkedIn, *"Data powers everything that we do,"* the chapter underscores that data is not merely an ingredient but the lifeblood of Autonomous Finance.

From data acquisition and governance to the alchemy of analytics, the discourse has illuminated how these elements drive decision-making, predictive accuracy, and operational efficiency. It has also grappled with the inherent challenges—data quality, privacy, and security—accompanying this data-centric journey. Yet, as the burgeoning field of data ethics suggests, these challenges are not insurmountable but call for a multifaceted approach involving regulation, third-party risk management, and continuous employee training.

As the narrative of this chapter culminates, it sets the stage for a future examination of 'Customer Experience,' a realm where data and analytics manifest their most tangible impact. The horizon is rife with possibilities: from hyper-personalization and seamless integrations to the inclusion of previously underserved populations, the next chapter promises to delve into how Autonomous Finance can serve as a force multiplier in enhancing human well-being.

In great physicist Richard Feynman's (1988) words, *"What I cannot create, I do not understand."* This chapter has sought to create a comprehensive understanding of the role of data and analytics in Autonomous Finance, paving the way for informed creation and innovation. The journey may be intricate, but it promises a future where every nuance adds to collective progress and well-being. The horizon beckons, but now, it is better understood, promising a journey of exploration and enlightened transformation.

10.12 Key Recommendations

a) **Invest in Effective Data Integration:** Invest in effective data integration systems that can handle the complexity and diversity of financial data, allowing for improved decision-making and service personalization.

b) **Continuous Improvement in AI and ML models is necessary:** Regularly update and calibrate AI and ML models to maintain their accuracy and effectiveness. Attention must be given to avoid overfitting and other pitfalls.

c) **Leverage advanced Data Visualization tools:** Enhance data visualization tools to make financial patterns and trends more comprehensible and intuitive for users and financial experts.

d) **Develop a comprehensive Data Security and Privacy approach:** Prioritize data security and privacy to maintain customer trust and avoid potential financial and reputational losses.

e) **Big Data and Analytics:** Develop and implement advanced data analytics skills and technologies to convert raw data into actionable insights, unlocking the potential of Big Data in Autonomous Finance.

10.13 References

Batini, C., & Scannapieco, M. (2016). *Data and information quality*. Cham, Switzerland: Springer International Publishing.

Benson, T., & Grieve, G. (2020). Principles of health interoperability*: SNOMED CT, HL7 and FHIR*. Springer.

Chen, H., Chiang, R. H., & Storey, V. C. (2012). Business Intelligence and Analytics: From Big Data to Big Impact. *MIS Quarterly, 36*(4), 1165-1188.

Davenport, T. (2014). *Big data at work: dispelling the myths, uncovering the opportunities*. Harvard Business Review Press.

Deming, W.E. (1982). *Out of the Crisis*. MIT Press

Few, S. (2009). *Now you see it: simple visualization techniques for quantitative analysis*. Analytics Press.

Feynman, R. P. (1988). What I cannot create, I do not understand [Quotation found on blackboard]. California Institute of Technology.

Gide, A. (1925). *Les faux-monnayeurs*. [The Counterfeiters]. NRF Gallimard.

Hale, G., Singh, N. Center for Analytical Finance (CAFIN). Leo, M., Sharma, S., & Maddulety, K. (2019). Machine learning in banking risk management: A literature review. *Risks, 7*(1), 29.

IPEA Australia Government [@ipeaAusGov]. (2021, April 8). Jeff Weiner: Here's an informative infographic on how to write a great research paper! #ResearchPaper #AcademicWriting [Image attached] [Tweet]. Twitter. https://twitter.com/ipeaAusGov/status/1380035861425741824

Jobs, S. (2005). *'You've got to find what you love,'* Jobs says. Stanford Report. Retrieved from https://news.stanford.edu/2005/06/14/jobs-061505/

Katal, A., Wazid, M., & Goudar, R. H. (2013, August). Big data: issues, challenges, tools and good practices. In *2013 Sixth international conference on contemporary computing (IC3)* (pp. 404-409). IEEE.

Khatri, V., & Brown, C. V. (2010). Designing data governance. *Communications of the ACM, 53*(1), 148-152.

Kirk, A. (2019). Data visualization: A handbook for data-driven design. *Data Visualization, 1*-328.

Ladley, J. (2019). *Data governance: How to design, deploy, and sustain an effective data governance program*. Academic Press.

Loshin, D. (2010). *The practitioner's guide to data quality improvement*. Elsevier.

Mehdi Khosrow-Pour, D.B.A. (2019). Robo-Advisor. In *Encyclopedia of Information Science and Technology (Fourth Edition)*, pp. 5456-5465. IGI Global.

Otto, B. (2011). Organizing data governance: Findings from the telecommunications industry and consequences for large service providers. *Communications of the Association for Information Systems, 29*(1), 3.

Rahm, E., & Do, H. H. (2000). Data cleaning: Problems and current approaches. *IEEE Data Eng. Bull., 23*(4), 3-13.

Redman, T. C. (1997). *Data quality for the information age*. Artech House, Inc.

Romdhane, R. B., Hammami, H., Hamdi, M., & Kim, T. H. (2019). A novel approach for privacy-preserving data aggregation in smart grid. In *2019 15th international wireless communications & Mobile Computing Conference (IWCMC)* (pp. 1060-1066). IEEE.

Shalimov, A. (2023). What are the Best Personal Finance Apps and How to Build One? *Eastern Peak - Technology Consulting & Development Company*. https://easternpeak.com/blog/personal-finance-app-development/

Siegel, E. (2013). *Predictive analytics: The power to predict who will click, buy, lie, or die*. John Wiley & Sons.

Wang, R. Y., & Strong, D. M. (1996). Beyond accuracy: What data quality means to data consumers. *Journal of management information systems, 12*(4), 5-33.

Weber, K., Otto, B., & Österle, H. (2009). One size does not fit all: A contingency approach to data governance. *Journal of Data and Information Quality (JDIQ)*, 1(1), 1-27. https://doi.org/10.1145/1515693.1515698

Weiner, J., (2012, June 18), fortune.com

Worldwide Big Data and Analytics Spending Guide. (2019). IDC: The Premier Global Market Intelligence Company. https://www.idc.com/getdoc.jsp?containerId=IDC_P33195

CHAPTER 11

Beyond Transactions

Crafting the Customer Experience in Autonomous Finance

"The customer's perception is your reality. What they think about your products MATTERS. If you don't put your customer's perception first, THE GAME IS OVER."

- SHARFARAZ AHMED (GLADCHUK, 2020)

11.0 Key Insights

- **Seamless Integration: Bridging Finance and Everyday Life:** The chapter emphasizes the growing integration of Autonomous Finance into lifestyle and business apps, contributing to a more streamlined and intuitive user experience. This evolution signifies a future where financial management becomes virtually indistinguishable from routine tasks.
- **Instant and Intuitive Support: AI's Role in Customer Service:** AI-driven support, including chatbots and predictive customer service, is significantly enhancing the user experience in Autonomous Finance. Future advancements could blur the lines between human and machine assistance, creating a more personalized and immediate support system.
- **Security and Privacy: Essential Pillars of Trust in Autonomous Finance:** The chapter underscores the importance of robust security measures and stringent privacy policies in Autonomous Finance. Strategies to maintain user trust include advanced encryption technology, ML for fraud detection, 'privacy-by-design' approaches, and biometrics.
- **Hyper-Personalization: AI's Promise for Future Customer Experiences:** The chapter anticipates a future where Autonomous Finance offers highly personalized financial advice, products, and services based on refined AI predictive analytics. Such hyper-personalization can lead to a financial management experience uniquely tailored to each individual.
- **Balancing Innovation with Responsibility: The Ethical Considerations:** While Autonomous Finance opens up endless opportunities for crafting unique and enriching customer experiences, it also brings complex regulatory and ethical considerations. The upcoming chapter is expected to delve into these challenges, exploring how Autonomous Finance navigates these intricate territories.

11.1 Introduction

In the intricate tapestry of finance, the warp and weft of transactions have historically been devoid of personalization, especially in the business-to-business (B2B) arena. Defined by cumbersome workflows, bureaucratic rigidity, and a one-size-fits-all approach, these interactions have often resembled a mechanistic assembly line rather than a dynamic, customer-centered experience. However, as dawn breaks in the age of Autonomous Finance, a transformative paradigm shift is unfolding.

Empowered by the avant-garde technologies of AI, ML, RPA, Blockchain, Big Data, Advanced Analytics, Cloud Computing, SaaS, and APIs, Autonomous Finance heralds an era where financial systems not only execute transactions but also comprehend, learn, and even anticipate the unique financial goals, needs, and behaviors of organizations. Imagine a financial landscape where systems transcend transactional functions to proactively deliver personalized financial insights, suggestions, and foresight. This notion, once the stuff of science fiction, is rapidly transitioning to a tangible reality.

Amid a technological renaissance, the world stands on the cusp of a financial revolution driven not by mere conjectures but by quantifiable transitions in market dynamics and consumer behaviors. According to a comprehensive analysis by Statista (2021), the digital payments sector has witnessed an astronomical rise, burgeoningly expanding its global transaction value from 4.1 trillion U.S. dollars in 2019 to an estimated 6.6 trillion U.S. dollars in 2021. This translates to a staggering growth of over 60% in just two years, underscoring a rapidly intensifying trend that showcases not merely an adoption but a burgeoning preference for digital financial mediums (Statista, 2021). In such a climate, the B2B domain is not a mere spectator but a dynamic participant, steering towards a horizon where Autonomous Finance is not a distant dream but an impending reality, pulsating with personalized and technologically enriched experiences.

As the B2C sector continues to set a high bar for enriched customer experiences, the B2B landscape is not far behind. The burgeoning array of features introduced by Autonomous Finance software vendors seeks to make B2B interactions as intuitive and user-friendly as their B2C counterparts. This chapter promises an in-depth exploration of the critical facets of customer experience in Autonomous Finance, from the allure of intuitive interface designs to the power of real-time, AI-driven financial advisories.

Yet, this transformative journey is not without its labyrinthine complexities and ethical conundrums, a dimension this chapter will explore with the acuity it deserves. Christine Lagarde, the President of the European Central Bank, once emphasized that *"The financial industry is a service industry. It should serve others before it serves itself"* (Inside Job (2010) - IMDb, n.d.). Drawing from this profound insight, this chapter endeavors to unfurl a financial panorama where customer service is paramount. The Autonomous Finance ecosystem, deeply rooted in futuristic technology, operates with a heart and conscience. It is driven to foster unparalleled customer value, traversing the dynamic landscapes of both B2B and B2C sectors.

So, brace for an exhilarating voyage through the uncharted waters of Autonomous Finance, where each chapter unveils a deeper layer of understanding. These unearthing insights could redefine customer experience in the financial cosmos. The expedition awaits.

11.2 Journey Beyond Transactions: The Paradigm Shift

In the traditional B2B financial setup, interactions have been primarily transactional, with a heavy emphasis on data entry, reconciliation, and record-keeping. With the world entering the era of Autonomous Finance, a substantial shift from a transaction-centric model to a relationship- and experience-centric one is emerging.

First, let us comprehend the essence of this paradigm shift. It transforms how businesses perceive and interact with their financial systems and processes. The focus has transitioned from mere transaction processing to designing a seamless, personalized, engaging business financial journey. As delineated in Figure 11.1, Autonomous Finance seeks to extend the user-friendly nature of B2C interactions into B2B transactions, turning financial management into an intuitive, empowering, and interactive experience.

This shift aligns with a broader trend observed across industries: Businesses are reorienting their strategies to be more customer-focused, aiming to deliver outstanding experiences that foster loyalty and satisfaction. According to a Salesforce report (Afshar, 2020),

84% of customers say the organization's experience is as important as its products and services, and this sentiment extends to B2B customers.

As experiences in the B2C sector impact expectations in the B2B sector, adapting financial systems to this new standard becomes critical. AI-powered Autonomous Finance systems are stepping up to the challenge. They offer intelligent, predictive, and personalized features that help businesses transact and interact with their financial data and gain valuable insights.

An example of this shift is evident in how software vendors are adding features like interactive dashboards, predictive analytics, and recommendation engines to their offerings. These features, reminiscent of those found in B2C apps, aim to make B2B financial operations more accessible and intuitive.

The potential benefits are enormous. By focusing on user experience and customer satisfaction, businesses can build stronger relationships with their clients, streamline their operations, and, ultimately, drive growth and profitability.

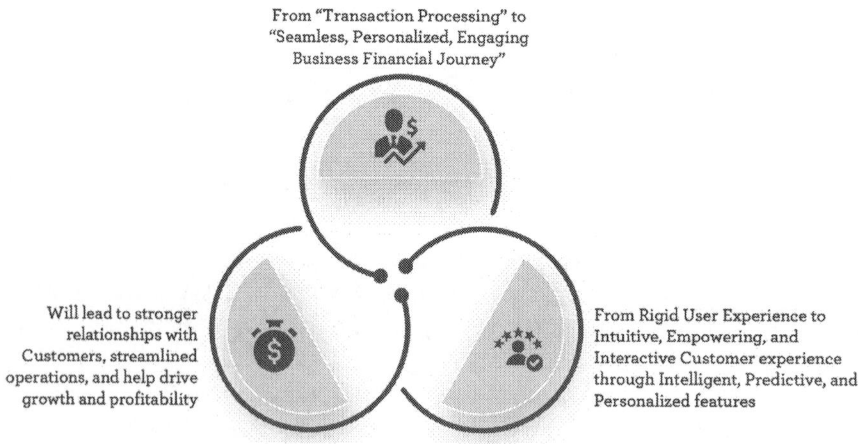

FIGURE 11.1: The Paradigm Shift

This paradigm shift toward an experience-centric model in Autonomous Finance represents a promising and exciting frontier in the financial landscape.

11.3 Fashioning Finance: The Power of Personalization

The key to enhanced user engagement in today's digital landscape is personalization, as well-proven by the B2C sector. It is time for the B2B world to embrace this revolution fully. Autonomous Finance is stepping up to this challenge, reshaping the user experience from a one-size-fits-all approach to a finely tailored interaction that caters to each unique user.

Historically, finance systems offered rigid user experiences narrowly tailored to specific roles or processes. Nonetheless, Autonomous Finance is revolutionizing this approach.

Through the power of AI, it is evolving to cater to individual user needs, refining experiences down to the level of single interactions.

For instance, let us consider how Autonomous Collections software solutions utilize AI. These tools examine multiple parameters, including invoice date, invoice amount, and historical payment behavior, to predict the most probable date when a customer will receive a payment. This level of tailored insight goes beyond standard transactional data, offering valuable predictive capabilities that streamline cash flow management.

Autonomous Finance does not stop at tailoring software features and processes—it is reimagining how interfaces look and feel for every user. The dashboards for a CFO and a financial analyst can vastly differ based on their respective needs. The ability of Autonomous Finance systems to present each user with a customized view enriches their interactions and supports better decision-making.

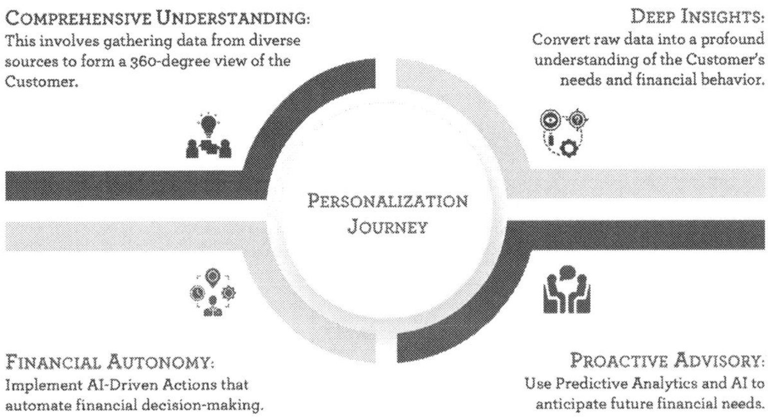

FIGURE 11.2: Tailored Approach to Customer Engagement and Decision-making

The personalization wave does not end at user interfaces and predictive capabilities—it is also changing how support systems function. AI-powered chatbots, for example, provide assistance tailored to individual needs and circumstances, enhancing the overall user experience.

In summary, Autonomous Finance is rewriting the user experience playbook, weaving a financial tapestry tailored to each individual's needs, as illustrated in Figure 11.2. By offering finely tuned experiences, Autonomous Finance enhances user satisfaction, encourages loyalty, and propels finance into a new era of personalized engagement.

11.4 Anytime, Anywhere: The Power of Accessibility

In a globally connected ecosystem that pulsates with activity around the clock, the contemporary individual demands instantaneous access to financial transactions, a need more pronounced now than ever before. Within this pulsating landscape, Autonomous Finance unveils itself as the herald of a new era where time and space constraints dissolve, offering financial transactions and insights at one's fingertips, irrespective of geographical or temporal boundaries. Figure 11.3 provides a multi-dimensional view into the architecture of financial ubiquity, mapping both the driving forces and the challenges, in a symphonic relation that reflects the dynamism inherent in the new era of Autonomous Finance.

Fintech companies stand at the forefront of this rapid evolution, steering the helm with a visionary approach that has redefined accessibility over the last decade. Mobile banking is a quintessential testimony to this transformation as the preferred conduit for financial engagements. A report delineating the paramount trends of 2023 by Accenture Banking underscores this phenomenon, highlighting that the contemporary banking customer predominantly engages through mobile platforms, with a scant 3% opting for face-to-face interactions (*Top 10 Banking Trends for 2023*, 2023).

Synchronically, contactless payments have undergone a remarkable growth, delineating the path of the future. The metamorphosis in the payment landscape is substantiated by a recent McKinsey analysis, which illuminated that from 2018 to 2021, the global compound annual growth rate (CAGR) of non-cash retail payment transactions ascended by 13%, a trajectory even more pronounced in emerging markets with a staggering 25% CAGR. This trend, which witnessed the fastest growth in select African and Asian markets, envisages a continued momentum, forecasting a CAGR of 15% from 2021 to 2026 (*Sustaining Digital Payments Growth: Winning Models in Emerging Markets*, 2022).

Additionally, in B2B Finance, the value of accessibility cannot be downplayed. As discussed in *Chapter 6: "Inside the Engine: The Tech that Powers Autonomous Finance,"* the rise of cloud computing and SaaS models is making finance-related software solutions increasingly accessible. For instance, ERP systems and financial management software that were once tethered to a specific location can now be accessed from anywhere, offering real-time financial insights and facilitating informed decision-making.

While these advancements are commendable, it is crucial to remain aware of the challenges accompanying increased accessibility. As discussed in *Chapter 10: "The Influence of Insight: Data and Analytics in Autonomous Finance,"* data security and privacy are significant concerns that need to be addressed as the realm of Autonomous Finance is explored further. For additional details, please refer to *Chapter 14: "The Digital Fortress: Cybersecurity in the Realm of Autonomous Finance."*

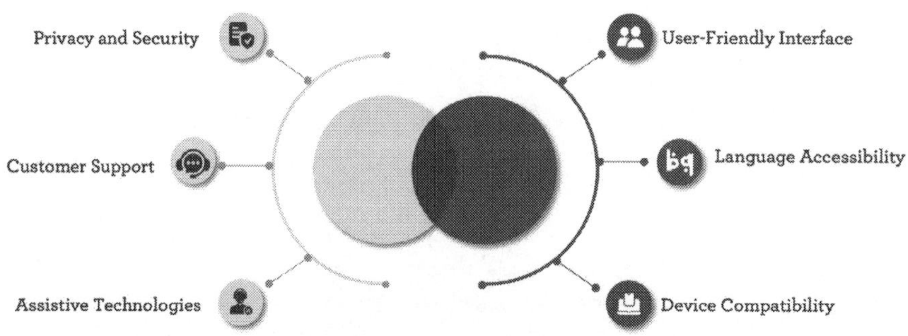

FIGURE 11.3: "Finance Anytime Anywhere" via Autonomous Finance

In conclusion, with the capability to engage with financial services and transactions anytime, anywhere, Autonomous Finance redefines how interactions with finances occur, propelling the world into a new era of financial accessibility.

11.5 A Step Ahead: Proactive Financial Management

In the era of Autonomous Finance, the concept of 'proactive' assumes a new significance. Autonomous Finance uses advanced technologies to identify potential risks, seize opportunities, and initiate actions before an issue becomes a problem. As depicted in Figure 11.4 below, this revolution allows businesses and individuals to stay ahead in their financial management.

For individuals, proactive financial management materializes through functionalities like predictive budgeting and personalized financial advice. For instance, personal finance tools like Mint and Quicken analyze users' spending habits and generate budgets, enabling informed financial decisions (Mint, 2020). Similarly, robo-advisors such as Betterment and Wealthfront leverage complex algorithms to create personalized investment portfolios, adjusting asset allocation based on market trends.

Autonomous Finance revolutionizes financial management for organizations by automating tasks and generating insights that promote proactive decision-making. One notable example is the evolution of ERP systems. These systems are transitioning from mere transactional tools to predictive instruments, allowing businesses to anticipate financial outcomes and take preventative actions.

For example, let us consider the domain of Autonomous Treasury Management. Advanced platforms utilize AI to optimize liquidity management, execute accurate cash forecasting, and mitigate financial risk. These tools can predict cash needs by analyzing vast data sets, enabling treasurers to make informed decisions about cash positioning and funding (HighRadius, 2023).

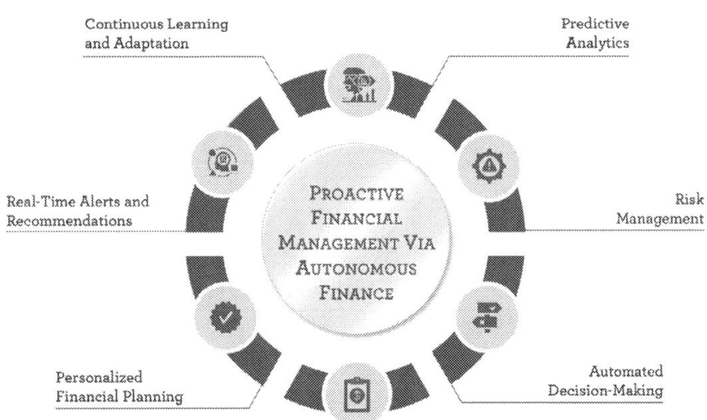

FIGURE 11.4: "Finance at your Fingertips" via Autonomous Finance

While embracing this shift toward proactive financial management, it is essential to remember the associated ethical and regulatory implications. As AI and ML become instrumental in decision-making processes, considerations around data privacy, algorithmic fairness, and transparency must be prioritized.

In summary, by harnessing technology to enable proactive financial management, Autonomous Finance empowers individuals and businesses to stay ahead in their financial journey.

11.6 Seamless Integration with Lifestyle Apps

In today's fast-paced, interconnected world, the boundaries between different sectors and services are becoming increasingly blurred. As a part of this transformation, Autonomous Finance is creating a more unified experience by integrating with lifestyle apps to make financial management a seamless part of everyday life.

One of the pivotal benefits of Autonomous Finance is its ability to connect with many other applications and platforms. From online shopping and food delivery apps to health and fitness trackers, Autonomous Finance solutions can be linked to myriad services, allowing users to manage their finances effortlessly within their lifestyle routines.

Take the case of super apps like WeChat and Grab. Initially launched as communication and ride-hailing apps, these platforms have evolved to include financial services such as payments, transfers, and investments. Their seamless integration of finance into the user's lifestyle represents a significant advancement in Autonomous Finance.

Similarly, popular social media platforms, such as Facebook and Instagram, have introduced in-app shopping and payment features, simplifying the process of browsing, purchasing, and paying. Users can transact without leaving the app, ensuring a frictionless shopping experience.

Let us consider the widespread adoption of Open Banking as an example of this integration. With APIs, finance teams can interact with third-party providers, resulting in a more holistic customer experience. For instance, linking a personal finance app with a food delivery platform can automatically categorize and track food-related expenses, thus simplifying budget management.

Similarly, on a B2B level, Autonomous Finance can enhance financial management by integrating with other business functions. For example, using APIs, modern ERP systems can seamlessly integrate with Autonomous Finance applications, enabling real-time data sharing and collaboration. This can significantly improve the accuracy of financial forecasts, facilitate informed decision-making, and foster agility.

Imagine a situation where an organization's supply chain management system is integrated with its Autonomous Payables system. Based on the insights generated from the supply chain (like when the inventory drops below a certain level), the Autonomous Payables system can automatically initiate transactions with the suppliers, ensuring business continuity.

While navigating the dynamic landscape of Autonomous Finance, the undeniable integration of this field with lifestyle and business apps is evident. By paving the way for a more intuitive, streamlined, and unified user experience, Autonomous Finance is subtly weaving itself into the fabric of our everyday activities. This evolutionary intertwining signifies a future where the demarcation between finance and routine tasks may be virtually indistinguishable.

11.7 Instant Assistance: AI Chatbots

In the age of Autonomous Finance, instant gratification is not just a trend—it is an expectation. Users do not want to wait for responses or resolutions. They need instant, efficient, and practical assistance. Enter AI-driven support—a game-changer in customer service, providing quick and accurate solutions to users' queries or issues, significantly enhancing the user experience in Autonomous Finance.

One of the most influential advancements in AI support is using chatbots, virtual assistants powered by AI that can understand and respond to human queries in real-time. These chatbots can understand context, gauge sentiment, and provide personalized responses through ML algorithms. For instance, J.P. Morgan's virtual assistant, COIN, processes over 12,000 documents every second, providing swift responses to queries from clients and internal teams, thereby vastly improving efficiency (Georgiou, 2017).

AI support also extends to Predictive Customer Service, a system where AI uses past data to predict future needs or issues. For example, an Autonomous Finance system could anticipate when users might encounter difficulties with a new feature based on their usage patterns. It could then proactively send instructions or offer assistance via the AI support system, mitigating potential problems before they arise.

AI support is not restricted to addressing current issues but focuses on continuous learning and improvement. By analyzing feedback and complaints, AI can identify patterns and areas of improvement, leading to a better and more seamless user experience. For example, an Autonomous Payables system could learn from the queries raised by the accounts payable team, refining its processes over time to minimize similar issues in the future.

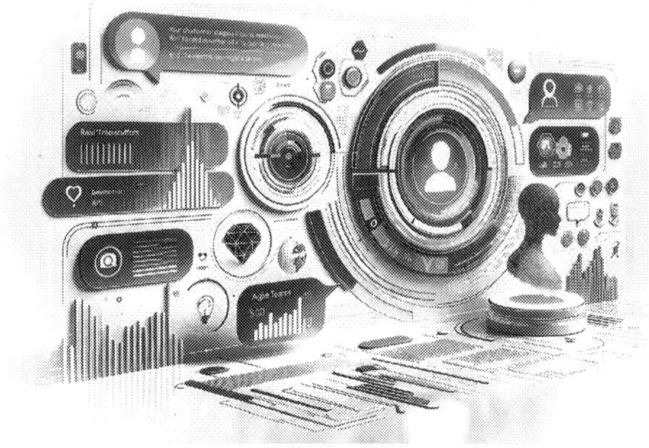

Finally, integrating AI support with voice recognition technology allows users to interact with the system through voice commands, making the process more intuitive and effortless. Amazon's Alexa and Google Assistant have set the trend in the consumer world, with B2B financial software vendors also beginning to incorporate voice-activated functionalities into their solutions (Moc & Moc, 2023).

The promise of AI support in Autonomous Finance lies in its potential to provide immediate, accurate, and personalized assistance, thereby significantly improving user experiences. As AI continues to evolve, so will its ability to support and engage with users, blurring the lines between human and machine assistance.

11.8 Security, Privacy, & Trust in Autonomous Finance

As users embark on their journey into Autonomous Finance, their expectations for robust security and stringent privacy measures increase. Whether in a B2B or B2C context, the sanctity of financial data remains paramount. Any potential compromise could lead to catastrophic repercussions, not just materially but also in terms of eroded trust and reputation. Thus, at the heart of Autonomous Finance lies a non-negotiable commitment to safeguarding user data—maintaining an unwavering focus on security and privacy.

Furthermore, Autonomous Finance systems can proactively identify potential fraudulent activities using ML algorithms. Unusual patterns and outliers in financial transactions can be identified in real-time, instantly alerting users and security teams. For instance, MasterCard's Decision Intelligence employs AI to assess transactional risk levels, flagging potential fraud before it occurs (Mastercard, 2023).

While technology bolsters security, privacy policies are essential for maintaining user trust. Organizations implementing Autonomous Finance must have transparent and user-centric privacy policies in place. GDPR in Europe and CCPA in California offer a regulatory framework for companies to build such policies, granting users more control over their data.

In an Autonomous Finance world, '*privacy by design*' is becoming the norm—incorporating privacy measures into the system's architecture from the outset (Medine, 2018). For instance, some Autonomous Treasury systems now allow individual user-level privacy settings, enabling users to decide what data they wish to share and with whom.

The advent of biometrics—face recognition, fingerprints, iris scanning—has bolstered security in Autonomous Finance (*Application of Biometric Face Identification Technologies in Financial Institutions | Biometric Update*, 2023). Companies like Apple and Samsung have popularized biometrics in the B2C world, and B2B vendors have followed suit, incorporating these technologies for user verification and access control.

Indeed, the narrative of Autonomous Finance continually underscores an enduring pledge to security and privacy. This evolution weaves a harmonious coexistence between trust and technology, shaping an ecosystem where financial innovation thrives under the robust security and privacy shield.

11.9 The Future of Customer Experience

Standing on the threshold of this technological revolution, the canvas of the future appears primed for Autonomous Finance to redraw the contours of customer experience. Propelled by AI's capabilities and driven by customer-centricity, the forthcoming innovations promise to take user experiences to unprecedented heights.

The initial brush stroke on this canvas is attributed to AI-driven hyper-personalization. As AI's predictive analytics become more refined and Autonomous Finance grows more capable of understanding and responding to individual user needs, the personalization of financial services will reach new depths (Correspondent & Correspondent, 2022). Imagine a world where financial advice, products, and services are so finely tuned to your needs that every interaction feels like it was designed just for you.

Subsequently, the expansion of integrating finance functions into everyday life will result in increasing seamlessness. As the blend between lifestyle and financial apps further intensifies, the line between finance and life will blur to indistinction. Autonomous Finance will permeate even more into daily routines, making financial management an effortless part of life.

Moreover, as progress continues, the potency of AI-powered assistance will be enhanced. The future holds the promise of AI assistants that can answer complex queries and predict and pre-empt user needs, offering solutions even before users recognize them. This will create a dynamic where finance becomes a companion rather than just a service ("How Artificial Intelligence Can Deliver Real Value to Companies," 2017).

Lastly, on the horizon of this future landscape is the consistent evolution of security measures. As Autonomous Finance's capabilities evolve, so will the methods to safeguard privacy and data security. Stringent safety measures will underpin every transaction and every decision in the future, fostering a more resilient and trust-oriented ecosystem ("How Artificial Intelligence Can Deliver Real Value to Companies," 2017).

These innovations will culminate in an Autonomous Finance landscape where user experience takes center stage. Financial management will achieve unprecedented levels of efficiency, insight, intuitiveness, engagement, and empowerment in the envisioned future. Such is the promise of Autonomous Finance— a realm where Finance serves individuals, understands their needs, and enables the pursuit of an optimal life.

11.10 Conclusion

In the evolving tableau of Autonomous Finance, the customer is no longer a passive actor but a central figure around whom the entire narrative revolves. This chapter has elucidated the monumental paradigm shift from a transaction-centric model to one where relationships and experiences are at the forefront. Enabled by the catalytic capabilities of AI and ML, this transformation promises personalized and hyper-personalized experiences tailored to the individual's financial DNA.

In a perceptive observation shared on a LinkedIn post, Ignacio Bracamonte brought to the fore a quotation attributed to Joe Sparano: *"Good design is obvious. Great design is transparent"* (Bracamonte, 2020). This principle echoes profoundly in the evolving landscape of Autonomous Finance, illustrating the zenith of aspiration - to engineer platforms that are not just functional and intuitive but transcendent in their design. The ultimate aspiration is to create functional, intuitive platforms and transcend design altogether. To create financial ecosystems so seamless, they become virtually invisible, embedded into the very fabric of our daily lives. The future vision manifests as a reality where financial management evolves from a task to an intrinsic part of our lived experience, informed and enriched by AI and ML algorithms that work silently in the background, optimizing, advising, and ensuring financial well-being.

However, this canvas of possibilities is not without its shadows. questions and challenges arise as we anticipate a future suffused with hyper-personalization, seamless integration, and impenetrable security. How can the pursuit of technological transcendence maintain a focus on ethical imperatives, data security, and regulatory compliance?

Autonomous Finance is not a static concept etched in stone but a dynamic, evolving entity. It responds, it learns, and it adapts. It manifests collective aspirations to make financial management more efficient and humane. As we segue into the subsequent chapter, an exploration of these ethical and regulatory landscapes awaits, offering a balanced perspective that marries technological innovation with social responsibility.

The palette and the canvas are set; the brush now lies in the hands of pioneers and visionaries. The opportunity beckons to shape a future where finance transcends its utilitarian boundaries to become a powerful catalyst for empowerment, inclusivity, and individual fulfillment. The voyage through the intricate landscape of Autonomous Finance continues, each chapter revealing another layer, each layer bringing us closer to a future replete with untold possibilities.

11.11 Key Recommendations

a) **Focus on Seamless Integration:** Businesses should continue to explore and leverage opportunities for integrating Autonomous Finance into lifestyle and business apps. This seamless integration would enhance the user experience by making financial management an intuitive part of everyday activities.

b) **Invest in AI-Driven Customer Support:** Companies should invest in AI-driven support systems like chatbots and predictive customer service to provide instant and personalized assistance to users. Continuous learning from feedback and complaints can help refine these systems and improve the overall customer experience.

c) **Prioritize Security and Privacy:** Organizations must prioritize data security and privacy in Autonomous Finance. Employing advanced encryption technologies, implementing 'privacy by design,' and adopting biometrics are some strategies to protect user data. Clear and user-centric privacy policies should also be formulated and communicated to the users.

d) **Pursue Hyper-Personalization through AI:** Autonomous Finance providers should strive to refine their AI algorithms to enable hyper-personalization. Tailoring financial advice, products, and services to individual user needs can result in a significantly enhanced user experience, fostering customer loyalty and satisfaction.

e) **Navigate Regulatory and Ethical Landscapes Responsibly:** As Autonomous Finance evolves, businesses must proactively understand and adhere to regulatory and ethical considerations. Balancing the drive for innovation with the necessity for responsible and compliant conduct is crucial in ensuring the sustainable growth of Autonomous Finance.

11.12 References

Afshar, V. (2020). New research uncovers big shifts in customer expectations and trust. *Salesforce*. https://www.salesforce.com/blog/digital-customers-research-blog/

Application of biometric face identification technologies in financial institutions | Biometric Update. (2023, February 20). Biometric Update |. https://www.biometricupdate.com/202302/application-of-biometric-face-identification-technologies-in-financial-institutions

Bracamonte, I. (n.d.). "Good Design is Obvious, Great Design is Transparent." *www.linkedin.com*. https://www.linkedin.com/pulse/good-design-obvious-great-transparent-ignacio-bracamonte-villaz%C3%B3n/

Correspondent, G., & Correspondent, G. (2022). Hyper-personalization: Future of banking. *Global Business Outlook*. https://www.globalbusinessoutlook.com/hyper-personalization-future-of-banking/

Georgiou, M. (2017). AI in Banking: A JP Morgan Case Study and How Your Business Can Benefit. *Imaginovation | Top Web & Mobile App Development Company Raleigh*. https://imaginovation.net/blog/ai-in-banking-jp-morgan-case-study-benefits-to-businesses/

Gladchuk, V. (2020, December 8). *Customer perception role and importance - SupportYourApp*. SupportYourApp Blog. https://supportyourapp.com/blog/customer-perception/

HighRadius. (2023, August 16). *Working Capital Management: A winning case for the CFO's office | Order to Cash Knowledge Center. HighRadius Resource Center*. https://www.highradius.com/resources/treasury/whitepapers/working-capital-management-for-cfos-office/

How artificial intelligence can deliver real value to companies. (2017). *In McKinsey & Company*. https://www.mckinsey.com/capabilities/quantumblack/our-insights/how-artificial-intelligence-can-deliver-real-value-to-companies

How Mint works. Mint. (2020). https://www.mint.com/how-mint-works

Inside Job (2010) - IMDb. (n.d.). Www.imdb.com. Retrieved September 9, 2023, from https://www.imdb.com/title/tt1645089/characters/nm1756660

Mastercard. (2023). Mastercard leverages its AI capabilities to fight real-time payment scams. *Mastercard Newsroom*. https://www.mastercard.com/news/press/2023/july/mastercard-leverages-its-ai-capabilities-to-fight-real-time-payment-scams/

Medine, D. (2018). Privacy by design for financial services | Mint. *Mint*. https://www.livemint.com/Opinion/1ShpKAOC59VlXiwgCkVv8O/Privacy-by-design-for-financial-services.html

Mint. (2020). How Mint works. Retrieved from https://www.mint.com/how-mint-works

Moc, & Moc. (2023). 10 Key Predictions for the Future Of Voice Assistants. *Master of Code Global*. https://masterofcode.com/blog/9-key-predictions-for-the-future-of-voice-assistants

Statista. (2021). *Digital Payments - Worldwide | Statista Market Forecast*. https://www.statista.com/outlook/dmo/fintech/digital-payments/worldwide

Sustaining digital payments growth: Winning models in emerging markets. (2022, October 13). McKinsey & Company. https://www.mckinsey.com/industries/financial-services/our-insights/sustaining-digital-payments-growth-winning-models-in-emerging-markets

Top 10 banking trends for 2023. (2023). Accenture. https://www.accenture.com/us-en/insights/banking/top-10-trends-banking

CHAPTER 12

Playing by the Rules

Regulatory & Ethical Landscapes in Autonomous Finance

"Ships are safe in harbor,
but that's not what ships are for."

- JOHN A. SHEDD (SHEDD, 1928)

12.0 Key Insights

- **Regulatory Compliance and Ethical Conduct:** These are cornerstones that underpin the stability and trustworthiness of Autonomous Finance, rather than mere checkboxes to be ticked off.
- **The Multidimensional Landscape of Autonomous Finance:** Its facets are shaped by data protection and privacy, financial regulations, AI ethics and fairness, cybersecurity, interoperability, and regulatory compliance.
- **The Influence of Cultural Values:** Perception and acceptance of Autonomous Finance are influenced by cultural values, necessitating a culturally sensitive approach to implementation.
- **Key Characteristics of Autonomous Finance:** Transparency, explainability, accountability, bias, fairness, and systemic risk play integral roles in shaping the design, implementation, and operation of Autonomous Finance systems.
- **Anticipating Future Shifts:** Anticipating future regulatory and ethical shifts is crucial for long-term success in Autonomous Finance.

12.1 Introduction

As the preceding chapter illuminated the nuances of crafting a superior customer experience in Autonomous Finance, the focus now shifts to another cornerstone: the intricate regulatory and ethical landscapes that govern this transformative field. The advent of Autonomous Finance, characterized by the ever-increasing role of AI, ML, and RPA in financial decision-making, inaugurates a new frontier laden with compliance and ethical complexities. This uncharted terrain, much like the vast oceans navigated by ancient mariners under the guidance of celestial bodies, demands an expert compass to traverse its intricacies.

Integrating AI into financial operations elevates the stakes of regulatory compliance and ethical conduct to levels never before witnessed. The repercussions of non-compliance or ethical missteps could destabilize the financial architecture and erode the very bedrock of consumer trust. However, there is a guide to navigating these tricky waters of data protection principles, privacy, financial regulations, and AI ethics. These principles, grounded in frameworks such as the EU's GDPR and the United States' Fair Credit Reporting Act (FCRA), illuminate the path through the labyrinthine landscape of Autonomous Finance.

While the lens of this chapter is global, its narrative is enriched by lessons predominantly drawn from the North American context. This chapter is written for a broad audience–from policymakers to business executives to ethical overseers, and it aims to untangle the complex regulatory and ethical dimensions that Autonomous Finance navigates. It elucidates the core principles that shape this realm, including but not limited to data protection, privacy, AI ethics, and cybersecurity.

This multifaceted narrative also explores the influences of geographic variances, industry-specific considerations, and financial sub-domains, offering a comprehensive analysis. It scrutinizes critical characteristics such as transparency, explainability, and fairness that influence Autonomous Finance systems' design, implementation, and operation. Moreover, the chapter peels back the layers of emerging regulatory shifts and outlines a concrete roadmap for achieving enduring compliance in this rapidly evolving field.

In summary, this chapter serves as a beacon, guiding the reader through the regulatory and ethical challenges and considerations in Autonomous Finance. With its nuanced insights and real-world case studies, it aims to equip readers for informed decision-making in this intricate landscape. Let this enlightening journey commence.

12.2 Navigating the Landscape: The Essential Lenses

An array of lenses offers nuanced perspectives on the Autonomous Finance landscape. These lenses illuminate potential obstacles and opportunities in the path of Autonomous Finance, enabling effective navigation. The ensuing subsections explore these lenses in-depth: Data Protection and Privacy, Financial Regulations, AI Ethics and Fairness, Cybersecurity, Interoperability, and Regulatory Compliance.

The Privacy Paradox: Balancing Data Protection and Utility

Autonomous Finance systems thrive on personal data, enabling personalized financial advice, automated transactions, and predictive analytics for market trends. However, this use of data can potentially infringe upon individuals' privacy rights, establishing a paradox: the balance of data utility with privacy protection.

A host of critical regulations govern this domain:

a) GDPR - EU

b) CCPA - US

c) Personal Information Protection and Electronic Documents Act (PIPEDA) - Canada

d) Data Protection Act - UK

e) The Personal Data Protection Act (PDPA) - Singapore

For instance, the GDPR in the European Union and the CCPA in the U.S. are notable illustrations of regulations with a laser focus on data privacy (Russell, 2020; Solove & Schwartz, 2022). These laws enforce stringent data collection, storage, processing, and sharing rules, shaping how Autonomous Finance systems manage personal data.

Organizations face a critical paradox: The more data they collect to improve their AI systems, the more they risk compromising user privacy. Organizations must embed *privacy-by-design* principles into their AI systems to navigate this paradox. Embedding *privacy-by-design* principles involves implementing robust data protection mechanisms, anonymizing data where possible, and securing informed consent from users before data collection (Cavoukian, 2010).

The Financial Maze: Navigating Regulatory Complexities

Financial regulations form a complex labyrinth that Autonomous Finance solutions must adeptly navigate. These regulations, varying by geography and specific financial areas, span aspects ranging from AML to fair lending practices and investor protection.

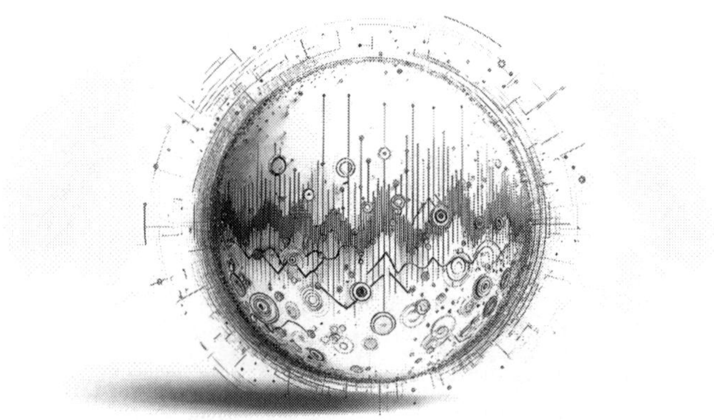

Key financial regulations in this space include

a) Bank Secrecy Act (BSA) - US

b) The FCRA - US

c) Basel III international regulatory framework - International

d) Dodd-Frank Wall Street Reform and Consumer Protection Act - US

e) Payment Services Directive (PSD2) - EU

To illustrate, the Bank Secrecy Act (BSA) in the U.S. obliges financial institutions to aid government agencies in detecting and preventing money laundering (*The Bank Secrecy Act | FinCEN.gov*, 2020). Conversely, the FCRA governs the collection and use of consumer credit information (*Fair Credit Reporting Act*, 2023). These laws, among others, shape how Autonomous Finance systems monitor transactions, utilize credit data, and make financial decisions.

The key to navigating this regulatory labyrinth lies in a comprehensive understanding of applicable laws, regularly monitoring regulatory updates and a proactive approach to compliance. AI-powered automated compliance solutions can be invaluable allies in this journey, capable of tracking regulatory changes in real-time and providing actionable insights (Cain, 2023).

The AI Ethical Dilemma: Striving for Fairness amid Bias

AI ethics and fairness present a critical lens in Autonomous Finance. They pose pivotal questions: How can fairness be ensured in AI financial decisions? How can these systems be prevented from perpetuating or amplifying societal biases?

Key regulations and guidelines in this domain include

a) Algorithmic Accountability Act - US (proposed),

b) Guidelines on Automated individual decision-making and Profiling - EU,

c) FCRA - US,

d) Equality Act - UK, and

e) Anti-Discrimination Act - Australia.

Biases in AI systems can stem from biased training data, biased algorithms, or a lack of diversity in AI development teams (Mehrabi et al., 2019). These biases can lead to unfair outcomes, such as discriminatory loan approvals or biased investment advice. Regulations like the Algorithmic Accountability Act proposed in the U.S. underscore the growing regulatory focus on mitigating AI biases (CaseGuard Video Redaction Software, 2022).

The pursuit of fairness necessitates the implementation of rigorous bias detection and mitigation mechanisms. This could involve conducting regular audits of AI systems, integrating fairness metrics into AI training, and promoting diversity in AI development teams (Simbeck, 2023).

The upcoming sections delve deeper into the remaining lenses: Cybersecurity, Interoperability, and Regulatory Compliance.

Cybersecurity: Safeguarding the Digital Treasure Trove

In Autonomous Finance, cybersecurity transcends mere digital protection; it is the bedrock of trust. A single breach can erode hard-won customer confidence and result in significant financial loss. Hence, cybersecurity regulations form an indispensable lens through which the landscape of Autonomous Finance is viewed.

Key cybersecurity regulations include

a) The Network and Information Systems Regulations (NIS) - EU,

b) New York's Cybersecurity Regulation (23 NYCRR 500) - US,

c) The Federal Information Security Management Act (FISMA) - US,

d) Payment Card Industry Data Security Standard (PCI DSS) - International, and

e) Health Insurance Portability and Accountability Act (HIPAA) - US (for Health Finance).

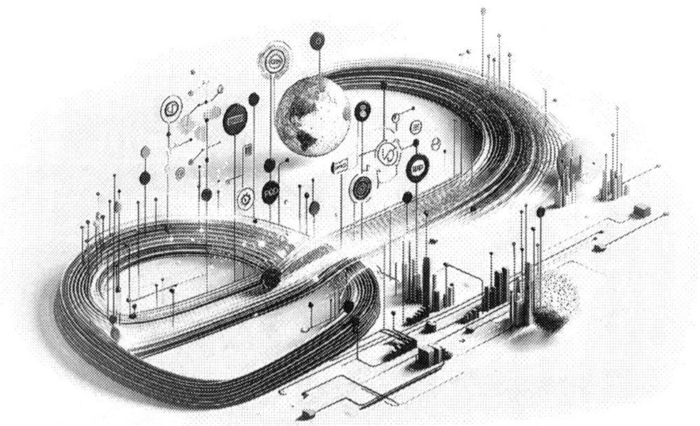

An example of stringent cybersecurity regulation is the New York Department of Financial Services Cybersecurity Regulation (23 NYCRR 500), which mandates a cybersecurity program to protect consumers' private data (Security & Security, 2023). The regulation reflects a broader global trend toward tightening cybersecurity controls, particularly in the financial sector.

Navigating the cybersecurity landscape requires a multifaceted approach that combines robust technological safeguards, continuous monitoring, employee training, and adherence to relevant regulations (Von Solms & Van Niekerk, 2013). Collaboration with regulatory bodies and adherence to global standards can foster a culture of continuous improvement in cybersecurity practices.

Interoperability: Bridging the Autonomous Islands

Interoperability refers to the ability of different Autonomous Finance systems to work together seamlessly. It is akin to bridges connecting islands, allowing for smooth traffic flow. Each system becomes an isolated island without interoperability, limiting collaboration and efficiency.

Essential standards and guidelines promoting interoperability include

a) The Financial Information eXchange (FIX) Protocol - International,

b) ISO 20022 Standard - International,

c) Open Banking Standard - UK,

d) PSD2 - EU, and

e) The Clearing House (TCH) Real-Time Payments (RTP) System - US.

For instance, the FIX Protocol has become the de facto messaging standard for pre-trade, trade, and post-trade communication, enabling a seamless exchange of financial information across platforms (FIXTrading, 2023). Similarly, the Open Banking Standard in the UK allows secure sharing of financial information, fostering innovation and competition (*What Is Open Banking? - Open Banking*, 2023).

The pursuit of interoperability necessitates embracing standards, adopting open APIs, and fostering stakeholder collaboration (Zachariadis et al., 2019). A proactive approach to interoperability can unlock new synergies, enhance efficiency, and pave the way for innovative financial products and services.

Compliance Conundrum: Traversing the Regulatory Jungle

Regulatory compliance is not merely a legal requirement but a strategic imperative. The regulatory landscape in Autonomous Finance is a dense jungle with myriad rules, guidelines, and standards. Traversing this jungle requires a sophisticated understanding of the regulations that govern various aspects of Autonomous Finance, from data protection to investor rights.

Critical aspects of regulatory compliance include

a) The Sarbanes-Oxley Act (SOX) - US,

b) AML directives - EU,

c) The Monetary Authority of Singapore (MAS) Guidelines on Fair Dealing - Singapore,

d) The Dodd-Frank Wall Street Reform and Consumer Protection Act - US, and

e) The Financial Conduct Authority (FCA) regulations - UK

For example, the Sarbanes-Oxley Act (SOX) in the US imposes strict audit requirements and financial disclosures for public companies to prevent corporate fraud (Anand, 2008). Likewise, AML directives in the EU aim to prevent the financial system from being used for illicit activities (Unger, 2009).

Navigating this regulatory jungle requires a comprehensive, agile, and proactive compliance strategy. This strategy must encompass continuous monitoring of regulatory changes, rigorous risk assessment, effective compliance measures, and a culture that emphasizes ethical conduct and adherence to laws (Arner et al., 2017).

These six lenses collectively provide a panoramic view of the Autonomous Finance landscape, offering unique insights and considerations. Together, they form a cohesive navigational guide essential for steering through the complex seas of Autonomous Finance. The following sections will further explore the dimensions of geography, industry, and the finance domain, providing a multidimensional perspective on the intricate world of Autonomous Finance.

12.3 The Multidimensional Regulatory Landscape

Autonomous Finance's complex regulatory terrain demands a nuanced understanding of the interplay between various geographies, industries, the finance domain, and governing ecosystems, as illustrated in Figure 12.1. Failure to navigate this intricately woven fabric can lead to severe consequences, as exemplified by real-world incidents.

FIGURE 12.1: A Multidimensional Mosaic of Regulations in Autonomous Finance

In 2018, HSBC faced allegations that its automated compliance systems failed to prevent money laundering activities, resulting in a forfeiture of $1.256 billion settlement (*HSBC Holdings Plc. And HSBC Bank USA N.A. Admit to Anti-Money Laundering and Sanctions Violations, Forfeit $1.256 Billion in Deferred Prosecution Agreement*, 2015). Similarly, PayPal's automated screening systems faced fines of $7.7 million in 2017 for failing to detect suspicious transactions across different countries, illustrating the challenges of meeting diverse anti-money laundering rules (Staff, 2015). These examples highlight the critical nature of understanding and complying with the multifaceted regulatory landscape.

The convergence of multiple dimensions, including geography, industry verticals, finance sub-domains, and governing ecosystems, adds further complexity. The following sections unravel this multi-dimensional mosaic, exploring key regulations and considerations for each dimension.

Geography: Mapping the Global Financial Terrain

The geography dimension uncovers various regulations across different regions, reflecting the varying legal, cultural, and socio-economic contexts.

Table 12.1 offers a curated snapshot of the regulatory landscape across three pivotal geographies—North America, Europe, and Asia Pacific—focusing exclusively on key regulations of critical importance to Autonomous Finance. This table does not aim for exhaustive completeness, but instead concentrates on regulations that significantly influence factors critical to Autonomous Finance landscapes. In doing so, it serves as a discerning guide for stakeholders vested in navigating the nuanced labyrinth of global regulatory compliance.

TABLE 12.1: Regulatory Landscape by Geography

ID	GEOGRAPHY	REGULATORY LANDSCAPE	KEY REGULATIONS
1.	**North America**	Emphasizes robust consumer protection, financial innovation, and stringent cybersecurity regulations	• Dodd-Frank Wall Street Reform Act • CCPA • Bank Secrecy Act (BSA) • Securities and Exchange Commission (SEC) guidelines • The FCRA
2.	**Europe**	Known for strong data protection laws and Open Banking initiatives	• GDPR • PSD2 • AML directives • Basel III international regulatory framework • European Banking Authority (EBA) guidelines
3.	**Asia Pacific**	Focused on financial inclusion, fintech innovation, and cross-border cooperation	• MAS Guidelines • China's Cybersecurity Law • Payment and Settlement Systems Act (Japan) • Reserve Bank of India (RBI) guidelines • Australian Privacy Act

Industry: The Vertical Spectrum

The industry horizons dimension reveals a spectrum of regulations influenced by the specific needs, challenges, and nature of different industry sectors. Recognizing and adapting to these regulations is a complex task but essential for building and maintaining trust, integrity, and compliance within Autonomous Finance.

Table 12.2 delineates the complex tapestry of key regulations across eight pivotal industries, each with unique regulatory intricacies. From Banking and Financial Services to Retail, this table emphasizes regulations with cardinal implications for each industry vertical. Though not exhaustive, the table is a focused compendium for stakeholders seeking to understand regulatory nuances germane to industries interfacing with Autonomous Finance.

TABLE 12.2: Regulatory Landscape by Industry Vertical

ID	INDUSTRY VERTICAL	KEY REGULATIONS
1.	**Banking and Financial Services:** The Banking and Financial Services industry is heavily regulated, owing to its critical role in the global economy	• Basel III • Dodd-Frank Wall Street Reform and Consumer Protection Act • PSD2 • FCA Regulations • MAS Guidelines on Fair Dealing
2.	**Consumer Product Goods (CPG):** The CPG industry vertical faces unique regulatory challenges centered around consumer protection and product safety.	• Federal Food, Drug, and Cosmetic Act • General Product Safety Directive (GPSD) • Consumer Product Safety Act (CPSA) • Competition and Consumer Act • Consumer Rights Act
3.	**Manufacturing:** The Manufacturing industry faces regulations encompassing many issues, from worker safety to environmental compliance.	• Occupational Safety and Health Act (OSHA) • Registration, Evaluation, Authorisation, and Restriction of Chemicals • Clean Air Act • Environmental Protection Law • Factory Act
4.	**Food & Beverage:** The Food and beverage industry vertical operates under stringent food safety, quality, and traceability regulations.	• Food Safety Modernization Act (FSMA) • Regulation (EC) No 178/2002 • The Alcoholic Beverage Labeling Act (ABLA) • Canada's Food and Drugs Act (FDA) • Japan's Food Sanitation Act
5.	**Healthcare:** Regulations in the Healthcare industry encompass a wide array of concerns, including patient rights, privacy, and quality of care	• HIPAA • Medical Device Regulation (MDR) • Affordable Care Act (ACA) • National Health Service Act • Pharmaceutical Affairs Law
6.	**Pharmaceuticals:** The Pharmaceutical industry vertical is subject to rigorous regulations to ensure the safety, efficacy, and quality of drugs	• Food and Drug Administration (FDA) Regulations • European Medicines Agency (EMA) Regulations • Good Manufacturing Practice (GMP) Guidelines • Drug Price Control Order (DPCO) • Therapeutic Goods Act
7.	**Insurance:** The Insurance industry vertical faces regulations related to consumer protection, solvency, and market conduct	• Solvency II • McCarran-Ferguson Act • Insurance Contracts Act • Insurance Act • China Insurance Regulatory Commission (CIRC) Regulations
8.	**Retail:** The Retail industry vertical is shaped by regulations concerning consumer rights, product safety, and competition	• Consumer Protection Act • Unfair Commercial Practices Directive (UCPD) • Competition Act • Consumer Guarantees Act • Fair Trading Act.

The regulatory terrain of Autonomous Finance is a multi-dimensional mosaic, reflecting the diversity and complexity of global commerce. By understanding the key regulations within each industry, organizations can navigate the regulatory landscape with greater confidence and compliance, aligning with legal obligations and industry best practices. The interplay between these regulatory frameworks and the previously discussed lenses provides a comprehensive understanding of this dynamic field's unique challenges and opportunities.

Functional Areas within Finance: The Niche Navigators

The intricacy of the regulatory terrain extends to the diverse functional areas within finance, each governed by distinct rules, standards, and practices. The specific functions and requirements of different financial activities shape the regulations within these functional areas.

Table 12.3 offers a curated compendium of key regulations shaping various finance areas today. While not exhaustive, this table crystallizes the most salient regulatory frameworks and guidelines, serving as an invaluable navigational aid for professionals steering through the complexities of modern financial governance, especially in the context of the journey to excellence in Autonomous Finance.

TABLE 12.3: Regulatory Landscape by Functional Areas within Finance

ID	FUNCTIONAL AREA	KEY REGULATIONS
1.	**Financial Accounting:** Financial accounting regulations guide the preparation and reporting of financial statements.	• GAAP • IFRS • SEC Regulations • Companies Act • Corporate Financial Reporting Standards (CFRS)
2.	**Management Accounting:** Management accounting focuses on internal decision-making.	• Institute of Management Accountants (IMA) Standards • Cost Accounting Standards (CAS) • European Management Accounting Guidelines • Activity-Based Costing Guidelines • Japanese Management Accounting Standards (JMAS)
3.	**Financial Planning & Analysis:** FP&A focuses on financial forecasting, planning, and analysis.	• Association for Financial Professionals (AFP) Guidelines • Financial Forecasting Best Practices • Sarbanes-Oxley Act (SOX) • Basel III • International Professional Practices Framework (IPPF)
4.	**Treasury Management:** Treasury management regulations governing cash, liquidity, and risk management	• Bank Secrecy Act (BSA) • European Market Infrastructure Regulation (EMIR) • Financial Services and Markets Act (FSMA) • Foreign Exchange Management Act (FEMA) • Liquidity Coverage Ratio (LCR) Guidelines
5.	**FRM:** FRM involves identifying, assessing, and mitigating financial risks	• Dodd-Frank Act • FRM Guidelines • Market Abuse Regulation (MAR) • Prudential Regulation Authority (PRA) Rules • Operational Risk Management Guidelines

ID	FUNCTIONAL AREA	KEY REGULATIONS
6.	**Corporate Finance:** Corporate Finance deals with capital structure, funding, and investment decisions.	• M&A Regulations • Capital Adequacy Requirements • Private Equity and Venture Capital Guidelines • Securities Act: • UK Takeover Code
7.	**Financial Systems and Technology:** Integrating technology in finance has fueled the need for specific regulations.	• PCI DSS • FinTech Regulations • Electronic Signature Regulations • Open Banking Standards • GDPR
8.	**Compliance and Regulatory Reporting:** Ensuring adherence to laws and regulations is paramount.	• Sarbanes-Oxley Act (SOX) • AML Directives • Dodd-Frank Wall Street Reform and Consumer Protection Act • Basel Committee on Banking Supervision Guidelines • MiFID II

These sub-domains further accentuate the complexity and diversity of the multi-dimensional regulatory terrain of Autonomous Finance. The interplay between these aspects highlights the intricacy and dynamism of this field, necessitating a profound understanding of the regulations and their implications. The ability to dissect and interpret these layers is crucial for organizations navigating the rapidly evolving landscape of Autonomous Finance.

Governing Ecosystems: The Overarching Framework

The governing ecosystems encompass the overarching frameworks and standards set by global and regional governing bodies. This includes international agreements, industry standards, and collective initiatives that shape the overall regulatory environment. For example:

a) **United Nations Sustainable Development Goals (SDGs):** These goals provide a global framework for sustainability and social responsibility.

b) **Financial Stability Board (FSB) Guidelines:** These guidelines aim at enhancing global financial stability by coordinating national financial authorities.

c) **International Organization of Securities Commissions (IOSCO) Principles:** A set of principles that guide securities regulation, focusing on investor protection, market integrity, and systemic risk reduction.

d) **Organization for Economic Co-operation and Development (OECD) Guidelines:** These guidelines offer responsible business conduct and governance principles.

e) **Basel Committee on Banking Supervision:** Established by central banks and banking regulators, this committee formulates standards to enhance financial stability, including Basel III regulations (BIS, 2010). It plays a crucial role in risk management and banking regulations.

f) **Financial Action Task Force (FATF):** An intergovernmental body that sets standards to combat money laundering and terrorist financing (The FATF, 1989). Its guidelines are vital in the compliance and regulatory reporting sub-domains.

g) **PCI Security Standards Council (SSC):** This global council develops standards to enhance payment card security, including the PCI DSS (Baykara, 2022). It is pivotal in the financial systems and technology sub-domains.

These governing ecosystems represent a unifying fabric that weaves together the multi-faceted regulatory landscape of Autonomous Finance. They offer a structured yet flexible framework for diverse geographies, industries, and finance sub-domains. Understanding these ecosystems is vital, as they shape the underlying principles, norms, and practices that guide the Autonomous Finance voyage.

In the ever-shifting sea of regulations, these governing bodies act as lighthouses, guiding the way and ensuring safe navigation. Recognizing their role and influence is paramount for those venturing into Autonomous Finance's complex and challenging waters. Their guidelines and standards steer the course and set the horizon for what lies ahead in this exciting and transformative field.

Summary of the Regulatory Terrain

The regulatory landscape of Autonomous Finance is a complex, multifaceted arena influenced by an array of diverse dimensions. Each dimension introduces unique rules, guidelines, and standards for Autonomous Finance. The intricate and dynamic interplay between such dimensions creates a challenging environment that demands cautious and informed navigation. Organizations considering embarking on the journey to implement Autonomous Finance solutions should keep this multi-layered regulatory complexity at the forefront of their strategic planning.

a) **Geographic Diversity:** Different countries and regions introduce varying legal frameworks, reflecting cultural values, economic priorities, and societal needs. From GDPR in Europe to CCPA in California, geographic regulations can significantly impact data protection, privacy, and other aspects of Autonomous Finance.

b) **Industry Verticals:** Industries like Banking, Insurance, Healthcare, and Retail bring specific regulatory considerations. The FCRA may govern credit decisions in banking, while HIPAA might influence financial handling in healthcare.

c) **Functional Areas within the Finance Domain:** Financial accounting, risk management, corporate finance, and more have unique regulations. Basel III may guide risk management, while IFRS impacts financial reporting across global boundaries.

d) **Governing Ecosystems:** International bodies and standards, such as FATF, FSB, and PCI SSC, knit together the regulatory fabric, setting global guidelines that transcend individual jurisdictions and sectors.

Navigating this multidimensional mosaic requires a deep understanding of the interconnections between these dimensions. It necessitates a flexible yet robust approach that adapts to the ever-evolving regulatory landscape. Recognizing the diversity and dynamism of this terrain is a critical step toward compliance, stability, and ethical conduct in the promising field of Autonomous Finance.

As the journey through this intricate regulatory terrain concludes, the focus now shifts to the defining characteristics that shape AI's ethical and responsible application in finance. These characteristics form the compass guiding the next segment of this exploration.

12.4 The Ethical Compass in Autonomous Finance

The journey through the intricate landscape of Autonomous Finance is guided by a set of defining characteristics or principles that form the ethical compass for this domain. These principles set the standards for responsible AI utilization and act as the navigational beacon guiding finance teams through the complex terrain.

Table 12.4 assembles a focused collection of nine guiding principles supported by primary key regulations that act as the ethical and regulatory linchpin for Autonomous Finance. It is crucial to note that this table is not an exhaustive compendium but highlights the most salient regulations today. These principles collectively serve as a dynamic, interconnected framework, affording crucial insights for discerning leaders, regulators, and operators committed to the responsible deployment of AI and ML in the fluid realm of Autonomous Finance.

TABLE 12.4: Guiding Principles for Ethical Compass in Autonomous Finance

ID	GUIDING PRINCIPLES	KEY REGULATIONS
1.	**Transparency:** Transparency is the cornerstone of trust in Autonomous Finance. Whether disclosing how AI models make decisions or being clear about data usage, transparency fosters confidence and promotes understanding among users and stakeholders (Roy, 2017).	• Right to Explanation under GDPR • U.S. Federal Trade Commission's (FTC's) Fair Information Practice Principles • SEC guidelines
2	**Explainability:** Explainability refers to interpreting and understanding AI decision-making. Both regulators and users must grasp how decisions are made, especially in critical areas like credit scoring or investment strategies (Doshi-Velez & Kim, 2017).	• EU's AI Ethics Guidelines for Trustworthy AI • Consumer Financial Protection Bureau (CFPB) regulations
3	**Accountability:** Accountability ensures that organizations are answerable for their AI systems' decisions and behaviors. It involves mechanisms to detect errors or biases and take corrective actions, promoting ethical behavior (Kroll et al., 2017).	• Sarbanes-Oxley Act • MAS Guidelines on Fair Dealing
4	**Bias and Fairness:** Bias can creep into AI systems through biased training data or algorithms. Ensuring fairness means actively detecting and mitigating these biases and preventing discrimination based on race, gender, or other prohibited characteristics (Barocas & Selbst, 2016).	• The FCRA • The Equal Credit Opportunity Act (ECOA)

ID	GUIDING PRINCIPLES	KEY REGULATIONS
5	**Ethics:** Ethics in Autonomous Finance goes beyond compliance, embodying respect, integrity, and social responsibility. Ethical guidelines form the moral fabric that guides AI use in finance (Hagendorff, 2020).	• FCA regulations • OECD Principles on AI
6	**Compliance:** Compliance ensures adherence to relevant laws and regulations, from data protection to AML directives. It is essential for maintaining the trust and avoiding legal pitfalls (Papantoniou, 2022).	• The Dodd-Frank Wall Street Reform and Consumer Protection Act • AML directives
7	**Stability:** Stability in Autonomous Finance refers to the system's robustness and resilience. Maintaining stability prevents cascading failures and financial crises (*Confronting Too Big to Fail*, 2009).	• Basel III international regulatory framework • The FSB guidelines
8	**Systemic Risk:** Systemic risk considers the potential for large-scale disruptions in the financial system. Understanding and mitigating systemic risk is vital for financial markets' overall health and integrity (Reinhart, 2011).	• The Volcker Rule • Financial Sector Assessment Program (FSAP)
9	**Interoperability:** Interoperability ensures that AI models can communicate and interact with other systems, promoting efficiency and collaboration across different platforms and institutions (Javaid, 2023).	• PSD2 • HIPAA • FIX Protocol

Together, these nine guiding principles form the ethical compass that steers the journey through the Autonomous Finance landscape. They are not isolated concepts but interwoven characteristics that collectively define the responsible development and deployment of AI in finance. They reflect the complex, multifaceted nature of the regulatory landscape and provide a roadmap for navigating the ethical challenges and opportunities in this rapidly evolving field.

12.5 Cultural Values and Autonomous Finance

Cultural diversity is the spice of life. It adds color, depth, and richness to our global village. However, it can also present challenges, especially regarding Autonomous Finance. The perception of privacy, data, and AI can differ greatly from one culture to another, affecting the acceptance and use of Autonomous Finance.

Privacy, data protection, and perception of AI can vary significantly across cultures (Floridi, 2014). In some cultures, technology is highly trusted and used in financial transactions. In others, there may be skepticism or even fear about using AI in finance. For instance, in some European countries, data privacy is highly valued, and there is strong support for regulations like the GDPR.

On the other hand, in some Asian countries, there may be a higher tolerance for data collection in exchange for the perceived benefits of technology. Understanding these cultural nuances is crucial when implementing Autonomous Finance systems in different regions.

Cultures also respond differently to ethical and regulatory challenges (Husted, 2000). Some may prioritize transparency and accountability, while others may place a greater emphasis on efficiency and effectiveness. For instance, in the U.S., there may be a stronger emphasis on individual rights and freedoms, while in China, there may be a stronger focus on collective benefit and social harmony. As a result, when dealing with ethical and regulatory challenges, companies need to consider the local cultural context and align their strategies accordingly.

One interesting case study uses Autonomous Finance in India. Given the country's vast population and significant financial inclusion challenges, Autonomous Finance has tremendous potential. However, the country also has a diverse culture with various perceptions of technology, privacy, and data.

In 2017, India's Supreme Court declared privacy a fundamental right (Sethi & Ramesh, 2017). This has implications for how Autonomous Finance systems collect and use personal data. Furthermore, the cultural emphasis on community and social connections could influence how AI systems are designed and implemented, particularly in rural areas where trust in technology may be lower. On the other hand, the Indian government has made significant strides in promoting digital transactions and reducing the reliance on cash. This has led to a rapid increase in the adoption of digital finance solutions, creating a fertile ground for adopting Autonomous Finance. This example illustrates how cultural values can challenge and facilitate the implementation of Autonomous Finance, underlining the importance of a culturally sensitive approach.

12.6 Roadmap to Compliance in Autonomous Finance

"In every conceivable manner, the family is a link to our past, a bridge to our future." This quote by Alex Haley may not directly speak about compliance or Autonomous Finance, but it perfectly encapsulates the significance of establishing a robust compliance framework. Like the family bonds that connect us to our heritage and propel us toward the

future, a robust compliance framework ties organizations to their ethical obligations. It paves the way for a sustainable future in Autonomous Finance.

This section provides a step-by-step roadmap for organizations to attain compliance with regulations and ethics (as shown graphically in Figure 12.2):

a) **Understanding the Regulatory Landscape:** Knowledge of the regulatory landscape is the cornerstone of compliance. Organizations must explore the terrain through the above mentioned lenses, including data protection, financial regulations, AI ethics, etc. Monitoring global, industry-specific, and domain-specific regulations ensures readiness to adapt to evolving requirements (Harrington & Niehaus, 2004).

b) **Risk Assessment:** Identifying potential risks is a pivotal step. This includes assessing risks related to data breaches, discriminatory practices, systemic vulnerabilities, and ethical dilemmas (Brunnermeier, 2009). The risk assessment must be comprehensive, considering the multidimensional aspects of regulations.

c) **Implementing Compliance Measures:** Tailoring compliance measures based on the risk assessment is essential. This encompasses creating robust data protection mechanisms, enhancing transparency, ensuring fairness in AI algorithms, and adhering to industry-specific regulations (*Ethical Algorithm Design Should Guide Technology Regulation | Brookings*, 2022).

d) **Training and Awareness:** Building a culture of compliance necessitates continuous education and alignment. Regular training sessions impart awareness of regulations, ethical standards, and the importance of adherence to all levels of the organization (Treleaven & Batrinca, 2017).

FIGURE 12.2: A Roadmap to Compliance in Autonomous Finance

e) **Audit and Review:** An ongoing process of audit and review ensures that compliance measures are effective. Regular audits detect potential issues, while periodic reviews allow for updates and refinements as regulations and business needs evolve (Power, 1997).

Compliance and ethics are intertwined and mutually reinforcing. Ethical considerations such as fairness, transparency, and accountability are often encoded in regulations. Conversely, compliance with legal requirements reinforces an organization's commitment to ethical conduct. Integrating ethical principles into compliance efforts fosters a culture where adherence to regulations is seen not just as an obligation but as a reflection of organizational values. This culture of integrity enhances customer trust and positions the organization as a responsible player in the Autonomous Finance landscape.

12.7 Future of Regulatory and Ethical Compliance

At the frontier of Autonomous Finance, a complex and rich tapestry of regulatory and ethical considerations is ready to be woven with innovation, foresight, and inclusivity threads. Standing at the precipice of a revolution fueled by AI and unprecedented technological advancements, the financial landscape is on the verge of a transformative metamorphosis, one where data transcends to become the new currency, and ethics emerge as the guiding compass.

This section embarks on a journey through visionary corridors, exploring potential future trajectories delineated by evolving regulations (see Figure 12.3). It ventures into the delicate equilibrium between globalization and localization, emphasizing the ethical imperatives that underpin financial evolution. The discourse further unravels the transformative potential of technology in redefining compliance and navigates the art of resilience in foreseeing the unforeseen. Moreover, it delineates the unique regulatory considerations incumbent upon diverse functional areas in this dynamic ecosystem. Herein, envision a future where the regulatory and ethical landscape metamorphoses into a dynamic entity, fostering a symphony of regulation harmonized with global trends and local wisdom, thus nurturing a financial ecosystem characterized by responsible innovation and ethical leadership.

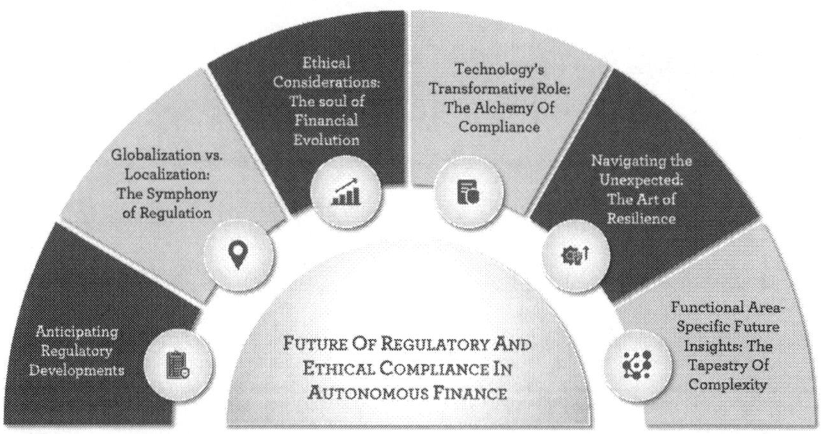

FIGURE 12.3: The Future of Regulatory and Ethical Compliance in Autonomous Finance

Anticipating Regulatory Developments

The insights gleaned through the six lenses previously explored offer an invaluable perspective. Here, the regulatory terrain is brought to life, revealing paths that may shape the future.

a) **Data Protection and Privacy:** In a world that breathes data, regulations like GDPR are more than mere guidelines; they manifest societal values. The future may witness nations crafting laws with a fine balance between individual rights and societal benefits, thus charting a new course in the global discourse on privacy (Greenleaf, 2013).

b) **Financial Regulations:** Digitalization is not merely changing financial services; it is revolutionizing them. A future guided by a unified approach to digital assets and fintech innovations may foster a new era of financial democracy and inclusivity (Feyen et al., 2021).

c) **AI Ethics and Fairness:** AI's power carries a profound responsibility. Future regulations may embody the principles of transparency, fairness, and empathy, transforming AI from a tool into a trusted partner in financial decision-making (Hagendorff, 2020).

d) **Cybersecurity:** The shadows of cyberspace hold untold risks. A future governed by robust cybersecurity regulations may herald a new age of trust, collaboration, and resilience in the face of evolving cyber threats (The White House, 2023).

e) **Interoperability:** The financial ecosystem is a symphony of diverse systems. Future standards ensuring seamless interaction could weave a tapestry of global financial integration, enriching lives across borders (Vanessa MOCK, 2017).

f) **Regulatory Compliance:** Given the nuanced and dynamic nature of compliance in a rapidly evolving field of Autonomous Finance, dynamic and adaptive systems guided by intelligent regulations could shape a future where compliance is not a burden but a catalyst for innovation and growth (Council on Foreign Relations, 2023).

Globalization vs. Localization: The Symphony of Regulation

The convergence and divergence of regulatory principles across jurisdictions present a complex paradox. On the one hand, globalization fosters a need for universal standards, promoting consistency and facilitating cross-border operations. For instance, the Basel III framework for banking supervision seeks to enhance global financial stability by setting minimum standards for banks worldwide (Tarullo, 2008).

On the other hand, localization recognizes the unique socio-economic, cultural, and legal landscapes of individual jurisdictions. Local regulations often respond to specific challenges, risks, and societal preferences. For example, China's focus on collective benefit and social harmony has led to distinct regulations governing data use and AI decision-making (Husted, 2000).

Navigating this symphony requires a deep understanding of global trends and local nuances. It requires a dynamic approach that can harmonize international best practices with local wisdom. A case in point is the GDPR's influence beyond European borders, inspiring similar data protection laws in countries like Brazil and Japan (Voigt & Von Dem Bussche, 2017).

Ethical Considerations: The Soul of Financial Evolution

Ethics transcends legal compliance, embodying principles of fairness, integrity, and social responsibility. The future of Autonomous Finance will be shaped not only by what is legally permissible but also by what is morally right.

Sustainable investing, for instance, has emerged as a vital ethical consideration. Investors increasingly look beyond financial returns, considering ESG factors (Friede et al., 2015). Similarly, AI models must be designed to avoid discriminatory outcomes, respecting principles of fairness and equality.

The rise of ethical banking practices, such as Islamic finance, which adheres to Sharia law, highlights how ethical considerations can define entire financial systems (El-Gamal, 2006). The future landscape will likely see greater integration of ethical principles into regulations, fostering a financial ecosystem that aligns with broader societal values.

Technology's Transformative Role: The Alchemy of Compliance

Compliance in the Era of Autonomous Finance is not a static obligation but a dynamic process. Technology is transforming the essence of compliance, enabling real-time monitoring, predictive analysis, and adaptive responses.

Blockchain technology, for example, is revolutionizing transparency and traceability in financial transactions (Tapscott & Tapscott, 2016). AI-powered tools enable organizations to monitor global regulatory changes and adapt compliance measures accordingly (Zetzsche, 2018). These technological advancements are not mere facilitators but transformative forces that redefine compliance from reactive obligation to proactive strategy. They empower organizations to anticipate risks, innovate responsibly, and contribute to shaping regulations that align with technological evolution.

Navigating the Unexpected: The Art of Resilience

The future regulatory landscape is uncertain. Economic upheavals, political changes, technological disruptions, and unforeseen global events like the COVID-19 pandemic can reshape regulations overnight.

Resilience in this context is not just about surviving unexpected shocks but about adapting and thriving. It involves building adaptive regulatory frameworks that respond to evolving risks and opportunities. The FSB's efforts to monitor and make recommendations about the global financial system exemplify this approach (*Evaluation of the Effects of Too-big-to-fail Reforms: Final Report*, 2021). Organizations must develop a flexible compliance strategy to navigate these shifting sands, leveraging technology, foresight, and collaboration to build a resilient financial ecosystem.

Functional Area-Specific Future Insights: The Tapestry of Complexity

Autonomous Finance is not monolithic; it is a complex tapestry woven from diverse functional areas such as FRM, corporate finance, FinTech, and compliance reporting. Each of these functional areas has its unique regulatory considerations. For example, in FRM, regulations focus on capital adequacy, stress testing, and risk reporting (Supervision & Settlements, 2010). In contrast, FinTech may be governed by regulations concerning innovation, consumer protection, and data security (*Annual Economic Report 2018*, 2018).

Understanding this complexity requires a nuanced approach that recognizes these functional areas' interconnected yet distinct nature. It demands a collaborative effort between regulators, industry experts, and technologists to create a cohesive regulatory landscape that fosters innovation while ensuring stability and integrity.

The future of regulatory and ethical compliance in Autonomous Finance is a multifaceted journey. It weaves together the global and local, the legal and ethical, the technological and human, the predictable and the unexpected. It demands a visionary approach that recognizes the complexity and dynamism of the evolving landscape.

In this journey, organizations have the opportunity to be more than mere participants; they can be pioneers, shaping the future through responsible innovation, ethical leadership, and strategic agility. The regulatory and ethical landscape is not just a set of rules; it is a living framework that can inspire a new era of financial excellence, guided by wisdom, enriched by diversity, and committed to creating a better world for all.

12.8 Conclusion

The labyrinthine matrix of regulations and ethics in Autonomous Finance represents not merely a complex backdrop but the actual stage upon which the future of global finance will unfold. This chapter has endeavored to dissect the nuanced interactions between technological innovation, regulatory governance, and ethical considerations, painting a vivid tapestry as intricate as it is illuminating.

Demystifying the regulatory terrain requires a comprehensive lens incorporating data protection, AI ethics, cybersecurity, and compliance, among other elements. The geographical, industrial, and functional dimensions, enriched by insights from the North American experience, reveal the symbiotic relationship between Autonomous Finance and its governing ecosystem.

The chapter's illumination through real-world case studies, such as HSBC and PayPal, serves not merely as academic touchpoints but as tangible beacons. These examples elucidate the compliance's gravity and labyrinthine nature, offering cautionary tales and operational wisdom.

As the future of Autonomous Finance unfurls, it carries with it a twofold promise: transformative innovation and ethical imperatives. The latter are not optional accouterments but foundational pillars for constructing an edifice of enduring impact and value in Autonomous Finance.

The roadmap to achieving and sustaining compliance is not etched in stone; it is a dynamic choreography that requires agility, vigilance, and ethical integrity. Here, technology is an enabler and a vigilant sentinel equipped to adapt to the shifting sands of regulatory landscapes.

As this chapter draws to a close, let it be known that the insights and foresight proffered herein are not mere intellectual exercises. They serve as the compass and sextant for those daring to navigate the tumultuous yet promising seas of Autonomous Finance. The choreography may be complex and the rhythms unpredictable, yet the dance can be graceful and transformative with the right blend of strategy, ethical resolve, and technological ingenuity.

The next chapter in this captivating narrative remains to be written. It promises to explore how this intricate dance can be orchestrated not merely with precision but with soul, ensuring that the Autonomous Finance of tomorrow is shaped by a harmonious blend of technology, ethics, and human insight—a dance as compelling as it is consequential.

12.9 Key Recommendations

a) **Holistic Understanding of the Multi-Dimensional Regulatory Terrain:** Organizations must develop an in-depth understanding of the complex regulatory landscape, encompassing not only data protection and financial regulations but also AI ethics, cybersecurity, interoperability, and regulatory compliance. This understanding should extend across geographies, industries, functional areas within finance, and governing ecosystems. Comprehensive risk assessments and real-time monitoring of regulatory changes are imperative.

b) **Strategic Implementation of Tailored Compliance Measures:** Beyond generic compliance protocols, organizations should design and implement robust measures tailored to their specific context and needs. This includes the creation of data protection mechanisms, fairness assurance in AI algorithms, and transparent protocols guided by insights derived from the multidimensional analysis of the regulatory landscape.

c) **Cultivating an Ethical Compass and Culture of Compliance:** Establishing a corporate culture where ethics and compliance are integral is non-negotiable. Regular training, awareness programs, and a commitment to transparency, accountability, and non-discrimination must be embedded into the organizational DNA. The ethical use of AI is both a regulatory requirement and a cornerstone of customer trust.

d) **Leveraging Technology for Adaptive Compliance and Auditing:** The dynamic nature of regulations demands continuous adaptation. Organizations should leverage AI and other advanced technologies for real-time monitoring, analysis, and response to regulatory changes. Regular audits powered by cutting-edge technology ensure compliance measures remain effective and aligned with the evolving landscape.

e) **Embracing Cultural Sensitivity and Global Responsiveness:** Autonomous Finance is not a one-size-fits-all solution. Organizations must be acutely aware of and responsive to the diverse cultural values, perceptions, and regulations across different regions. Localization strategies, culturally sensitive designs, and an understanding of global responses to ethical and regulatory challenges are vital.

12.10 References

Anand, S. (2008). *The Sarbanes-Oxley Body of Knowledge SOXBOK: An Introduction.* Van Haren.

Annual Economic Report 2018. (2018). https://www.bis.org/publ/arpdf/ar2018e.htm.

Arner, D. W., Barberis, J., & Buckley, R. P. (2017). FinTech, RegTech, and the reconceptualization of financial regulation. *Northwestern Journal of International Law and Business, 37*(3), 371. http://hub.hku.hk/bitstream/10722/235796/1/content.pdf.

Barocas, S., & Selbst, A. D. (2016). Big Data's Disparate Impact. *California Law Review, 104*(3), 671–732. http://www.jstor.org/stable/24758720.

Baykara, S. (2022). What is PCI DSS and PCI Compliance? PCI DSS GUIDE. https://pcidssguide.com/what-is-pci-dss-and-pci-compliance/#.

Brunnermeier, M. K. (2009). Deciphering the liquidity and credit crunch 2007–2008. *Journal of Economic Perspectives, 23*(1), 77–100. https://doi.org/10.1257/jep.23.1.77.

Cain, D. (2023). Bias in Artificial Intelligence: Charting a path forward. *www.linkedin.com*. https://www.linkedin.com/pulse/bias-artificial-intelligence-charting-path-forward-david-cain/.

CaseGuard Video Redaction Software. (2022). Algorithmic Accountability and Law in the U.S. *CaseGuard*. https://caseguard.com/articles/the-algorithmic-accountability-act-bias-and-tech/.

Cavoukian, A. (2010). Privacy by Design: The 7 Foundational Principles Implementation and Mapping of Fair Information Practices. https://iab.org/wp-content/IAB-uploads/2011/03/fred_carter.pdf.

Confronting too big to fail. (2009). Board of Governors of the Federal Reserve System. https://www.federalreserve.gov/newsevents/speech/tarullo20091021a.htm.

Council on Foreign Relations. (2023). *What will the future of AI regulation look like?* [Video]. YouTube. https://www.youtube.com/watch?v=kFDf1wekVEc.

Doshi-Velez, F., & Kim, B. (2017). Towards A Rigorous Science of Interpretable Machine Learning. *arXiv (Cornell University)*. https://doi.org/10.48550/arxiv.1702.08608.

El-Gamal, M. A. (2006). *Islamic finance: Law, economics, and practice.* Cambridge University Press.

Ethical algorithm design should guide technology regulation | Brookings. (2022). Brookings. https://www.brookings.edu/articles/ethical-algorithm-design-should-guide-technology-regulation/

Evaluation of the effects of too-big-to-fail reforms: Final Report. (2021). Financial Stability Board. https://www.fsb.org/2021/03/evaluation-of-the-effects-of-too-big-to-fail-reforms-final-report/.

Fair Credit Reporting Act. (2023). Federal Trade Commission. https://www.ftc.gov/legal-library/browse/statutes/fair-credit-reporting-act.

Feyen, E., Frost, J., Gambacorta, L., Natarajan, H., & Saal, M. (2021). Fintech and the digital transformation of financial services: implications for market structure and public policy. In *Bank for International Settlements.* https://www.bis.org/publ/bppdf/bispap117.pdf

FIXTrading. (2023). *What is FIX? • FIX Trading Community.* FIX Trading Community v1.9. https://www.fixtrading.org/what-is-fix/.

Floridi, L. (2014). *The Fourth Revolution: How the infosphere is reshaping human reality.* http://ci.nii.ac.jp/ncid/BB16630552.

Friede, G., Busch, T., & Bassen, A. (2015). ESG and financial performance: Aggregated evidence from more than 2000 empirical studies. *Journal of Sustainable Finance & Investment, 5*(4), 210-233. https://doi.org/10.1080/20430795.2015.1118917.

Greenleaf, G. (2013). Sheherezade and the 101 Data Privacy Laws: Origins, significance and global trajectories. *Social Science Research Network.* https://doi.org/10.2139/ssrn.2280877.

Hagendorff, T. (2020). The Ethics of AI Ethics: An Evaluation of Guidelines. *Minds and Machines, 30*(1), 99–120. https://doi.org/10.1007/s11023-020-09517-8.

Harrington, S. E., & Niehaus, G. R. (2004). Risk management and insurance. Tata McGraw-Hill Education.

Hendricks, D., Kambhu, J., & Mosser, P. (2006). *Systemic Risk and the Financial System Background Paper NAS-FRBNY Conference on New Directions in Understanding Systemic Risk Federal Reserve Bank of*

New York. https://www.newyorkfed.org/medialibrary/media/newsevents/events/research/2006/0518-background.pdf

HSBC Holdings Plc. and HSBC Bank USA N.A. Admit to Anti-Money Laundering and Sanctions Violations, Forfeit $1.256 Billion in Deferred Prosecution Agreement. (2015). https://www.justice.gov/opa/pr/hsbc-holdings-plc-and-hsbc-bank-usa-na-admit-anti-money-laundering-and-sanctions-violations.

Husted, B. W. (2000). The impact of national culture on responsible corporate behavior. Journal of International Business Studies, 31(4), 529-545. https://doi.org/10.1023/a:1006250203828

Javaid, S. (2023). Data Interoperability & Machine Learning in 2023 & Beyond. *AIMultiple.* https://research.aimultiple.com/data-interoperability/

Kroll, J., Huey, J., Barocas, S., Felten, E., Reidenberg, J., Robinson, D., & Yu, H. (2017). Accountable Algorithms. *University of Pennsylvania Law Review, 165*(3), 633. https://scholarship.law.upenn.edu/penn_law_review/vol165/iss3/3

Mehrabi, N., Morstatter, F., Saxena, N., Lerman, K., & Galstyan, A. (2019). A survey on Bias and Fairness in Machine Learning. *arXiv (Cornell University).* https://doi.org/10.48550/arxiv.1908.09635.

Papantoniou, A. A. (2022). Regtech: steering the regulatory spaceship in the right direction? *Journal of Banking and Financial Technology, 6*(1), 1–16. https://doi.org/10.1007/s42786-022-00038-9

Power, M. (1997). *The Audit Society: Rituals of Verification.* Oxford University Press.

Roy, M. J. (2017). Cathy O'Neil. Weapons of Math Destruction: How big data increases inequality and threatens democracy. New York: Crown Publishers, 2016. 272p. Hardcover, $26 (ISBN 978-0553418811). *College & Research Libraries, 78*(3), 403. https://doi.org/10.5860/crl.78.3.403

Security, R., & Security, R. (2023). Breaking Down The Requirements for 23 NYCRR 500. *RSI Security.* https://blog.rsisecurity.com/breaking-down-the-requirements-for-23-nycrr-500/#.

Sethi, A. & Ramesh, A. (2017). The Right to Privacy in the Supreme Court of India. https://www.law.upenn.edu/live/news/7450-the-right-to-privacy-in-the-supreme-court-of-india/news/international-blog

Shedd, J. A. (1928). *Salt from my attic.* The Mosher Press.

Simbeck, K. (2023). They shall be fair, transparent, and robust: auditing learning analytics systems. *AI And Ethics.* https://doi.org/10.1007/s43681-023-00292-7.

Solove, D., & Schwartz, P. (2022). *Privacy Law Fundamentals.*

Staff, R. (2015). PayPal to pay $7.7 million in U.S. Treasury sanctions case. *U.S.* https://www.reuters.com/article/us-usa-treasury-ebay/paypal-to-pay-7-7-million-in-u-s-treasury-sanctions-case-idUSKBN0ML28620150325.

Supervision, B. C. O. B., & Settlements, B. F. I. (2010). *Basel III: A Global Regulatory Framework for More Resilient Banks and Banking Systems.* https://www.bis.org/publ/bcbs189.pdf

Tapscott, D., & Tapscott, A. (2016). *Blockchain revolution: how the technology behind bitcoin is changing money, business, and the world.* Penguin.

Tarullo, D. K. (2008). *Banking on Basel: The future of international financial regulation.* Peterson Institute for International Economics.

The Bank Secrecy Act | FinCEN.gov. (2020). https://www.fincen.gov/resources/statutes-and-regulations/bank-secrecy-act#.

The FATF. (1989). https://www.fatf-gafi.org/en/the-fatf.html.

The White House. (2023). *NATIONAL CYBERSECURITY STRATEGY.* https://www.whitehouse.gov/wp-content/uploads/2023/03/National-Cybersecurity-Strategy-2023.pdf

Treleaven, P., & Batrinca, B. (2017). Algorithmic regulation: Automating financial compliance monitoring and regulation using AI and blockchain. *Journal of Financial Transformation, 45*, 14-21.

Unger, B. (2009). Money laundering: A newly emerging topic on the international agenda. *Review of Law & Economics, 5*(2), 807-819. https://doi.org/10.2202/1555-5879.1417.

Vanessa MOCK, L. (2017). Payment Services Directive (PSD2): Regulatory Technical Standards (RTS) enabling consumers to benefit from safer and more innovative electronic payments. *Journal of Payment Systems, 10*(3), 200-215.

Voigt, P., & Von Dem Bussche, A. (2017). *The EU General Data Protection Regulation (GDPR): A Practical guide.* https://dl.acm.org/citation.cfm?id=3152676

Von Solms, R., & Van Niekerk, J. (2013). From information security to cyber security. *Computers & Security, 38*, 97–102. https://doi.org/10.1016/j.cose.2013.04.004.

What is open banking? - Open Banking. (2023). Open Banking. https://www.openbanking.org.uk/what-is-open-banking/.

Zachariadis, M., Hileman, G., & Scott, S. V. (2019). Governance and control in distributed ledgers: Understanding the challenges facing blockchain technology in financial services. *Information and Organization, 29*(2), 105–117. https://doi.org/10.1016/j.infoandorg.2019.03.001.

Zetzsche, D. A., Buckley, R. P., Arner, D. W., & Barberis, J. N. (2018). Regulating a Revolution: From Regulatory Sandboxes to Smart Regulation. *Fordham Journal of Corporate & Financial Law, 23*(1), 31-103.

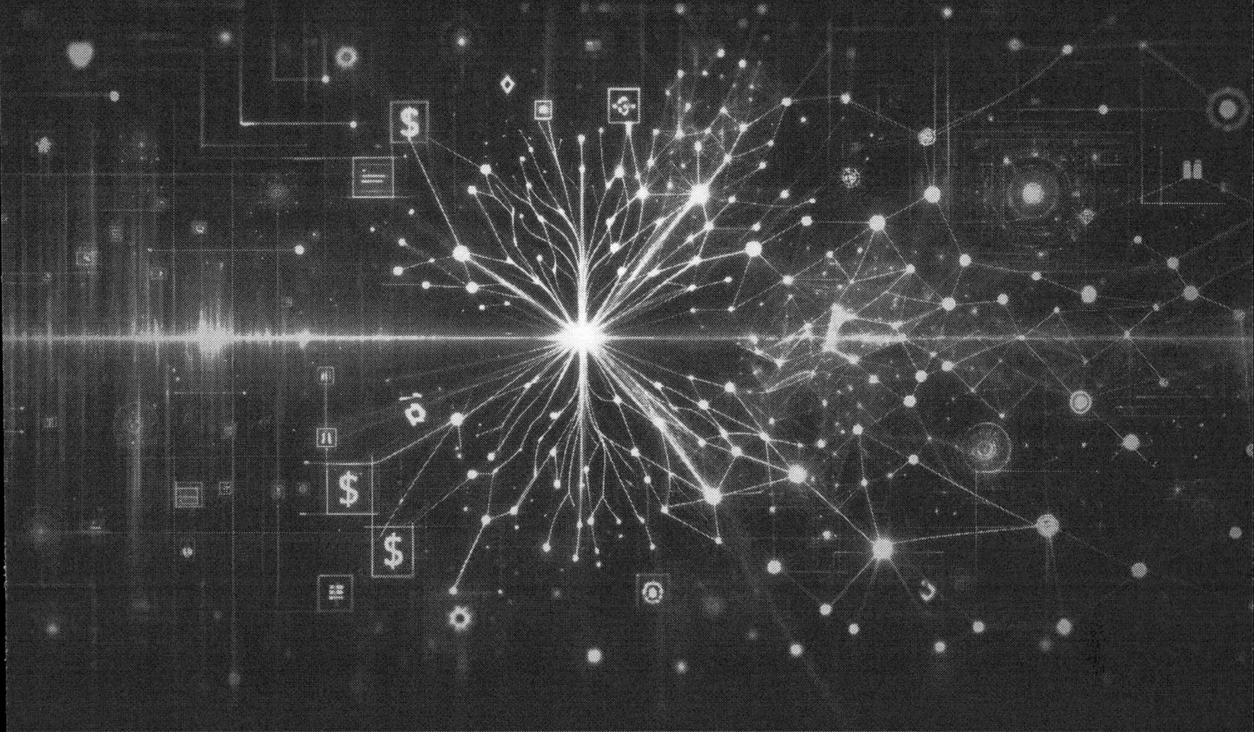

CHAPTER 13

Unveiling the Mystique

Ethical & Explainable AI & ML in Autonomous Finance

"We must ensure that the
yearning for the vast possibilities
of the digital age with AI
does not blind us to the potential downsides."

- SATYA NADELLA (2018)

13.0 Key Insights

- **Importance of Explainability:** AI and ML models are often called *"black boxes"* due to their complex inner workings. Users and stakeholders must understand how decisions are made in these models, which is crucial for ensuring fairness and building trust.
- **Need for Bias Mitigation:** AI and ML models can inadvertently learn and replicate existing societal biases, leading to unfair outcomes. It is essential to have strategies and processes to identify, mitigate, and monitor bias in these models.
- **Significance of Governance:** Governing AI and ML involves defining clear roles and responsibilities, establishing guidelines and procedures, and using tools for monitoring and management. Effective governance is key to ensuring the responsible use of AI and ML.
- **Role of Regulations:** The regulatory landscape for AI and ML is evolving, with an increasing emphasis on data protection, privacy, and user rights. Compliance with these regulations is legally necessary and helps foster trust with customers and stakeholders.
- **Impact of Real-world Cases:** Real-world cases highlight the practical challenges and opportunities in implementing ethical, explainable, and well-governed AI and ML. They serve as valuable lessons for organizations embarking on the Autonomous Finance journey.

13.1 Introduction

Embarking on the labyrinthine journey through the realms of AI and ML necessitates a meticulous charting of a course that integrates not only the learnings from the previous chapter but also the ground realities illuminated by recent studies. In this endeavor, the pivotal role played by ethical and explainable AI and ML stands at the forefront, a harbinger of a new era in Autonomous Finance grounded in trust, ethics, and transparency.

As one navigates from the regulatory and ethical landscapes elucidated in the preceding chapter — a rich tapestry woven with multidimensional regulatory mosaics and guiding principles for ethical compass in Autonomous Finance — it is imperative to delve deeper, to unravel the potent forces of AI and ML, forces that hold the promise of reshaping the Autonomous Finance ecosystem, yet brim with challenges that demand astute governance.

Drawing upon the insights from the Cisco 2023 Data Privacy Benchmark Study, one observes a remarkable ascension in the significance of privacy and data protection in recent years. Over 70% of organizations acknowledged deriving "significant" or "very significant" benefits in critical areas such as operational efficiency, building trust with customers, and mitigating losses from data breaches from their investments in privacy (Ikeda, 2023). This accentuates the economic rationale behind privacy investments and underscores the intrinsic link between privacy and trust. This bond is essential in fostering an Autonomous Finance ecosystem that is resilient and ethical.

Furthermore, 95% of respondents in the study emphasized the essentiality of equipping "all of their employees" with the knowledge to protect data privacy, echoing the sentiment of collective responsibility in safeguarding the sanctity of data (Ikeda, 2023). As this chapter unravels the roles of data scientists and AI ethics officers, it becomes clear that these guardians of ethics stand as the custodians of trust, weaving a tapestry of ethicality encompassing every stratum of the organizational hierarchy.

In navigating this complex landscape, it becomes pertinent to heed the insights of Donald Knuth, a computer science luminary who astutely observed that *"while AI has mastered the realms that require thinking, it still grapples with tasks that human beings undertake without thinking"* (*Donald Knuth Quotes*, n.d.). This observation serves as a compass, guiding this chapter in its exploration of the untrodden paths where the aspiration is to foster AI and ML systems that are not just intelligent but intuitively ethical, capable of actions that resonate with innate human sensibilities.

With a foundation grounded in such deep insights, this chapter endeavors to forge ahead, demystifying the complex realms of AI and ML, unraveling the nuances of ethical and explainable technologies, and envisioning a future where Autonomous Finance is not just a beacon of innovation but a bastion of trust, ethics, and humane values.

Let this chapter serve as a guide, navigating through the intricate pathways of AI and ML with a discerning lens, fostering a harmonious symphony of technology and ethics, and steering towards a horizon where the Autonomous Finance ecosystem embodies the pinnacle of ethical evolution, grounded in justice, trust, and the higher principles of human values.

13.2 Decoding AI & ML: From Complex to Clear

AI and ML are the twin engines propelling finance into a new era of automation and efficiency. Nevertheless, it is crucial to understand how these technologies work and make decisions to navigate this new landscape effectively.

AI, at its core, is a field of computer science that aims to mimic human intelligence. Whether recognizing patterns in data or predicting future trends, AI systems can perform tasks that usually require human intelligence (Russell & Norvig, 2010). On the other hand, ML, a subset of AI, involves the design of algorithms that allow computers to learn from and make decisions based on data (Goodfellow et al., 2016).

However, what is the source of this data, and how do these models learn? The key lies in the training data - a dataset used to teach the ML model about the task it needs to perform. The quality, diversity, and representativeness of this training data can significantly impact the performance and fairness of the AI and ML models (Gebru et al., 2021).

In Autonomous Finance, these models can be trained to perform various tasks, such as predicting stock market trends and detecting fraudulent transactions. However, while these models can be compelling, they are not infallible. Biases in the training data, opaque decision-making processes, and the lack of oversight can lead to unethical and potentially harmful outcomes.

This brings us to the crux of our discussion - the need for ethical, explainable AI and ML in Autonomous Finance. The following section will dive deeper into explainability and why it is crucial in finance.

13.3 Explainability: The Key to Trust in AI and ML

In the age of data-driven decision-making, AI and ML models have proven to be powerful allies, able to sift through vast amounts of information and extract valuable insights. However, as these models become more complex, they become more opaque. This so-called *"black box"* problem poses significant challenges in Autonomous Finance, where transparency and accountability are paramount.

Explainability in AI and ML refers to the ability to understand and interpret the decisions made by these models. In other words, it is about opening the *"black box"* and understanding the logic behind its outputs (Doshi-Velez et al., 2017).

Why is explainability so important? For one, it builds trust. According to a study by Accenture (2021), 74% of consumers are more likely to trust a company if it offers more transparency in its AI-based decisions. Moreover, explainability enables accountability. If an AI or ML model makes a decision that leads to an undesirable outcome, it is essential to understand why that decision was made and who (or what) should be held accountable (Mittelstadt et al., 2019).

There are various techniques for achieving explainability in AI and ML, ranging from more straightforward methods like decision trees and rule lists to more complex ones like LIME (Local Interpretable Model-agnostic Explanations) and SHAP (Shapley Additive Explanations). These methods provide different ways of interpreting and explaining the decisions of AI and ML models (Molnar, 2020).

However, explainability is just one piece of the puzzle. To truly ensure ethical AI and ML in Autonomous Finance, the issue of bias needs to be addressed, which is the primary topic of discussion in the next section.

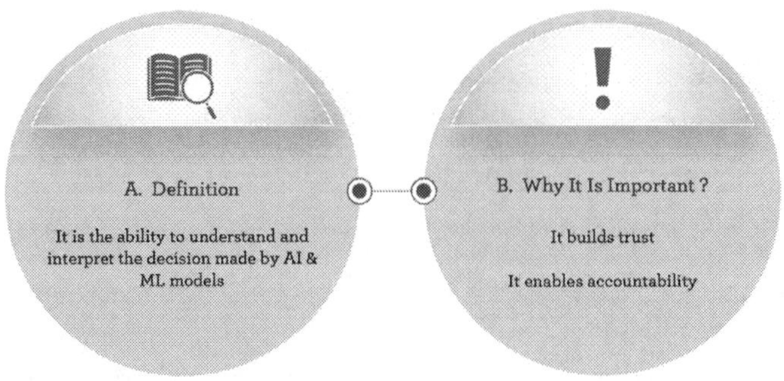

FIGURE 13.1: Explainability in Autonomous Finance

13.4 Battling Bias: Ensuring Fairness in AI and ML

Progressing further into the intricate labyrinth of AI and ML, the formidable adversary of bias is encountered. Even with their objective algorithms, AI and ML models are often found to reflect or magnify human biases, leading to unfair consequences (Buolamwini & Gebru, 2018). A stark example of this is a loan approval model that could unjustly refuse loans to specific population segments upon being trained on skewed data.

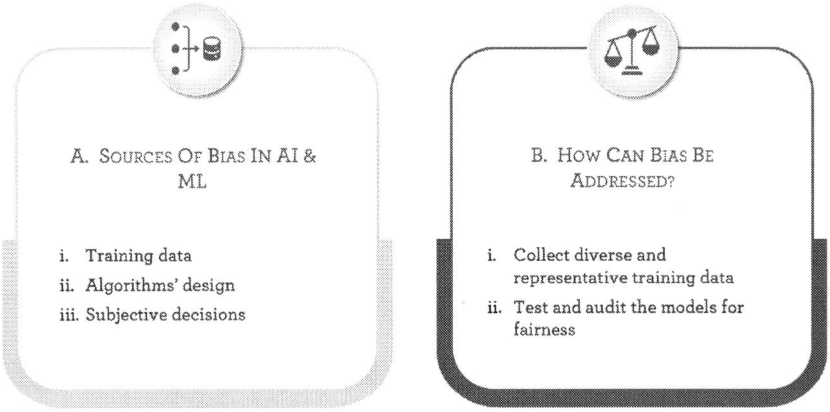

FIGURE 13.2: Battling Bias in Autonomous Finance

As shown in Figure 13.2, bias in AI and ML models can stem from various sources, with biased training data being one of the most common culprits. If the data used to train the model reflects existing prejudices, the model will likely learn and perpetuate these biases (Barocas & Selbst, 2016). Furthermore, bias can creep in through the algorithms' design or the subjective decisions made during the model development.

Tackling bias is critical to ensuring ethical AI and ML in Autonomous Finance. Figure 13.2 outlines two of the most common strategies organizations leverage to address bias in AI and ML models. One popular method for checking bias is the *"fairness through unawareness"* approach, which omits sensitive attributes (like race or gender) from the data. However, this method has limitations because bias can still occur indirectly through other correlated attributes (Dwork et al., 2012). The other popular method is to audit AI and ML models for fairness by scrutinizing the outcome, process, data, and context. Such an audit is not a one-off exercise but a recurring ritual in the lifecycle of the AI/ML models powering Autonomous Finance. These audits become integral to the AI/ML governance architecture with real-time monitoring and dynamic calibration mechanisms that adjust the AI/ML model for emergent biases.

Organizations must cultivate an environment that encourages recognition and responsibility towards bias, endorsing routine audits of AI and ML models and fostering transparency in their decision-making mechanisms. This leads to the critical subject governing AI and ML models, which will be dissected in the subsequent section.

13.5 Steering AI & ML Towards Good Governance

Traversing the domain of ethical, explainable AI and ML in Autonomous Finance, one reaches a crucial milestone - governance. Governance encompasses the mechanisms, processes, and principles that ensure the conscientious application of AI and ML (Bostrom & Yudkowsky, 2018). It acts as the helm, guiding these influential technologies appropriately.

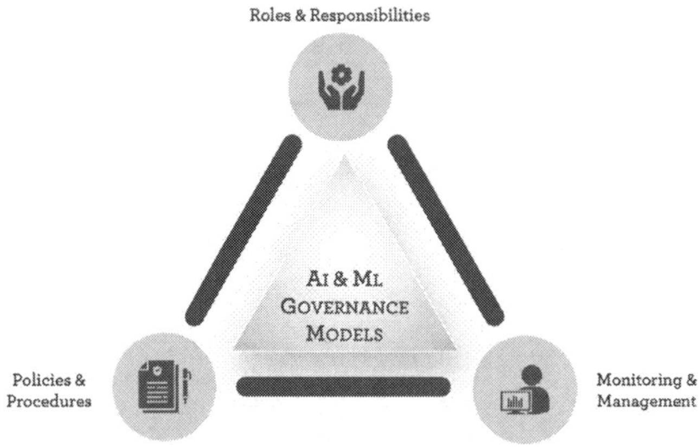

FIGURE 13.3: Battling Bias in Autonomous Finance

AI and ML Governance model includes various components, as illustrated in Figure 13.3. First and foremost, it requires clear roles and responsibilities, which may involve setting up a dedicated team or appointing an AI Ethics Officer to oversee the organization's use of AI and ML (Jobin et al., 2019).

Secondly, it requires policies and procedures. These could include guidelines for data collection and use, procedures for testing and auditing the models, and policies for handling any ethical issues that arise (Hagendorff, 2020).

Last but not least, it involves monitoring and management. Management of the AI and ML Governance model could involve using tools to track the performance and fairness of the models, conducting regular audits, and continually updating the models and policies based on these audits (Galindo et al., 2021).

Effective governance ensures ethical, explainable AI and ML in Autonomous Finance. It not only helps to tackle bias and enhance explainability but also fosters trust among customers and stakeholders. The following section will delve into the regulatory landscape surrounding AI and ML, providing a broader context for our discussion.

13.6 Compliance with AI and ML Rules

The exploration of the AI and ML realm uncovers that it is not an unregulated wilderness. On the contrary, it is a landscape continuously being molded by regulations. These current and prospective regulations strive to ensure the ethical utilization of AI and ML, safeguarding consumer rights while endorsing fairness and transparency (Cath et al., 2018).

For example, the EU's GDPR has set high data protection and privacy standards. The GDPR includes provisions related to automated decision-making and profiling, and it emphasizes the right to explanation – a user's right to know how a decision that affects them was made by an algorithm (Goodman & Flaxman, 2017).

In the United States, there is a growing call for AI regulation. While there are no comprehensive AI laws at the federal level, various states have passed or proposed AI-related laws:

a) The FTC has the authority to regulate unfair or deceptive practices involving AI under Section 5 of the FTC Act. In 2020, the FTC released guidance stating that AI systems must be transparent, fair, and accountable. The FTC has brought enforcement actions related to harms caused by algorithmic systems.

b) Sector-specific federal agencies like the FDA, Federal Aviation Administration (FAA), and CFPB have provided guidance around AI use in their regulated industries. For example, the FDA has issued procedures for AI/ML-based software as a medical device.

c) In 2019-2020, the White House released reports and draft guidance promoting AI principles such as transparency, fairness, and accountability. However, these are not binding regulations.

d) Several proposals for federal AI regulations, such as algorithmic accountability laws, have not been enacted yet. Lawmakers proposed the Algorithmic Accountability Act of 2022 to require impact assessments for specific automated systems.

e) AI used by federal government agencies is subject to some standards under policies like the OMB A-130 Circular and NIST risk management framework. However, there are calls for more robust federal procurement regulations.

So, in summary, there are currently no comprehensive federal statutes or regulations governing the Private sector use of AI. However, government guidance indicates increasing scrutiny and expectations around responsible AI practices. Significant legal and policy developments are likely in the coming years as AI use expands.

Complying with these regulations can be challenging, given the complex and evolving nature of AI and ML. However, it is an essential part of AI and ML governance. It helps organizations avoid hefty penalties and fosters trust among customers and stakeholders.

The following section will discuss the role of data scientists and AI Ethics Officers - the professionals on the front lines of ethical, explainable, and well-governed AI and ML.

13.7 The Guardians of AI Ethics

As the course for ethical, explainable AI and ML in Autonomous Finance is mapped out, it is imperative to acknowledge the crucial individuals who serve as compass bearers and guardians - Data Scientists and AI Ethics Officers. These professionals are responsible for designing and applying AI and ML models.

Data scientists, with their deep understanding of AI and ML models, play a key role in mitigating bias, enhancing explainability, and ensuring the overall ethical use of these models (Donoho, 2017). They are involved in every step of the model development process, from collecting and preprocessing the data to training, testing, and deploying the models.

On the other hand, AI Ethics Officers (or similar roles) serve as the stewards of ethical AI and ML usage within an organization. They are responsible for setting up and overseeing the governance mechanisms, ensuring compliance with regulations, and promoting a culture of awareness and accountability toward AI and ML ethics (Jobin et al., 2019).

These professionals form the backbone of ethical, explainable AI and ML in Autonomous Finance. Their work ensures the responsible use of powerful technologies, fostering trust among customers and stakeholders.

The next section will explore real-world examples illustrating the importance of ethical, explainable, and well-governed AI and ML models.

13.8 Lessons from the Field: Ethics in AI and ML

Exploring the intricacies of ethical, explainable AI and ML offers no better way to understand the importance of these concepts than through real-world examples. Although many of these case studies do not directly pertain to Autonomous Finance, they provide an essential foundation for its future trajectory. They offer valuable predictions about the directions that businesses, governments, and society might take in grappling with the ethical use, explainability, and governance of AI and ML. These case studies reflect the past and indicate the future, leading the way to an era where AI and ML are used responsibly, ethically, and beneficially in Autonomous Finance.

a) In 2018, Amazon scrapped an AI recruiting tool after finding it was biased against women. The system penalized resumes with words like *"women's"* and down ranked graduates of all women's colleges (Dastin, 2018).

b) In 2019, Apple apologized and updated its credit card algorithm after complaints that it gave higher spending limits to men than women. This incident raised questions about gender bias in black-box credit approval models (*TechCrunch Is Part of the Yahoo Family of Brands*, 2021).

c) Uber's ML system for driver rewards was found to be unfairly denying rewards to drivers based on unexplainable factors outside their control. Uber agreed to pay drivers over $100 million to settle FTC charges (Metz, 2022).

d) A Stanford study in 2022 found TikTok's recommendation algorithm could steer users toward content promoting eating disorders or depression when trained on those viewing patterns. This incident raised concerns about platforms' AI ethics (Cheng & Li, 2023).

e) Google fired ethics researchers and allegedly manipulated research findings that were critical of risks in AI designs. This action of Google exemplified business pressures around transparent AI research (Wakabayashi & Metz, 2022).

f) After receiving consumer complaints about Apple's lack of transparency in data usage for personalization, the EU reviewed Apple's use of personalized ads in 2021. New EU regulations restrict the use of specific AI applications (Chee & Kar-Gupta, 2023).

g) Some banks and lenders have created *"explainable AI"* dashboards to increase transparency in their credit risk models and comply with laws requiring explainability (Townson, 2020).

Scrutinizing case studies about AI and ML (see Figure 13.4 below) reveals that these instances are veritable treasure troves of crucial insights. Challenges common to applying AI and ML models across diverse functional areas and processes in Finance surface, offering priceless lessons for Autonomous Finance.

Firstly, these case studies highlight the tendency of AI and ML systems to reflect and amplify existing societal biases inadvertently. This tendency underscores the critical necessity for ongoing audits, ensuring these systems do not unintentionally propagate inherent human prejudices.

The focus on audits should not eclipse the equally important principles of transparency, explainability, and accountability. These principles are fundamental in building trust and minimizing potential harm from AI systems. In addition, a crucial lesson from these case studies is the importance of diversity in AI and ML design processes. Incorporating a wide spectrum of stakeholders can contribute to creating inclusive AI and ML systems that reflect societal needs. However, implementing these principles may sometimes collide with business interests that favor the opacity of *"black box"* models.

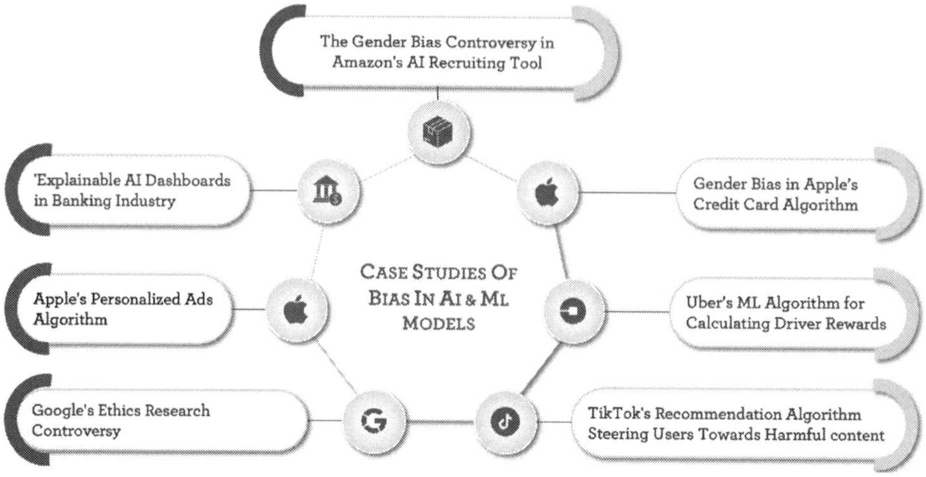

FIGURE 13.4: Case Studies of Bias in AI & ML in Autonomous Finance

The case studies also acknowledge the need for regulation and oversight in the AI domain. Nevertheless, the transition from principles to effective practices is not seamless, and the extent of government involvement in AI regulation remains a complex, unresolved question.

Unexpected outcomes of AI systems, irrespective of their initial design intentions, underline the essential preparedness for unanticipated consequences, a testament to the complex nature of AI and ML. The case studies highlight the intricate balancing act between accuracy, efficiency, transparency, and other objectives in AI and ML systems objectives. This intricate balancing act underlines that there are no simple, clear-cut solutions in AI and ML but rather a series of delicate compromises to navigate. While technical interventions such as explainable AI offer potential solutions, the obstacles related to organizational and governance issues present substantial hurdles to the development of accountable AI.

13.9 Conclusion

As Chapter 13 draws close, it illuminates an inflection point in the Autonomous Finance narrative. The labyrinthine complexities of AI and ML, technologies as potent as they are enigmatic, have been unraveled. Yet, the dualistic nature of these advanced systems is laid bare: they possess the capacity to revolutionize efficiency and perpetuate societal inequities should ethical considerations and governance be disregarded.

Anchored by ethical application, explainability, and governance, the chapter confronted exigent questions at the heart of responsible technology use. The insights shared are not merely theoretical abstractions but serve as navigational waypoints for stakeholders, illuminating the imperatives of ethical governance, transparency, and a commitment to fairness.

In 2006, British mathematician Clive Humby articulated a vision where he likened data to oil, emphasizing that, like oil, data harbors its true value not in its crude form but in its potential to be refined and transformed into a resource of invaluable insights. He stated, "*Data is the new oil*" (Humby et al., 2008). This potent parallel finds a resonant echo in the contemporary discourse on explainability in AI and ML. Just as data transitions from being a latent asset to a wellspring of insights through a meticulous refining process, the voluminous data maneuvered by AI and ML systems awaits its true actualization. This data transcends its latent state through explainability, morphing into a transparent and accountable entity, thereby unveiling its transformative potential to foster innovations and steer informed decision-making.

The discourse on ethical and explainable AI and ML has delineated the urgent imperatives for stringent governance and regulatory compliance. These are the bedrock of a trust-imbued, socially equitable Autonomous Finance ecosystem. As the narrative pivots toward the next frontier—cybersecurity—one finds oneself at the cusp of a new era. The ethical scaffolding erected in this chapter serves as a sentinel, vigilantly preparing us for the challenges and promise of the labyrinthine world ahead.

In closing, the journey toward an Autonomous Finance future is as multifaceted as it is inevitable. Armed with the invaluable insights of this chapter, the invitation is thus extended to continue this journey, to navigate a future shaped not merely by technological prowess but by a relentless commitment to ethical integrity and social equity. As philosopher Immanuel Kant wisely noted, *"He who is cruel to animals becomes hard also in his dealings with men. We can judge a man's heart by his treatment of animals"* (Kant, 1997). May this serve as a gentle reminder that the measure of our ethical commitment will indelibly imprint on the world we seek to create.

13.10 Key Recommendations

a) **Invest in Explainability:** Organizations must invest in technologies that leverage AI and ML and emphasize explainability. Understanding how these models make decisions can help audit their performance, ensure fairness, and build customer trust.

b) **Mitigate Bias:** Bias in AI and ML models can lead to unfair outcomes. Organizations must adopt practices for identifying and mitigating bias in their AI and ML models, including diverse data collection, rigorous testing for bias, and ongoing monitoring.

c) **Establish Governance Mechanisms:** Organizations should establish precise governance mechanisms for using AI and ML. The governance mechanism should include defining roles and responsibilities, setting up guidelines and procedures, and implementing tools for monitoring and management.

d) **Comply with Regulations:** As regulations around the use of AI and ML continue to evolve, organizations must stay informed and ensure compliance. Ensuring and enforcing compliance will help avoid legal repercussions and foster trust with customers and stakeholders.

e) **Promote Ethical Practices:** Lastly, organizations should promote a culture of ethical AI use, including employee training, stakeholder engagement, and setting up an ethics board or appointing an AI Ethics Officer.

13.11 References

Barocas, S., & Selbst, A. D. (2016). Big Data's Disparate Impact. *California Law Review, 104*(3), 671–732. http://www.jstor.org/stable/24758720.

Bostrom, N., & Yudkowsky, E. (2018). The ethics of artificial intelligence. *Cambridge Handbook of Artificial Intelligence, 1*, 316-334.

Buolamwini, J., & Gebru, T. (2018). Gender shades: Intersectional accuracy disparities in commercial gender classification. *Proceedings of Machine Learning Research, 81*, 1-15.

Cath, C., Wachter, S., Mittelstadt, B., Taddeo, M., & Floridi, L. (2018). Artificial intelligence and the 'good society': the US, EU, and UK approach. *Science and engineering ethics, 24*(2), 505-528. https://doi.org/10.1007/s11948-017-9901-7

Chee, F. Y., & Kar-Gupta, S. (2023, February 28). EU antitrust regulators narrow charges against Apple. *Reuters*. https://www.reuters.com/technology/eu-antitrust-regulators-narrow-case-against-apple-2023-02-28/

Cheng, Z., & Li, Y. (2023). Like, Comment, and Share on TikTok: Exploring the Effect of Sentiment and Second-Person View on the User Engagement with TikTok News Videos. *Social Science Computer Review*, 089443932311786. https://doi.org/10.1177/08944393231178603

Dastin, J. (2018). Amazon scraps secret AI recruiting tool that showed bias against women. *U.S.* https://www.reuters.com/article/us-amazon-com-jobs-automation-insight-idUSKCN1MK08G

Donald Knuth Quotes. (n.d.). https://www.goodreads.com/author/quotes/14228100.Donald_Knuth

Donoho, D. (2017). 50 years of data science. Journal of Computational and Graphical Statistics, 26(4), 745-766. https://doi.org/10.1080/10618600.2017.1384734

Doshi-Velez, F., & Kim, B. (2017). Towards a rigorous science of interpretable machine learning. *ArXiv, abs*/1702.08608.

Dwork, C., Hardt, M., Pitassi, T., Reingold, O., & Zemel, R. (2012). Fairness through awareness. *Proceedings of the 3rd Innovations in Theoretical Computer Science Conference*, 214-226.

Galindo, L., Perset, K., & Sheeka, F. (2021). An overview of national AI strategies and policies, *OECD Going Digital Toolkit Notes*, No. 14, OECD Publishing, Paris, https://doi.org/10.1787/c05140d9-en

Gebru, T., Morgenstern, J., Vecchione, B., Vaughan, J. W., Wallach, H., Daumeé III, H., & Crawford, K. (2021). Datasheets for Datasets. *ArXiv, abs*/1803.09010. https://doi.org/10.48550/arXiv.1803.09010

Goodfellow, I., Bengio, Y., & Courville, A. (2016). *Deep Learning*. MIT Press.

Goodman, B., & Flaxman, S. (2017). European Union regulations on algorithmic decision-making and a right to explanation. AI Magazine, 38(3), 50-57. https://doi.org/10.1609/aimag.v38i3.2741

Hagendorff, T. (2020). The ethics of AI ethics: An evaluation of guidelines. *Minds and Machines, 30*, 99-120. https://doi.org/10.1007/s11023-020-09517-8

Humby, C., Hunt, T., & Phillips, T. (2008). *Scoring points: How Tesco Continues to Win Customer Loyalty.* Kogan Page Publishers.

Ikeda, S. (2023, February 10). Cisco 2023 Data Privacy Benchmark Report: Customer trust hinges on transparency, strong desire to opt out . . . *CPO Magazine*. https://www.cpomagazine.com/data-privacy/cisco-2023-data-privacy-benchmark-report-customer-trust-hinges-on-transparency-strong-desire-to-opt-out-of-ai/

Jobin, A., Ienca, M., & Vayena, E. (2019). The global landscape of AI ethics guidelines. *Nature Machine Intelligence, 1*(9), 389-399. https://doi.org/10.1038/s42256-019-0088-2

Kant, I. (1997). Lectures on Ethics. In *Cambridge University Press eBooks*. https://doi.org/10.1017/cbo9781107049512

Metz, C. (2022, September 12). Uber agrees to pay N.J. $100 million in dispute over drivers' employment status. *The New York Times*. https://www.nytimes.com/2022/09/12/technology/uber-new-jersey-settlement.html

Mittelstadt, B., Russell, C., & Wachter, S. (2019). Explaining Explanations in AI. Proceedings of FAT* '19: Conference on Fairness, Accountability, and Transparency, 279-288. https://doi.org/10.48550/arXiv.1811.01439

Molnar, C. (2020). *Interpretable Machine Learning*. Lulu.com.

Russell, S., & Norvig, P. (2010). *Artificial Intelligence: A Modern Approach.* Pearson.

TechCrunch is part of the Yahoo family of brands. (2021). https://techcrunch.com/2021/03/23/new-yorks-department-of-financial-services-says-apple-card-program-didnt-violate-fair-lending-laws/

Townson, S. (2020). *AI can make bank loans more fair*. Harvard Business Review. https://hbr.org/2020/11/ai-can-make-bank-loans-more-fair

Wakabayashi, D., & Metz, C. (2022, May 2). Another firing among Google's A.I. brain trust, and more discord. *The New York Times*. https://www.nytimes.com/2022/05/02/technology/google-fires-ai-researchers.html

CHAPTER 14

The Digital Fortress

Cybersecurity Imperatives in Autonomous Finance

"There are only two types of companies: those that have been hacked and those that will be. Even that is merging into one category: those that have been hacked and will be again,"

- ROBERT S. MUELLER, III ("COMBATING THREATS IN THE CYBER WORLD: OUTSMARTING TERRORISTS, HACKERS, AND SPIES," 2012)

14.0 Key Insights

- **The Emergence of Autonomous Finance Unleashes New Cybersecurity Threats to Tackle:** The rise of Autonomous Finance opens a new frontier of opportunities but also introduces a myriad of potential cybersecurity threats, including data breaches, AI/ML manipulation, API vulnerabilities, and others.
- **The Imperative of Robust Cybersecurity Risk Management in Autonomous Finance:** A comprehensive and robust cybersecurity risk management is crucial, which includes a systematic cycle of identifying, assessing, mitigating, and monitoring risks. Implementing a solid cybersecurity framework that encompasses governance, technological measures, and incident response is pivotal.
- **Creating a Cyber-Literate Workforce is Vital for Tackling Autonomous Finance Challenges:** Promoting cybersecurity awareness and education within the organization is essential. Continuous, targeted training can build a cyber-literate workforce able to address the specific challenges of Autonomous Finance.
- **Adherence to International Cybersecurity Standards Bolsters Trust in Autonomous Finance:** Adhering to globally recognized cybersecurity standards, such as ISO/IEC 27001, offers a trustworthy set of controls and bolsters confidence among stakeholders, including customers, regulators, and partners.
- **Innovative, Scalable Cybersecurity Strategies are Key to Safeguarding the Future of Autonomous Finance:** In the face of Autonomous Finance's rapid growth, cybersecurity strategies need to be scalable and innovative, utilizing AI, ML, and automation for threat detection and security tasks, while embracing cloud-based solutions for broader protection. Successfully navigating these challenges ensures the resilience and reliability of Autonomous Finance systems.

14.1 Introduction

As the odyssey into the transformative world of Autonomous Finance continues—a realm where AI and automation meld to rewrite the annals of financial history—a new frontier emerges on the horizon. This is not merely a tale of technological mastery and unparalleled efficiency; it is an epic saga that unfolds in a realm fraught with newfound complexities and vulnerabilities. Welcome to *'Cybersecurity Imperatives in Autonomous Finance.'*

Autonomous Finance heralds a seismic shift, transforming the very architecture of financial operations. No longer are investment strategies, risk management, or financial planning solely entrusted to human intuition. Instead, they are orchestrated by intricate algorithms, operating in real-time to decipher market complexities.

Yet, as the technological tapestry of Autonomous Finance grows increasingly elaborate, it unveils a latent landscape teeming with sophisticated cybersecurity threats. While these threats are a global concern, transcending national borders, the experience and insights drawn upon in this chapter derive primarily from North America's regulatory and business environment.

This chapter endeavors to dissect the multi-dimensional cybersecurity challenges intrinsic to Autonomous Finance. It delves into the ever-morphing cyber threat landscape and its far-reaching implications—from economic repercussions to systemic instabilities and breaches of ethical norms. The discourse extends to vital areas such as regulatory compliance, secure infrastructure design, data protection, and privacy.

As this exploration continues, the narrative will be enriched with seminal case studies, comprehensive analyses, and expert commentaries. The aim is to give the reader a reactive set of tools and a visionary roadmap for proactive cybersecurity. This strategic approach anticipates, mitigates, and neutralizes threats before they burgeon into crises.

Furthermore, this chapter recognizes that the fortress of cybersecurity is buttressed not only by technological advancements but also by human acumen and ethical responsibility. Strategies to fortify this digital stronghold will thus integrate technological solutions with human behavior insights, culminating in a multi-disciplinary approach to cybersecurity in Autonomous Finance.

The expedition into this challenging yet exhilarating domain of Autonomous Finance remains underway. This chapter invites its readers to unravel complexities, fathom risks, and seize the unparalleled opportunities this technological finance renaissance avows.

14.2 Cybersecurity Threats in Autonomous Finance

Charting the course in the vast sea of Autonomous Finance, it is vital to understand the shifting cybersecurity landscape that mirrors the unpredictable nature of an ocean. While the advantages offered by this new realm are clear and compelling, it has also increased cyber risks, thus making the seas significantly rougher for the financial vessels navigating it.

The widening of the attack surface, primarily due to the increasing use of AI and ML, makes for the first major wave of challenges in the cybersecurity ocean. AI, while a crucial component of Autonomous Finance, brings the vulnerability of potential exploitation. Cyber adversaries with the right knowledge and tools could attempt to manipulate these AI systems, corrupting the decision-making algorithms and compromising financial transactions.

The massive volume of data accumulated, processed, and stored in Autonomous Finance is the second substantial wave. For instance, customer data, transaction details, and risk assessment information, all crucial for delivering personalized and efficient financial services, are potentially attractive targets for cybercriminals. This wealth of data poses a cybersecurity challenge unseen in traditional finance. The constant threat of data breaches, data leaks, and misuse of information makes it imperative for organizations to prioritize data security.

Thirdly, Autonomous Finance's dependency on various third-party entities adds another complexity to the cybersecurity challenge. These third-party entities, from cloud service providers to fintech partners, contribute to the functionality and efficiency of the financial ecosystem. However, they also introduce new points of vulnerability that may not be entirely within the control of the primary organization, extending the cybersecurity responsibility into the ecosystem of partnerships.

Moreover, just as the tools and technologies used in Autonomous Finance are evolving, so are the methods employed by cyber attackers. Cyber adversaries are becoming more sophisticated and audacious, continually upgrading their attacking strategies to find and exploit the slightest weaknesses in defenses. This phenomenon and the widespread adoption of digital technologies in finance significantly amplify the potential risks.

14.3 The Autonomous Finance Threat Landscape

Mirroring a complex and ever-changing battlefield, the threat landscape of Autonomous Finance is multifaceted and teeming with potential threats and challenges. The myriad of threats and the increasing dependency on technology in finance amplify the potential risks exponentially. To effectively formulate potent cybersecurity strategies, it is vital to have an in-depth understanding of this terrain, marked by its diverse sources of threats, the nature of these threats, and the implications of a potential breach. For a graphical representation of the threat landscape in Autonomous Finance, please refer to Figure 14.1.

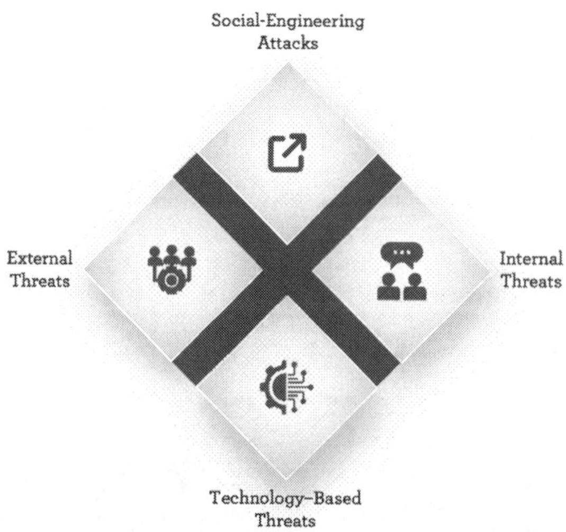

FIGURE 14.1: The Threat Landscape in Autonomous Finance

a) **Externally Sourced Threats:** The first dimension of this intricate landscape is constituted by threats originating from external sources. These external threats often come in the form of cybercriminals or state-sponsored hackers who employ advanced strategies and tools to exploit vulnerabilities in financial systems (Choo, 2011). They target data centers, networks, applications, and even individual users to gain unauthorized access to critical information. These cyber adversaries continually learn and adapt, using the latest technologies and techniques to launch increasingly sophisticated and damaging attacks. State-sponsored attacks, in particular, pose a significant risk. Countries with advanced cyber capabilities often engage in cyber warfare, targeting finance teams to destabilize economies, gather intelligence, or create havoc. With its high digital dependence, Autonomous Finance becomes an attractive target for such entities, exponentially increasing the risk factor for finance teams.

b) **Internal Threats,** on the other hand, emerge from within the organization and can sometimes be more damaging than external threats. They can be intentional or unintentional, stemming from disgruntled employees, simple human error, or even insufficient knowledge of cybersecurity best practices. Intentional internal threats often involve disgruntled employees or insiders with malicious intent who exploit their access privileges to cause damage to the organization. This could involve data leaks, system tampering, or other activities compromising the organization's cybersecurity measures. Unintentional threats often result from employees unknowingly creating vulnerabilities, such as by falling for phishing scams, misconfiguring settings, or using weak passwords.

c) **Technology-Based Threats:** These threats often target the very heart of Autonomous Finance - its digital infrastructure. This includes hardware, software, databases, and networks integral to the functioning of Autonomous Finance systems. Attackers exploit vulnerabilities in the system architecture, flawed algorithms, or loopholes in the security configurations to gain unauthorized access to data, disrupt operations, or even launch more extensive attacks. For instance, vulnerabilities in application code can lead to software exploits, where attackers gain access to systems or data. Similarly, hardware can be targeted with physical attacks or malware that exploits hardware vulnerabilities. The increasing complexity of the technologies used in Autonomous Finance escalates these risks, as attackers can target more potential points of failure.

d) **Social Engineering Attacks** present a growing and considerable threat in the landscape. These attacks exploit the weakest link in the cybersecurity chain - the human element (*DBIR Report 2023 - Introduction | Verizon Business*, 2023). Social engineering involves manipulating individuals into divulging sensitive information or performing actions compromising cybersecurity. Techniques such as phishing, pretexting, baiting, and quid pro quo are commonly used to trick employees into falling for the scams. In the context of Autonomous Finance, social engineering attacks can have severe consequences, given the amount of sensitive financial information involved. It underlines the importance of technical measures in cybersecurity and robust security awareness training for staff.

Each dimension of the threat landscape presents unique challenges, and an effective cybersecurity strategy must address each comprehensively. As shown in Figure 14.2, the sources of cybersecurity threats in the era of Autonomous Finance range from individuals to organized crime groups, from state-sponsored cyber attackers to internal employees. Here is an elaboration on the sources of cybersecurity threats in the Autonomous Finance landscape:

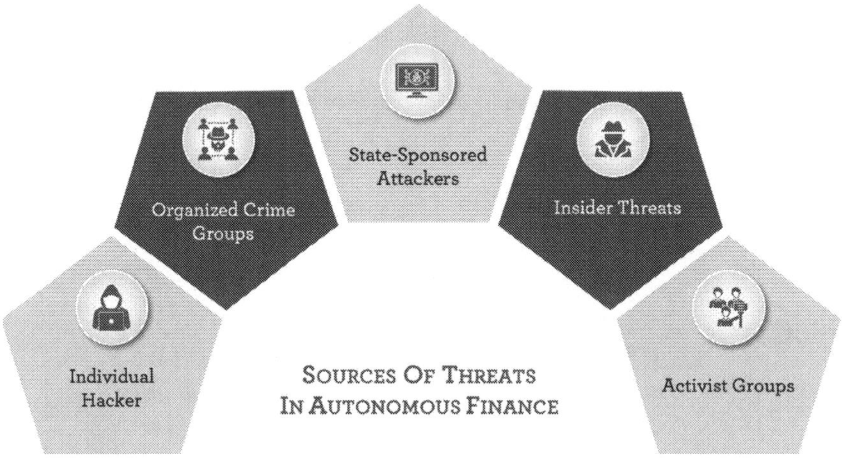

FIGURE 14.2: Sources of Threats in Autonomous Finance Landscape

a) **Individual Hackers** or 'lone wolves' often conduct their operations independently. They might be motivated by personal financial gain, the thrill of hacking, or the intellectual challenge it poses. They might employ tactics like hacking, malware distribution, identity theft, or phishing to exploit vulnerabilities within Autonomous Finance systems.

b) **Organized crime groups** pose a more severe threat as they often have more resources, sophisticated hacking tools, and a higher degree of organization (Chen et al., 2012). Their operations are usually driven by substantial financial gain, making the lucrative Finance domain a prime target. They often deploy advanced persistent threats (APTs), engage in large-scale data breaches, and employ ransomware attacks.

c) **State-sponsored attackers** are backed by national governments and engage in cyber espionage, cyber warfare, and economic sabotage (Hjortdal, 2011). Their targets often include critical infrastructures, including Finance teams, making Autonomous Finance a potential target. Their operations are characterized by high sophistication, vast resources, and extended durations, often undetected for long periods.

d) **Insider threats** come from individuals with legitimate access to the organization's systems—such as employees, contractors, or business partners—and misuse that access to carry out malicious activities or inadvertently cause security incidents (Probst et al., 2010). Given their inherent access, insider threats can be especially damaging and challenging to mitigate in Autonomous Finance.

e) **Hacktivist groups**, such as Anonymous, use hacking to promote or advance political or social causes. They may target finance teams to protest corporate actions, influence policy, or bring attention to specific issues. While their actions are generally less about financial gain, they can cause significant disruption and reputational damage.

Understanding the source of threats allows institutions to anticipate potential attack vectors, motivations, and tactics, informing a more targeted and effective defense strategy in the Autonomous Finance landscape.

Lastly, let us examine the various areas of vulnerability within the Autonomous Finance landscape (as shown in Figure 14.3), ranging from hardware and software applications to data, AI and ML models, APIs, people and processes, and third-party vendors. Each of these areas represents a distinct point of potential exploitation for malicious actors, and thus, they need to be safeguarded with specialized strategies and solutions. The following areas are particularly vulnerable in Autonomous Finance:

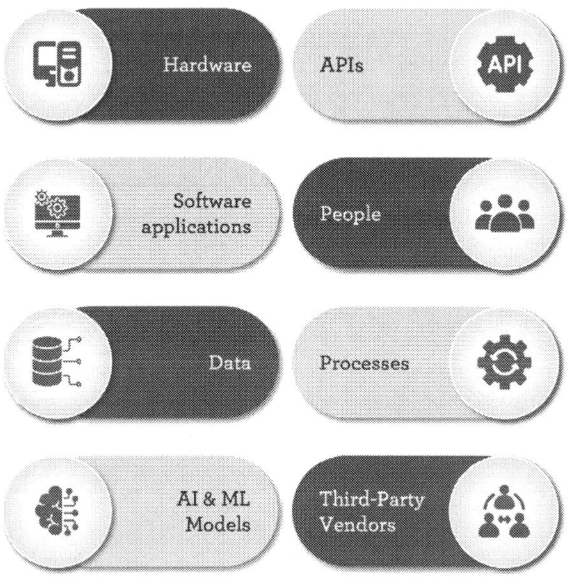

FIGURE 14.3: Areas of Vulnerabilities in Autonomous Finance Landscape

a) **Hardware and Infrastructure:** Hardware forms the foundation on which all software runs, including cloud-based infrastructure. Any vulnerabilities at the hardware level can be exploited to gain unauthorized access or disrupt services. For example, a well-known hardware vulnerability is the Spectre and Meltdown flaws in CPU architecture that potentially allow malicious programs to steal data processed on the computer (Kocher et al., 2020).

b) **Software Applications:** Financial applications and systems, including those using AI and ML, can have vulnerabilities that attackers may exploit. This can range from insecure code to flawed security configurations.

c) **Data:** Data is a prime target for cybercriminals. Data at rest and in transit can be vulnerable to attacks. Adopting Autonomous Finance typically involves using a large amount of sensitive financial and personal data, making it a lucrative target for cyberattacks.

d) **AI and ML Models:** As the backbone of Autonomous Finance, AI and ML models can be targeted in unique ways. For instance, attackers may attempt to poison the data used to train these models, leading to skewed results and faulty decision-making (Biggio & Roli, 2018).

e) **APIs:** APIs allow communication between different software components. Attackers can potentially exploit vulnerabilities in APIs to access sensitive information or disrupt services.

f) **People and Processes:** Even the most secure systems can be rendered vulnerable by human error or social engineering. Processes that involve human interaction are particularly susceptible.

g) **Third-Party Vendors:** As finance teams increasingly rely on third-party services for various tasks, the security posture of these vendors becomes a potential vulnerability. A breach in a vendor's system can potentially expose the financial institution's and its customers' data.

Mitigating these vulnerabilities requires a comprehensive cybersecurity strategy that spans all areas of an organization's Autonomous Finance ecosystem, from the hardware and software to people and processes.

14.4 The Top 10 Cybersecurity Threats

Having explored the Autonomous Finance threat landscape, attention now turns to comprehending the specific threats that could disrupt an Autonomous Finance ecosystem. Using insights from numerous real-world incidents and expert knowledge, the top 10 cybersecurity threats, particularly prevalent and damaging Autonomous Finance, have been identified and are presented. This section seeks to bolster understanding these critical threats and highlight malicious actors' tactics and techniques. It aims to provide finance teams and stakeholders with a strategic perspective essential for proactive threat anticipation and mitigation.

Table 14.1 offers a critical dissection of the top ten cybersecurity threats imperiling the realm of Autonomous Finance, each elucidated with a case study, its business ramifications, and corresponding mitigation strategies. This table should not be construed as an exhaustive inventory but rather as a concentrated analysis of the most pressing threats as they currently manifest, offering critical insights for safeguarding the technological and financial ecosystems in the Autonomous Finance era.

TABLE 14.1: Top 10 Cybersecurity Threats in Autonomous Finance

NO.	THREAT	DESCRIPTION	CASE STUDY	BUSINESS IMPACT	MITIGATION STRATEGY
1.	**Data Breaches:**	Unauthorized access to, use, or disclosure of personal and financial data.	Equifax, 2017: Exposure of the personal information of 147 million people (Equifax Data Breach Settlement, 2022).	Financial loss, reputational damage, regulatory fines, and loss of customer trust.	Implementing robust access controls, encrypting sensitive data, regular audits, and employee training.
2.	**Phishing Attacks**	are A method of gathering sensitive information by pretending to be trustworthy.	Lithuanian man, 2016: Tricked two tech companies into wiring over $100 million (Gatlan, 2019).	Financial loss, loss of sensitive data, damage to brand reputation.	Employee training to identify phishing attempts, use of email filtering software, and two-factor authentication.
3	**AI-Based Attacks**	Cyberattacks leveraging AI to automate or enhance malicious activities.	DeepLocker, 2018: An AI-powered malware that avoided detection until a specific victim was identified (Harford, 2022).	Compromise of AI systems, advanced persistent threats, and large-scale data breaches.	AI-driven threat detection, updating and patching AI software, and encrypting sensitive data.
4	**Cloud-Based Attacks**	Attacks on data stored in the cloud.	Capital One, 2019: Hacker exploited a configuration vulnerability in the cloud, affecting 106 million people (2019 Capital One Cyber Incident | What Happened | Capital One, 2019/2022).	Data loss, business interruption, reputational damage.	Secure cloud configuration, robust access control, and encryptions.
5	**Ransomware Attacks**	A type of malicious software designed to block access to a computer system until a payment is made..	Colonial Pipeline, 2021: Hackers took systems offline, affecting fuel supply for days (Kerner, 2022).	Business interruption, financial loss, and potential harm to the public.	Regular backups, staff training, robust access controls, and up-to-date systems.
6	**API vulnerabilities**	Exploit the way different software applications communicate with each other.	Facebook, 2018: An API bug exposed the private photos of nearly 6.8 million users (Barrett, 2018).	Data breaches, loss of sensitive data, and unauthorized operations.	Regular testing, patching, access control, and encryption (Open Web Application Security Project, 2021).

NO.	THREAT	DESCRIPTION	CASE STUDY	BUSINESS IMPACT	MITIGATION STRATEGY
7	**Supply Chain Attacks**	Attacks that target less-secure elements in the network of suppliers.	SolarWinds, 2020: Attackers compromised software updates to breach the networks of numerous organizations (Kerner, 2023).	Data loss, unauthorized access to internal systems, and operational disruption.	Vendor risk management, network segmentation, and threat hunting (Cybersecurity & Infrastructure Security Agency, 2020).
8	**Insider Threats**	Malicious activities conducted from within the organization.	Twitter, 2020: Employee was manipulated into providing access to administrative tools, affecting high-profile accounts (Statt, 2020).	Data breaches, operational disruption, and financial loss.	Employee training, access control, and activity monitoring.
9	**Malware Infections**	Malicious software designed to cause harm to systems.	WannaCry, 2017: Ransomware attacks affected systems worldwide, causing billions in damages (What Is WannaCry Ransomware?, 2023).	Data loss, system damage, financial loss.	Regular system updates, antivirus software, and employee training.
10	**Distributed Denial of Service (DDoS)**	An attack that overwhelms systems, services, or networks with traffic, causing a denial of service.	GitHub, 2018: A DDoS attack spiked incoming traffic at 1.35 terabits per second (The Hacker News, 2018).	Service interruption, loss of availability, potential data loss.	Firewalls, traffic filtering, redundancy, and scalability.

14.5 Proactive Strategies for Cybersecurity

The dawn of Autonomous Finance ushers in an age of unprecedented potential and profound complexities. As financial transactions and decision-making processes become increasingly automated, the threat landscape evolves in parallel, demanding robust and proactive strategies to tackle potential cyber threats. Here are four critical proactive strategies organizations must consider in Autonomous Finance:

- Risks due to Regulatory and Industry Standards,
- Risks due to Data Protection and Privacy Threats,
- Risks to Infrastructure and Secure Transactions, and
- Risks with leveraging Advanced Technologies.

In the following sections, risk mitigation strategies for each threat listed above will be discussed in detail.

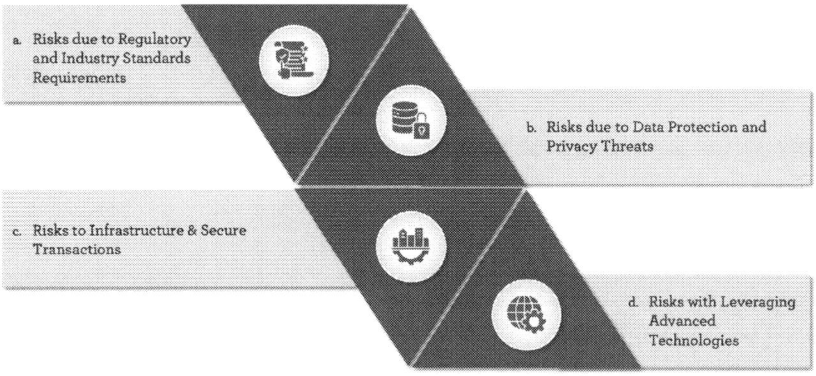

FIGURE 14.4: Proactive Strategies for Cybersecurity in Autonomous Finance Landscape

Regulatory and Industry Standards Requirements

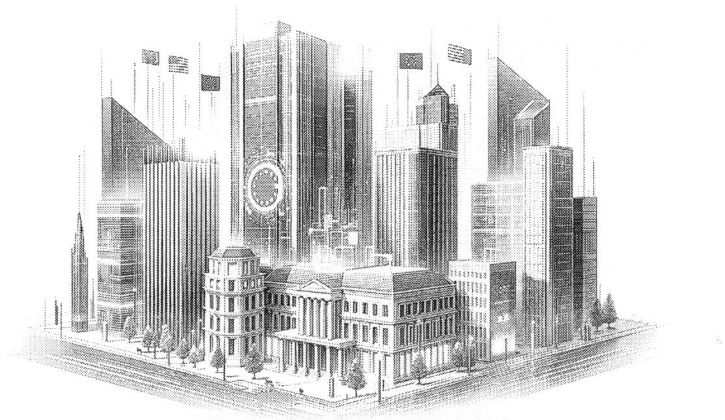

Regulatory adherence and industry standards form the cornerstone of any cybersecurity strategy. These frameworks provide a blueprint for creating a secure environment and enhancing consumer trust in Autonomous Finance initiatives. However, the rapidly evolving nature of Autonomous Finance requires an equally dynamic regulatory response.

The EU's GDPR and the U.S.'s Gramm-Leach-Bliley Act (GLBA) are significant in Autonomous Finance. GDPR, for example, enforces strict data protection and privacy rules, thereby compelling organizations to prioritize cybersecurity in their operations (European Union, 2018). GLBA mandates Finance teams to explain their data collection and sharing practices, thereby fostering a culture of transparency and accountability (*Financial Privacy Rule*, 2020).

Navigating the complex regulatory landscape can be daunting. Multiple jurisdictions, continuous updates, and the heterogeneous nature of regulatory frameworks may create compliance challenges. Organizations should focus on achieving 'Regulatory Intelligence' by leveraging technologies such as AI and ML to stay updated about regulatory changes, understand their impact, and adjust their cybersecurity strategies accordingly (Alston & Bird, 2022).

Industry standards, like the National Institute of Standards and Technology (NIST) Cybersecurity Framework, offer valuable guidelines for ensuring cybersecurity. The NIST framework's five pillars—Identify, Protect, Detect, Respond, and Recover—provide a comprehensive methodology to address cybersecurity threats (*Understanding the NIST Cybersecurity Framework*, 2022). Implementing these industry standards helps organizations build robust cybersecurity systems, foster a culture of cybersecurity awareness, and enhance their resilience against cyber threats.

Data Protection and Privacy Threats

Data is a crucial asset in Autonomous Finance. Finance teams leverage vast data volumes to fuel their AI and ML models, offering personalized services and enhancing operational efficiency. However, the reliance on data brings its own set of challenges, particularly concerning data protection and privacy.

Autonomous Finance systems, by their nature, collect, process, and store a wide range of sensitive customer data. The threat of data breaches and unauthorized access looms large. High-profile data breaches, such as the one at Capital One in 2019, highlight the importance of robust data protection strategies and the implications of inadequate data security measures (Murphy, 2022).

To safeguard data, organizations should consider a multi-faceted approach. Encryption protocols for data-at-rest and data-in-transit can help prevent unauthorized access. Regular security audits and vulnerability assessments can ensure continuous monitoring and early detection of potential threats. Moreover, technologies such as Differential Privacy can help organizations use sensitive data while minimizing the risk of privacy violations (Sweeney, 2023).

But ensuring data protection is only half the battle. Upholding data privacy is crucial, particularly in a global context with varying data privacy regulations. Privacy-by-design, a concept endorsed by the GDPR, encourages organizations to incorporate privacy measures right from the design phase of products or services (Cavoukian, 2011). Employing such a proactive approach helps address privacy concerns effectively and builds customer trust.

Risks to Infrastructure & Secure Transactions

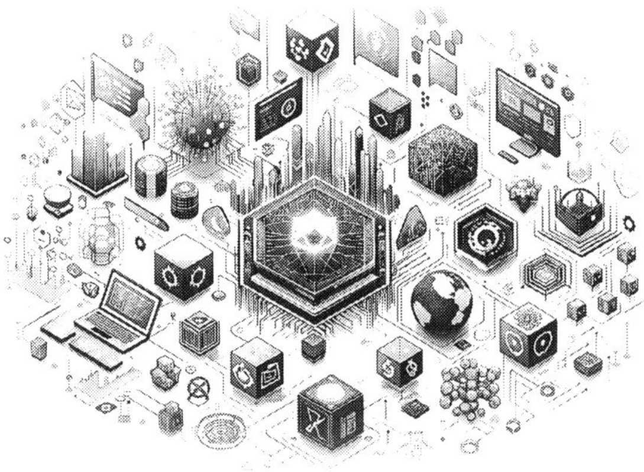

Building a resilient infrastructure is integral to the cybersecurity strategy of any Autonomous Finance system. Such systems should be designed to withstand cyber threats and ensure operational continuity even under adverse conditions.

Infrastructure resilience extends beyond technology; it covers organizational culture, processes, and governance mechanisms. Resilience should be ingrained in every aspect of the organization, from top-level strategy to operational activities (Bodeau et al., 2017). Regular risk assessments, disaster recovery planning, and incident response mechanisms are critical for achieving infrastructure resilience.

In parallel, the integrity of financial transactions is of utmost importance in Autonomous Finance systems. Security vulnerabilities can lead to fraudulent transactions, causing financial loss and damaging customer trust. In 2019, First American Financial Corp.'s data exposure incident served as a reminder of the consequences of a security flaw in transaction mechanisms (*First American Financial Corp. Leaked Hundreds of Millions of Title Insurance Records*, 2019).

Implementing secure transaction mechanisms involves several components. Secure coding practices can help prevent software vulnerabilities that could be exploited for fraudulent transactions. Furthermore, tokenization and cryptography utilization can enhance transaction security by rendering the data useless to hackers (TechMagic, 2023).

Risks with Leveraging Advanced Technologies

The cybersecurity landscape continuously evolves, with new threats emerging regularly. Autonomous Finance initiatives must embrace cutting-edge technologies like AI, ML, RPA, Big Data, Advanced Analytics, Cloud Computing, SaaS, and APIs.

AI and ML offer promising solutions for enhancing cybersecurity. These technologies can detect and predict cyber threats in real-time by analyzing patterns and anomalies in vast datasets. Early threat detection allows organizations to respond promptly, minimizing potential damage (Buczak & Guven, 2016).

With its decentralized and immutable nature, Blockchain technology provides a secure and transparent platform for conducting financial transactions. Its use in Autonomous Finance can increase transparency, reduce fraud, and streamline transactions, enhancing overall system security. Companies like Ripple are leading the way by leveraging blockchain for secure cross-border transactions (Mougayar, 2016).

In conclusion, adopting proactive strategies is vital for managing cybersecurity risks in the rapidly evolving landscape of Autonomous Finance. By understanding and aligning with regulatory frameworks, prioritizing data protection and privacy, building resilient infrastructure, and leveraging advanced technologies, organizations can navigate the path of Autonomous Finance with enhanced security and confidence.

14.6 Implementing a Cybersecurity Framework

Navigating the complex and evolving terrain of Autonomous Finance necessitates a comprehensive cybersecurity strategy. This is where a well-articulated cybersecurity framework comes into play. A robust framework helps organizations manage and mitigate risks and fosters resilience and adaptability in the face of emerging threats. This section outlines the key elements of such a framework and provides practical insights into its implementation.

Key Elements of a Robust Cybersecurity Framework

Autonomous Finance requires a comprehensive, adaptable, robust, and proactive cybersecurity framework. Figure 14.5 highlights that such a framework must encompass all areas of an organization and be ingrained in its culture, strategies, and operations. This subsection presents and expands on ten critical elements of a practical cybersecurity framework.

a) **Proactive Risk Management:** Autonomous Finance systems are not immune to vulnerabilities, and organizations can stay ahead of potential threats by proactively identifying and managing these risks. This involves performing regular risk assessments and audits to uncover and address potential weaknesses (ISO/IEC, 2018). It is an ongoing process that requires continual reevaluation as new threats and vulnerabilities emerge (NIST, 2022).

b) **Advanced Technical Controls:** The complexity of Autonomous Finance systems calls for advanced technical controls. Encryption, intrusion detection systems, multi-factor authentication, and real-time monitoring are some of the technical measures that can help protect against cyber threats (NIST, 2022). AI and ML can further enhance the ability to detect and respond to threats (Columbus, 2023).

c) **Secure Infrastructure Design and Deployment:** The architecture and deployment of an Autonomous Finance system play a pivotal role in its security. Security considerations must be integral to the system's design and deployment processes (OECD, 2021). This means embedding security measures into the infrastructure from the onset, including network segmentation, secure gateways, and security monitoring tools.

d) **Data Protection and Privacy:** With vast amounts of sensitive financial data, robust data protection and privacy measures are crucial. These include data encryption, secure data storage practices, and privacy-preserving computation techniques (Gupta & Shukla, 2023). Regulatory compliance with standards such as GDPR and CCPA is vital in ensuring data privacy (Tuteja, 2021).

e) **Secure Transactions and Processes:** Transactions in Autonomous Finance must be secured against various threats like fraud, identity theft, and data breaches. Secure transactions can be achieved through encryption, secure communication protocols, fraud detection systems, and secure payment gateways (*Visa Security | Securing the World's Payments*, 2023).

f) **Authentication and Authorization Frameworks:** Proper authentication and authorization mechanisms prevent unauthorized access to systems and data. Such measures include multi-factor authentication, biometric verification, and role-based access control systems (NIST, 2023).

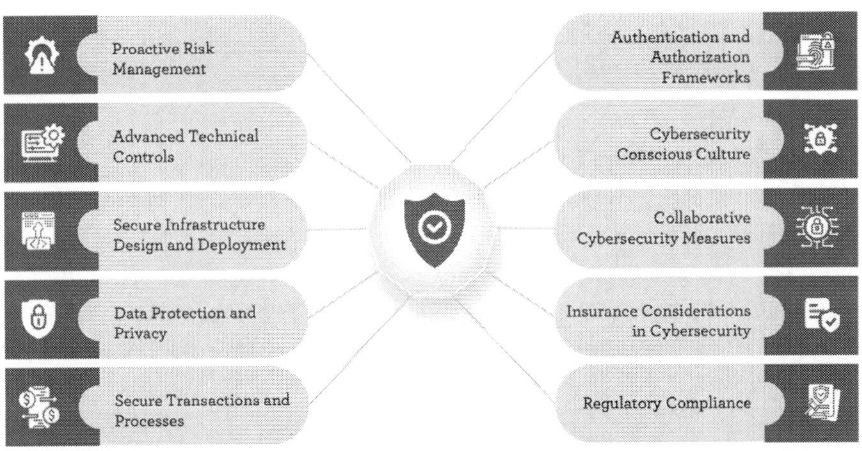

FIGURE 14.5: Cybersecurity Framework in Autonomous Finance Landscape

g) **Cybersecurity Conscious Culture:** The human factor remains critical to cybersecurity. Fostering a cybersecurity-conscious culture within the organization is essential (Americas, 2021). This involves comprehensive cybersecurity awareness and training for all employees and creating an environment where security is a shared responsibility. For additional details, please refer to section 9.7 below.

h) **Collaborative Cybersecurity Measures:** Organizations must adopt a collaborative approach to cybersecurity. This could involve partnerships with other organizations, industry bodies, and governments to share threat intelligence, best practices, and resources. Collaborative measures can significantly enhance an organization's cybersecurity resilience and effectiveness. For additional details, please refer to section 9.8 below.

i) **Insurance Considerations in Cybersecurity:** As part of a robust cybersecurity strategy, organizations should consider investing in cybersecurity insurance. Such insurance can provide critical financial support in a breach and incentivize proactive management of cybersecurity risks (*Perspectives on Transforming Cybersecurity*, 2017).

j) **Regulatory Compliance:** Compliance with industry standards and regulations is vital to any cybersecurity framework. Regulations such as GDPR and CCPA are particularly relevant. Compliance ensures legal protection and promotes customer trust (*2022 State of Cybersecurity - US | CompTIA*, 2022).

Organizations that invest in a comprehensive cybersecurity framework that covers these elements will be best positioned to navigate the evolving cyber threat landscape.

Implementing the Cybersecurity Framework

Implementing a robust cybersecurity framework is a strategic process that requires careful planning, resource allocation, and continuous monitoring and evaluation. The following steps, as outlined in Figure 14.6 below, provide a detailed approach to effectively operationalize the critical elements of the cybersecurity framework in Autonomous Finance.

a) **Conduct a Comprehensive Security Audit and Risk Assessment:** The first step is to conduct a thorough security audit and risk assessment to identify potential vulnerabilities and threats. This involves mapping out the organization's infrastructure, data flows, and security measures currently in place and testing them for weaknesses. The ISO/IEC 27001:2018 standard guides conducting risk assessments and audits (*ISO/IEC 27001 Standard – Information Security Management Systems*, 2018). The goal is to identify areas where security measures are lacking or insufficient and then prioritize remediation efforts based on the risk level.

b) **Implement Advanced Technical Controls and Secure Infrastructure:** Once the areas of improvement have been identified, the next step is to implement advanced technical controls and secure infrastructure design. This could involve deploying encryption technologies, setting up intrusion detection systems, network segmentation, secure gateways, and implementing real-time monitoring. Specific technologies will depend on the organization's unique requirements and risk profile. The NIST Cybersecurity Framework provides a useful guide for selecting and implementing appropriate controls (Understanding the NIST Cybersecurity Framework, 2022).

c) **Enhance Data Protection and Secure Transactions:** Organizations should establish stringent data protection measures to safeguard sensitive financial data. This can include advanced encryption techniques, secure data storage, and privacy-preserving computation techniques. Similarly, secure transactions can be ensured through secure communication protocols, fraud detection systems, and payment gateways. Compliance with data protection regulations like GDPR and CCPA is also essential (*Cyber and Data: Regulatory Challenges*, 2022).

d) **Establish Authentication and Authorization Frameworks:** The fortress of financial systems and data demands vigilant guardianship. Through the deployment of robust authentication and authorization frameworks, a secure barrier is erected, permitting entry only to those bearing the rightful credentials. This security architecture may encompass multi-factor authentication, biometric verification, and role-based access control systems, each layer adding to the resilience of the defense. Guided by the standards and insights from the NIST (NIST, 2023), these frameworks are the sentinels of integrity and trust within Autonomous Finance.

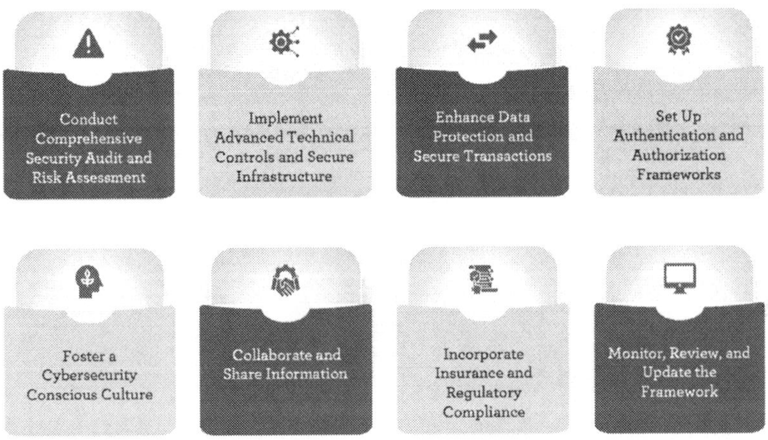

FIGURE 14.6: Implementing a Cybersecurity Framework in Autonomous Finance

e) **Foster a Cybersecurity Conscious Culture:** This step involves embedding cybersecurity awareness and training into the organization's culture. Regular briefings on the latest threats, phishing simulations, and interactive online training can be instrumental. Leadership should consistently communicate the importance of cybersecurity to all staff members and encourage a culture where security is a shared responsibility. For additional details, please refer to section 14.7 below.

f) **Collaborate and Share Information:** In cybersecurity, isolation is a vulnerability; collaboration is a strength. An organization transcends its capabilities by forging alliances with fellow organizations, industry guardians, and sovereign institutions. It becomes part of a collective intelligence, sharing insights into lurking threats, exchanging best practices, and marshaling resources. This confluence of minds and efforts fortifies the ramparts of cybersecurity and redefines resilience and efficacy. The fabric of this collaborative model is further unraveled in section 14.8, where readers can delve into the intricate tapestry of interconnected cybersecurity.

g) **Incorporate Insurance and Regulatory Compliance:** Implementing cybersecurity insurance can provide financial support in the event of a breach and incentivize proactive management of cybersecurity risks. Ensuring regulatory compliance provides legal protection and promotes customer trust (*Cyber and Data: Regulatory Challenges*, 2022).

h) **Monitor, Review, and Update the Framework:** Implementing the framework is not a one-time event; it requires ongoing monitoring and evaluation. Regular security audits and risk assessments should be conducted to identify new threats and vulnerabilities and update the framework accordingly. Keeping the framework adaptable and up-to-date is vital to maintaining robust cybersecurity in the face of evolving cyber threats.

Through careful planning and execution, a cybersecurity framework can provide a strong foundation for cybersecurity in Autonomous Finance, enabling organizations to navigate the risks of this rapidly evolving field confidently.

14.7 Fostering a Cybersecurity-Conscious Culture

Despite technological advances in Autonomous Finance, human factors remain integral to cybersecurity. Studies have consistently highlighted the critical role of human error in cybersecurity breaches. Consequently, fostering a cybersecurity-conscious culture is imperative to cybersecurity strategy in Autonomous Finance. This section explores the importance of cybersecurity awareness and training, and Figure 14.7 depicts the primary methods organizations leverage to foster a culture of security awareness.

FIGURE 14.7: Fostering a Cybersecurity-Conscious Culture in Autonomous Finance

Cybersecurity Awareness and Training for Autonomous Finance

As Autonomous Finance grows in complexity and influence, there is a corresponding need for comprehensive cybersecurity awareness and training. As it stands, 95% of cybersecurity breaches are due to human error, illustrating that technology alone cannot fully safeguard against threats (Cybint, 2022). Therefore, employees at all levels of an organization should have a foundational understanding of their role in preventing cybersecurity threats.

Effective cybersecurity training programs include regular briefings on the latest threats, phishing simulations, and interactive online training. These should be tailored to different organizational roles to ensure relevance and engagement. For example, IT staff may need more technical training, while executives may need training on strategic risk management.

One example of practical training is IBM's Cyber Range, a physical facility where organizations can simulate real-world cyber threats in a controlled environment. This hands-on, immersive approach ensures that training is engaging and practical (*IBM X-Force Cyber Range*, 2022).

Importance of Cybersecurity Education and Training

Cybersecurity education and training is not a one-time event but a continual process that adapts to the evolving threat landscape. As per a report by ISACA, a professional association for IT governance, risk, and cybersecurity, there will be a global shortage of 2 million cybersecurity professionals by 2023, highlighting the critical importance of ongoing education.

Organizations often face challenges in keeping cybersecurity training relevant and engaging. Gamification, where training is presented as a game or challenge, can increase participation and knowledge retention. In addition, professional certifications, such as the Certified Information Systems Security Professional (CISSP) or Certified Ethical Hacker (CEH), provide a clear framework for skills development and are highly valued in the industry.

Fostering a Culture of Security Awareness

Creating a culture of security awareness goes beyond training; it involves creating an environment where security is a shared responsibility and is integrated into all aspects of the organization. Employees should feel comfortable reporting potential threats, and there should be clear policies and procedures for handling security incidents.

A survey by EY found that organizations with a strong security culture had 70% fewer security incidents than those without (Amy, 2023). One successful case is that of Barclays, which implemented a security awareness program called 'The Digital Eagles,' which empowered employees to take personal responsibility for cybersecurity.

Companies can foster a culture of security awareness through regular communication from leadership about the importance of cybersecurity, incentives for following security protocols, and the celebration of 'security champions' who go above and beyond in their attention to cybersecurity issues.

Insurance Considerations in Cybersecurity

As part of a robust cybersecurity strategy, organizations should consider investing in cybersecurity insurance. With the constant rise in the magnitude and frequency of cyber threats, cybersecurity insurance serves as a critical financial lifeline in the aftermath of a breach.

Such insurance can cover a range of risks, from data breaches to business interruptions caused by cyber-attacks. However, choosing a cybersecurity insurance policy is a complex task that depends on the organization's risk profile, the sensitivity of the data it handles, and its overall cybersecurity posture.

Lower premiums are often available to organizations with robust cybersecurity measures in place. This incentivizes proactive management of cybersecurity risks and can lead to significant cost savings. For example, retail giant Target's insurance covered $90 million of the $162 million in expenses incurred from its infamous 2013 data breach (*Target Breach Costs: $162 Million*, 2015).

However, organizations should know that insurance is not a substitute for effective cybersecurity practices. It should be seen as a component of a comprehensive cybersecurity strategy that includes proactive measures such as risk assessments, training, and robust technical controls.

It is also essential to understand the fine print of any cybersecurity insurance policy, as coverage may not extend to all types of cyber incidents or costs. Legal advice may be beneficial when assessing policies.

In conclusion, while technology is vital in securing Autonomous Finance initiatives, the human element is equally important. Organizations that invest in comprehensive training programs, foster a culture of cybersecurity and leverage insurance as a part of their overall cybersecurity strategy will be best positioned to mitigate the risks of the evolving cyber threat landscape.

14.8 Collaborative Measures in Cybersecurity

As Autonomous Finance continues to expand and evolve, so does the threat landscape accompanying it. Traditional cybersecurity models that operate in isolation are increasingly insufficient in the face of advanced, persistent threats. The need for a more holistic, collaborative approach has never been more crucial. This section will explore the significance and potential of collaborative measures in Autonomous Finance cybersecurity, including the role of partnerships, collective resilience, and the power of information sharing.

Collaboration and Partnerships in Cybersecurity

In the cyber world, it is not an exaggeration to say that there is strength in numbers. Just as threats can emerge from numerous sources, so can defenses be built and strengthened through collective efforts. Collaboration and partnerships between organizations, industry bodies, regulatory entities, and governments can be powerful tools for mitigating cybersecurity risks, as suggested in Figure 14.8.

An example of successful collaboration can be seen in the Financial Services Information Sharing and Analysis Center (FS-ISAC), a global finance domain partnership focused on reducing cybersecurity risk through the sharing of threat intelligence (FS-ISAC, 2023). Such collaborations enable member organizations to stay informed of emerging threats and effective countermeasures, leading to a more robust cybersecurity posture.

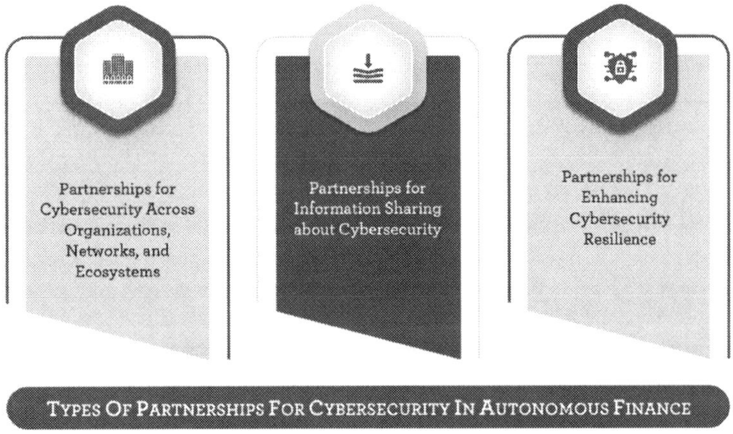

FIGURE 14.8: Collaborative Measures in Autonomous Finance Cybersecurity

Collaborative Approach to Enhancing Cybersecurity Resilience

An organization's ability to prepare for, respond to, and recover from cyber attacks is called Cybersecurity Resilience. Collaboration is vital in enhancing resilience, enabling organizations to pool resources, share best practices, and learn from each other's experiences.

Organizations can foster collaboration for enhanced resilience by participating in cross-industry cybersecurity exercises. For example, Quantum Dawn, a series of cybersecurity exercises conducted by the Securities Industry and Financial Markets Association (SIFMA), simulates large-scale cyber attacks on the finance domain to assess and improve collective response capabilities (SIFMA, 2023).

Partnerships and Information Sharing for Effective Cybersecurity

Sharing threat intelligence is a powerful way of strengthening defenses against cyber attacks. By sharing information about threats, vulnerabilities, and breaches, organizations can preemptively strengthen their defenses and respond more effectively when breaches occur.

One successful model for such sharing is the Cyber Threat Alliance (CTA), a group of cybersecurity providers and other organizations that share threat intelligence to improve defenses against advanced cyber adversaries. CTA members share real-time, actionable threat data, which helps all members respond more swiftly to emerging threats.

In conclusion, collaboration and partnerships are vital tools in the cybersecurity toolkit for Autonomous Finance. Organizations can significantly enhance their cybersecurity resilience and effectiveness by joining forces, sharing information, and pooling resources.

14.9 Conclusion

The dawning of Autonomous Finance marks an epochal shift that promises to rewrite the financial lexicography with terms such as 'hyper-efficiency,' 'personalization,' and 'universal financial inclusion.' Yet, this narrative of unprecedented promise lurks the shadow of equally unprecedented complexity—a complexity manifesting itself most acutely in cybersecurity.

This chapter navigated the labyrinthine corridors of cybersecurity within Autonomous Finance, dissecting a pantheon of threat actors, vulnerabilities, and potential attack vectors. In doing so, it aimed to arm financial institutions, technology partners, and regulators with a strategic arsenal that transcends reactive responses in favor of proactive threat anticipation and mitigation.

The observation by Elon Musk, *"A failure is an option here. If things are not failing, you are not innovating enough"* (Satara, 2018) serves as a philosophical lodestar as we envisage a future saturated with Autonomous Finance systems. It is a call to arms, an invocation for the cybersecurity community within Autonomous Finance to embrace an ethos of relentless innovation, even if that entails confronting and learning from failures.

To realize this vision, the chapter delineated a robust cybersecurity framework, encompassing compliance with evolving regulatory landscapes, intricate data protection mechanisms, and the deployment of next-generation cybersecurity technologies. It further urged the critical necessity for a cybersecurity-conscious organizational culture, emphasizing that the last line of defense in cybersecurity is often not a line of code but a human being equipped with both technological tools and ethical guidelines.

In the grand theater of Autonomous Finance, autonomous systems and humans are not mere actors but co-authors of an unfolding narrative. While the machine contributes computational prowess and real-time analytics, the human adds layers of ethical discernment, creative problem-solving, and societal context. This harmonious confluence will be the cornerstone of the ensuing chapter, exploring how this delicate balance of strengths could reshape the financial world.

14.10 Key Recommendations

a) **Secure-by-Design Approach:** Business and IT leaders should ensure that cybersecurity is integrated into every Autonomous Finance system development and operations stage. This approach should start from the initial design phase, integrating security protocols and controls within the architecture of the product or system instead of as an afterthought.

b) **Continuous Employee Training and Awareness:** Given the critical role that human error plays in cybersecurity breaches, continuous training and awareness programs are essential. This involves regular briefings on the latest threats, simulations, and other interactive training tailored to different organizational roles. Creating a security-conscious culture, where cybersecurity is viewed as a shared responsibility, should be a priority.

c) **Invest in Cybersecurity Insurance:** Cybersecurity insurance can provide critical financial support in the event of a breach and can incentivize proactive risk management. However, it should be treated as a component of a comprehensive cybersecurity strategy rather than a substitute for effective cybersecurity practices.

d) **Establish Collaborative Partnerships:** By collaborating with other organizations, industry bodies, and governments, companies can enhance their ability to prepare for, respond to, and recover from cyber-attacks (Connelly et al., 2017). Sharing threat intelligence and best practices can significantly strengthen an organization's cybersecurity posture and resilience.

e) **Regular Monitoring and Updating of the Framework:** Cybersecurity is a continuous process, not a one-time event. Regular security audits, risk assessments, and updates to the cybersecurity framework are necessary to identify and address new threats and vulnerabilities. This ensures the framework remains adaptable and robust in an evolving threat landscape.

14.11 References

2019 *Capital One Cyber incident | What happened | Capital One*. (2019/2022). Capital One. https://www.capitalone.com/digital/facts2019/

2022 State of Cybersecurity - US | CompTIA. (2022). CompTIA.org. https://www.comptia.org/content/research/cybersecurity-trends-research

Al-Rubaie, M., & Chang, J. M. (2019). Privacy-preserving machine learning: Threats and solutions. *IEEE Security & Privacy*, *17*(2), 49-58.

Alston & Bird. (2022). AI Regulation in the U.S.: What's Coming, and What Companies Need to Do in 2023. *www.alston.com*. https://www.alston.com/en/insights/publications/2022/12/ai-regulation-in-the-us

Americas, E. (2021). *Global Information Security Survey (GISS)*. https://www.ey.com/en_us/giss

Amy. (2023). *10 facts about the benefits of Security Awareness Training | ID Agent*. ID Agent. https://www.idagent.com/blog/10-facts-about-the-benefits-of-security-awareness-training/

Barrett, B. (2018). Facebook Exposed 6.8 Million Users' Photos to Cap Off a Terrible 2018. *WIRED*. https://www.wired.com/story/facebook-photo-api-bug-millions-users-exposed/

Biggio, B., & Roli, F. (2018). Wild patterns: Ten years after the rise of adversarial machine learning. *Pattern Recognition, 84*, 317-331. https://doi.org/10.48550/arXiv.1712.03141

Bodeau, D. J., Graubart, R., Picciotto, J., & Mcquaid, R. (2017). *Cyber Resiliency Engineering Framework*. MITRE Corporation.

Buczak, A.L. & Guven, E. (2016). "A Survey of Data Mining and Machine Learning Methods for Cyber Security Intrusion Detection," in IEEE Communications Surveys & Tutorials, vol. 18, no. 2, pp. 1153-1176, https://ieeexplore.ieee.org/document/7307098.

Cavoukian, A. (2011). *Privacy by Design: The 7 Foundational Principles*. Information and Privacy Commissioner of Ontario, Canada.

Chen, H., Chiang, R. H., & Storey, V. C. (2012). Business Intelligence and Analytics: From Big Data to Big Impact. *MIS Quarterly, 36*(4). https://doi.org/10.2307/41703503

Choo, K. K. R. (2011). The cyber threat landscape: Challenges and future research directions. Computers & Security, 30(8), 719-731. https://doi.org/10.1016/j.cose.2011.08.004

Columbus, L. (2023). IBM study reveals how AI automation protects enterprises against data breaches. *VentureBeat*. https://venturebeat.com/security/ibm-study-reveals-how-ai-automation-protect-enterprises-against-data-breaches/

Combating threats in the cyber world: outsmarting terrorists, hackers, and spies. (2012). In *FBI*. RSA Cyber Security Conference, San Francisco, CA, United States of America. https://archives.fbi.gov/archives/news/speeches/combating-threats-in-the-cyber-world-outsmarting-terrorists-hackers-and-spies

Connelly, E. B., Allen, C. R., Hatfield, K., Palma-Oliveira, J. M., David, •, Woods, D., & Linkov, I. (2017). Features of resilience. Environment Systems and Decisions, 37(1), 46–50.

Cyber and data: Regulatory challenges. (2022). KPMG. https://advisory.kpmg.us/articles/2022/ten-key-regulatory-challenges-2022-cyber-data.html

Cybersecurity *& Infrastructure Security Agency*. (2020). Supply chain compromise. https://www.cisa.gov/supply-chain-com.

Cybint. (2022). Cybersecurity Facts, Figures & Statistics for 2022. *Cybint.com*. https://www.cybintsolutions.com/cyber-security-2.

DBIR *Report 2023 - Introduction | Verizon Business*. (2023). Verizon Business. https://www.verizon.com/business/resources/reports/dbir/2023/introduction/

Deloitte. (2020). *Regulatory Technology* (RegTech). Deloitte.

Equifax *Data breach settlement*. (2022). Federal Trade Commission. https://www.ftc.gov/enforcement/refunds/equifax-data-breach-settlement

European Union. (2018). Regulation (EU) 2016/679 of the European Parliament and of the Council of 27 April 2016 on the protection of natural persons with regard to the processing of personal data and on the free movement of such data, and repealing Directive 95/46/EC (General Data Protection Regulation). *Official Journal of the European Union, L 119*, 1-88. https://eur-lex.europa.eu/legal-content/EN/TXT/?uri=CELEX%3A32016R0679.

Financial Privacy rule. (2020, March 6). Federal Trade Commission. https://www.ftc.gov/legal-library/browse/rules/financial-privacy-rule

First American Financial Corp. leaked hundreds of millions of title insurance records. (2019). https://krebsonsecurity.com/2019/05/first-american-financial-corp-leaked-hundreds-of-millions-of-title-insurance-records/

Gatlan, S. (2019, March 20). Lithuanian pleads guilty to stealing $100 million from Google and Facebook. *BleepingComputer*. https://www.bleepingcomputer.com/news/security/lithuanian-pleads-guilty-to-stealing-100-million-from-google-facebook/

Gupta, P., & Shukla, S. (2023). *Privacy-Preserving Techniques in Cloud Computing*. IGI Global.

Harford, I. (2022). Does AI-powered malware exist in the wild? Not yet. *Security.* https://www.techtarget.com/searchsecurity/tip/Does-AI-powered-malware-exist-in-the-wild-Not-yet

Hjortdal, M. (2011). China's use of cyber warfare: Espionage meets strategic deterrence. *Journal of Strategic Security, 4*(2), 1-24. http://dx.doi.org/10.5038/1944-0472.4.2.1

IBM X-Force Cyber Range. (2022). https://www.ibm.com/services/xforce-cyber-range

ISO/IEC 27001 Standard – Information Security Management Systems. (2018). ISO. https://www.iso.org/standard/27001

Kerner, S. M. (2022). Colonial Pipeline hack explained: Everything you need to know. *WhatIs.com*. https://www.techtarget.com/whatis/feature/Colonial-Pipeline-hack-explained-Everything-you-need-to-know

Kerner, S. O. S. M. (2023). SolarWinds hack explained: Everything you need to know. *WhatIs.com*. https://www.techtarget.com/whatis/feature/SolarWinds-hack-explained-Everything-you-need-to-know

Kocher, P., Horn, J., Fogh, A., Genkin, D., Gruss, D., Haas, W., ... & Yarom, Y. (2020). Spectre attacks: Exploiting speculative execution. arXiv preprint *arXiv:1801.01203*. https://doi.org/10.48550/arXiv.1801.01203

Mougayar, W. (2016). *The Business Blockchain: Promise, Practice, and Application of the Next Internet Technology*. Wiley.

Murphy, D. (2022). Capital One data breach highlights the importance of Data-Centric security. *Lepide Blog: A Guide to IT Security, Compliance and IT Operations*. https://www.lepide.com/blog/capital-one-data-breach-highlights-importance-of-data-centric-security/

NIST. (2023). Guide to role-based access control (RBAC). National Institute of Standards and Technology. https://csrc.nist.gov/projects/role-based-access-control

OECD. (2021). Artificial Intelligence, Machine Learning and Big Data in Finance - *OECD. www.oecd.org*. https://www.oecd.org/finance/artificial-intelligence-machine-learning-big-data-in-finance.htm

Open Web *Application Security Project*. (2021). API Security. https://owasp.org/www-project-api-security/

Perspectives on transforming cybersecurity. (2017). McKinsey & Company. https://www.mckinsey.com/capabilities/risk-and-resilience/our-insights/perspectives-on-transforming-cybersecurity

Probst, C. W., Hunker, J., Gollmann, D., & Bishop, M. (2010). Aspects of Insider Threats. In *Insider Threats in Cyber Security* (pp. 1-15). Springer, Boston, MA.

Satara, A. (2020). In 2 sentences, Elon Musk explains why the key to success is failure. *Inc.com*. https://www.inc.com/alyssa-satara/in-2-sentences-elon-musk-explains-why-key-to-success-is-failure.html

SIFMA. (2023). *Cybersecurity Exercise: Quantum Dawn V - SIFMA*. https://www.sifma.org/resources/general/quantum-dawn-v/.

Statt, N. (2020). Twitter reveals that its employee tools contributed to an unprecedented hack. *The Verge*. https://www.theverge.com/2020/7/15/21326656/twitter-hack-explanation-bitcoin-accounts-employee-tools

Sweeney, M. (2023). The future of AdTech lies in Privacy-Enhancing technologies (PETs). *Clearcode | Custom AdTech and MarTech Development*. https://clearcode.cc/blog/adtech-future-privacy-enhancing-technologies/

Target breach costs: $162 million. (2015). BankInfoSecurity. https://www.bankinfosecurity.com/target-breach-costs-162-million-a-7951

TechMagic. (2023). Payment tokenization: How to do it properly? *www.linkedin.com.* https://www.linkedin.com/pulse/payment-tokenization-how-do-properly-techmagic/

The Hacker News. (2018). *Biggest-ever DDoS attack (1.35 tbs) hits GitHub website.* https://thehackernews.com/2018/03/biggest-ddos-attack-github.html

Tuteja, A. (2021). Protecting the privacy frontier. *KPMG.* https://kpmg.com/xx/en/home/insights/2021/11/protecting-the-privacy-frontier.html

Understanding the NIST cybersecurity framework. (2022). *Federal Trade Commission.* https://www.ftc.gov/business-guidance/small-businesses/cybersecurity/nist-framework

Visa Security | Securing the world's payments. (2023). Visa. https://usa.visa.com/run-your-business/visa-security.html

What is WannaCry ransomware? (2023). *usa.kaspersky.com.* https://usa.kaspersky.com/resource-center/threats/ransomware-wannacry

CHAPTER 15

The Symbiotic Dance

Autonomous Finance and the Future of Work

"For good ideas and true innovation, you need human interaction, conflict, argument, debate."

- MARGARET HEFFERNAN (HEFFERNAN, 2012)

15.0 Key Insights

- **The Mutual Dance: Autonomous Finance and Future of Work:** The relationship between Autonomous Finance and the Future of Work is not one-directional. They shape each other, creating a dance of mutual influence. This reciprocal relationship will shape the future of organizations.
- **The Changing Landscape: Roles and Responsibilities:** Autonomous Finance will cause significant shifts in roles and responsibilities in finance teams. As AI takes over routine tasks, humans must focus on higher-level tasks requiring judgment, creativity, and strategic thinking.
- **The Skills Revolution: New Skills for the New Era:** The rise of Autonomous Finance will necessitate a skills revolution. As AI integrates into finance operations, employees must develop new technical skills, data literacy, and soft skills such as change management.
- **Guarding the Fort: Cybersecurity in the Age of AI:** As finance operations become more automated and digital, cybersecurity will become a key concern. Ensuring the security of financial transactions and customer data will be crucial in an era of Autonomous Finance.
- **AI Governance:** The Need for Transparency and Fairness - With the increasing use of AI in finance, the importance of AI governance cannot be overstated. Companies must ensure their AI systems are transparent, fair, and accountable to maintain trust and compliance.

15.1 Introduction

In the wake of the previous chapter's exploration into the complexities of cybersecurity within Autonomous Finance, it becomes evident that this emergent paradigm is a multifaceted entity. It is not merely an innovation in financial technologies but a transformative wave synchronizing with other societal shifts, notably the Future of Work. This intricate ballet of mutual influence and adaptation forms the cornerstone of the current chapter. This choreography is continually reshaping professional milieus, modulating methodologies, and redefining the very essence of finance.

Pioneered by a constellation of progressive technologies—AI, ML, RPA, Big Data, SaaS, Cloud Computing, Advanced Analytics, and APIs—Autonomous Finance heralds a new era. It transcends traditional financial paradigms to catalyze a financial ecosystem that is self-driving, self-learning, and self-adapting, capable of real-time operations and services.

Parallel to the rise of Autonomous Finance, the Future of Work is also experiencing a seismic shift. Conventional roles are metamorphosing into specialized positions like AI Systems Trainers, Data Ethics Analysts, and Cybersecurity Specialists. According to the World Economic Forum, by 2025, automation will displace 85 million jobs while creating 97 million new roles (World Economic Forum, 2020b). This landscape is not one of dystopian joblessness but rather a dynamic vision of job transition, transformation, and creation.

However, this transformative era is fraught with challenges. A widening skill gap calls for an unprecedented scale of reskilling and upskilling. The specter of cybersecurity, as elaborated in the preceding chapter, demands robust and agile protective measures. An ever-evolving regulatory framework necessitates adaptive compliance strategies, while data privacy emerges as a non-negotiable aspect requiring stringent governance.

This chapter aims to provide a panoramic view of the interplay between Autonomous Finance and the Future of Work. It dissects this complex dance through the multifaceted perspectives of industry leaders, policy-makers, and the workforce. Moreover, it concretizes these discussions through a case study focused on the A/R function, offering tangible insights into how roles, skills, and organizational structures are evolving.

Invoking the wisdom of computer scientist Alan Kay, *"We cannot predict the future, but we can invent it"* (Gabor, 1963), this chapter invites you. The future landscape of Autonomous Finance and the Future of Work remains a blank canvas, awaiting your innovative strokes and choreographic mastery. As you peruse the insights and recommendations herein, you are invited to join this dance, resonating with the tempo of change to carve out a future imbued with vision, wisdom, and ethical integrity.

15.2 Autonomous Finance Reshaping the Future of Work

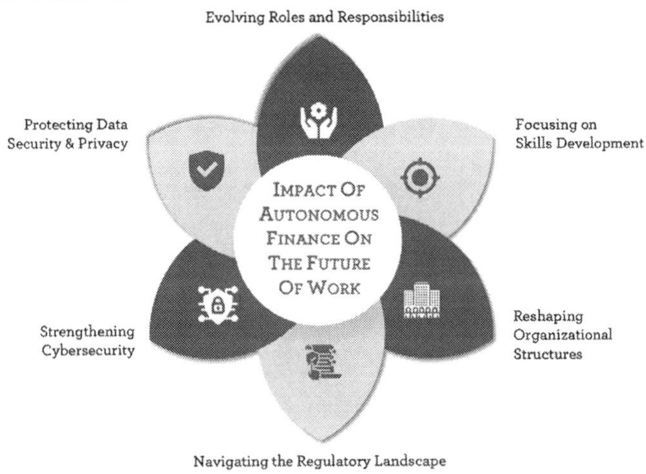

FIGURE 15.1: The Impact of Autonomous Finance on the Future of Work

With its AI-driven capabilities, Autonomous Finance already significantly impacts the Future of Work, as shown in Figure 15.1. The first and most noticeable impact is the evolution of roles and responsibilities. Routine tasks that human employees once performed are now being automated. This trend has led to new roles, such as AI Trainer and AI Ethics Officer, while traditional roles are redefined. According to the World Economic Forum, 75 million current job roles may be displaced by the shift in the division of labor between humans and machines by 2022, but 133 million new job roles may emerge simultaneously (World Economic Forum, 2018).

The second impact of Autonomous Finance on the Future of Work is in the area of skill development. As AI takes over routine tasks, the demand for soft skills such as creativity, critical thinking, and complex problem-solving is increasing. LinkedIn's 2020 Workplace Learning Report indicates that 57% of senior leaders today say soft skills are more critical than hard skills (LinkedIn, 2020).

Thirdly, Autonomous Finance is reshaping organizational structures. Traditional hierarchical structures give way to network-based structures with cross-functional teams collaborating on complex tasks. Microsoft's CEO Satya Nadella stated, *"As work becomes more digital, the flow of digital information becomes more important"* (Nadella, 2018b). This shift necessitates new forms of collaboration and communication facilitated by AI.

The fourth impact is on the regulatory landscape. Autonomous Finance, with its potential for rapid decision-making and execution, often operates at the cutting edge of what is legally permissible. This rapid evolution requires regulatory agility, with laws and regulations needing to keep pace with technological innovation.

The fifth impact is on cybersecurity. Advanced cybersecurity protocols are paramount with increased data flow and interconnectivity between systems. A report from the Capgemini Research Institute reveals that 69% of organizations acknowledge that they cannot respond to critical threats without AI (Capgemini, 2019).

Lastly, the rise of Autonomous Finance has brought data privacy and security into sharper focus. As AI systems handle more sensitive financial data, safeguarding this data becomes crucial. This is reflected in Tim Cook's statement that *"Privacy is a fundamental human right"* (Cook, 2018).

In conclusion, Steve Jobs once said, *"Innovation distinguishes a leader from a follower"* (Jobs, 2005). In the dance between Autonomous Finance and the Future of Work, those who can navigate these impacts with innovation and foresight will lead the way.

15.3 The Future of Work Steering Autonomous Finance

It is evident from Figure 15.2 that the Autonomous Finance ecosystem is being shaped by the Future of Work. The rise of remote work, accelerated by recent global events, is a significant driver. A Gartner survey reveals that 82% of company leaders plan to allow employees to work remotely for at least part of the time after the pandemic (Gartner, 2020). This trend pushes businesses toward greater automation and digitalization of financial processes to support distributed workforces.

As work becomes increasingly flexible and digital, the demand for 24/7 financial services is growing. Robo-advisors, capable of managing financial portfolios round the clock, are expected to manage $4.6 trillion by 2022. This expectation is driving the expansion of Autonomous Finance capabilities to provide anytime, anywhere financial services.

The Future of Work is also prompting a stronger emphasis on cybersecurity. With employees accessing financial systems from various locations and devices, the risk of cyber threats increases. This risk is met with AI-enhanced cybersecurity measures, further integrating Autonomous Finance into the work landscape (Capgemini Research Institute, 2019).

Moreover, the shift toward a digital-first work environment is leading to a surge in data, which Autonomous Finance systems can leverage for better predictive modeling, risk assessment, and decision-making. Satya Nadella, the CEO of Microsoft, has described data as *"the new electricity"* (Nadella, 2018b), highlighting its central role in powering AI systems.

Conversely, the Future of Work also highlights the importance of data privacy and security in Autonomous Finance. Mark Zuckerberg noted that *the future is private* (Zuckerberg, 2019), underlining the need for Autonomous Finance systems to prioritize data protection as they handle increasingly sensitive financial data.

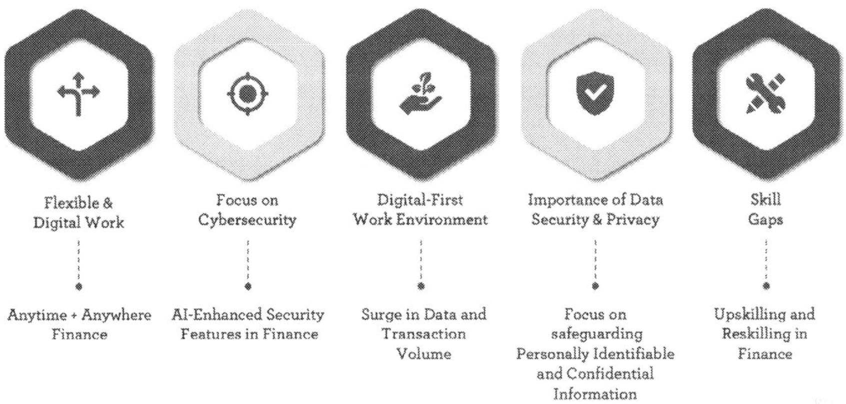

FIGURE 15.2: The Impact of the Future of Work on Autonomous Finance

Finally, as the Future of Work evolves, so do the skills needed in finance. The World Economic Forum predicts that by 2022, 54% of all employees will require significant re- and upskilling (World Economic Forum, 2018). As Autonomous Finance takes over routine tasks, the focus for human workers is shifting toward skills that complement AI, such as oversight, ethical decision-making, and complex problem-solving.

As Steve Jobs once said, *"Technology is nothing. What's important is that you have faith in people, that they're good and smart, and if you give them tools, they'll do wonderful things with them"* (Jobs, 2005). In this context, the Future of Work and Autonomous Finance is not just about technology but about empowering people to do wonderful things with new tools.

15.4 The Human-Machine Dance in Autonomous Finance

In the grand performance of Autonomous Finance and the Future of Work, the harmonious collaboration of humans and machines takes center stage. As the curtain rises, a symphony of unique strengths and abilities is witnessed, each enhancing and complementing the other. This extraordinary fusion of humans and AI is not a mere experiment but a carefully choreographed dance guided by principles that ensure a graceful and ethical performance. These principles are shown in Figure 15.3, and the strengths of Humans and AI in each are outlined in Figure 15.4.

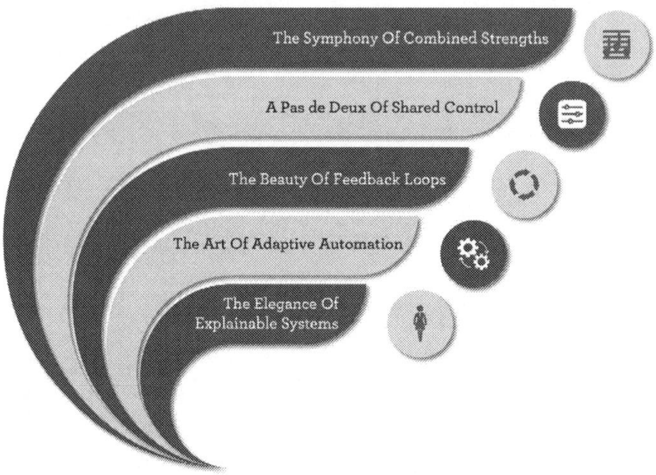

FIGURE 15.3: The Human-Machine Dance in Autonomous Finance

The Symphony of Combined Strengths

As the performance unfolds, the dancers – humans and AI – move together ideally. With their strategic insight, humans' profound understanding of business context and ability to provide oversight share the stage with AI's impressive ability to rapidly analyze vast volumes of financial data, detect patterns, and execute tasks with unprecedented speed and scale. Each dancer enhances the other, creating an efficient and effective performance. In the words of visionary entrepreneur Elon Musk, *"Humans and AI will be symbiotes, all linked via a neural lace"* (Carmichael, 2016).

A Pas de Deux of Shared Control

As the dance progresses, the principle of shared control is apparent. The human and AI dancers move together, guiding and responding to each other. In this partnership, AI systems handle routine tasks while humans oversee the overall strategy and ensure ethical considerations are met. A robo-advisor, for example, might handle the task of rebalancing a portfolio. At the same time, a human advisor guides the overall investment strategy and ensures the client's best interests are at the forefront.

The Beauty of Feedback Loops

The dance of humans and AI is not a one-time performance. It is an ongoing evolution, a series of movements and counter-movements continually refining choreography. This dance is characterized by feedback loops, where humans provide corrections and additional context to improve automated systems. AI systems use this feedback to enhance their performance and adapt to the changing rhythm of the finance domain and teams.

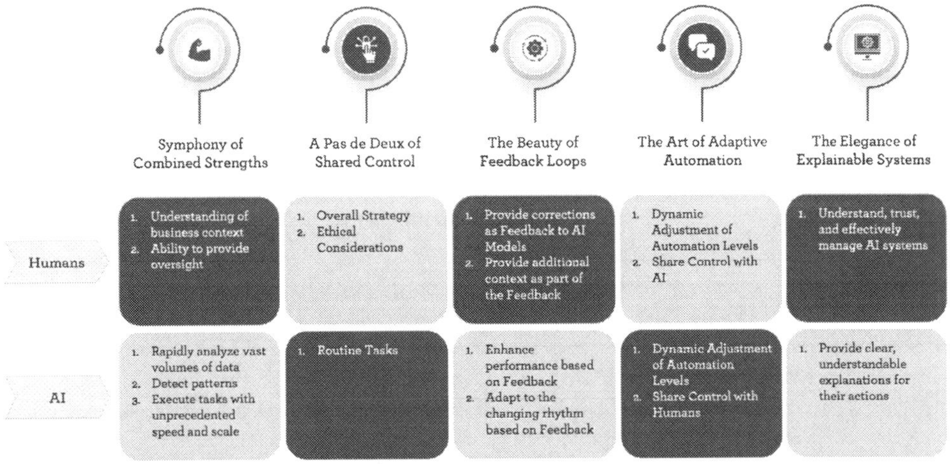

FIGURE 15.4: Guiding Principles for the Human-Machine Dance in Finance

The Art of Adaptive Automation

The dance of Autonomous Finance and the Future of Work is not a rigidly choreographed performance. It is fluid and adaptive, responding to the changing rhythms and dynamics of the financial world. Automation levels are adjusted dynamically based on the situation, with humans and AI systems sharing control and responding to each other's moves. This adaptive automation ensures that the dance stays fresh, relevant, and responsive to the ever-evolving financial landscape.

The Elegance of Explainable Systems

Transparency and interpretability are central to the dance of humans and AI. As the AI systems move and make decisions, they don't do so in a cloud of mystery. Instead, they provide clear, understandable explanations for their actions – a principle known as explainability. This ensures that humans can understand, trust, and effectively manage AI systems, ensuring a harmonious and successful performance.

In this grand performance of Autonomous Finance and the Future of Work, these guiding principles ensure a graceful, efficient, and ethical dance. They ensure that the fusion of humans and AI is not just a spectacle but a powerful, transformative force in finance.

15.5 Navigating the Cross-Roads: Challenges & Strategies

In the intricate dance between Autonomous Finance and the Future of Work, there are several crossroads that organizations must navigate. These intersections pose unique challenges that require careful choreography and strategic maneuvering. Table 15.1 thoroughly explores these challenges and presents a roadmap of strategies to address each. The aim is to provide a practical guide for organizations to ensure a smooth and harmonious dance between humans and machines.

TABLE 15.1: Challenges in the Human-Machine Dance & Strategies to Address Them

THE CHALLENGES	STRATEGY TO ADDRESS THE CHALLENGE
Changes in Roles and Responsibilities The dance of Autonomous Finance and the Future of Work brings to the fore the dynamic and rapidly evolving landscape of roles and responsibilities. A report by McKinsey suggests that AI and automation technologies could affect 375 million jobs by 2030. As AI becomes more prevalent in finance, traditional roles will likely be redefined, and entirely new roles will emerge. As Klaus Schwab, Founder of the World Economic Forum, aptly puts it, *"We stand on the brink of a technological revolution that will fundamentally alter the way we live, work, and relate to one another"* (Schwab, 2016). The challenge lies in ensuring a smooth transition, maintaining human value within the workforce, and integrating humans and AI into a symbiotic relationship.	**Embrace the Change** As Autonomous Finance continues to evolve, roles and responsibilities within the financial industry are bound to change. Automation will likely eliminate some tasks, but it will also create opportunities for new roles that require uniquely human skills. Individuals and organizations must be flexible and adaptive to navigate these shifts successfully. Organizations can ease this transition by fostering a culture of continuous learning and maintaining an open dialogue about the changing nature of work. This might involve regular training sessions, mentorship programs, or even opportunities for job rotation. By helping employees understand and adapt to their changing roles, organizations can ensure a smoother transition to Autonomous Finance. (World Economic Forum, 2020a)
The Evolution of Skills and the Gap Dilemma With the evolution of roles comes the evolution of skills. The World Economic Forum's Future of Jobs Report indicates that by 2025, machines will perform more current work tasks than humans, compared to 71% being performed by humans (World Economic Forum, 2020b). In such a scenario, hard skills like data analysis might decrease demand, while soft skills such as creativity, critical thinking, and emotional intelligence will become increasingly valuable. However, this transition presents a significant challenge: the skills gap. As the demand for new skills outpaces the growth in their supply, organizations and workers will need to navigate a potentially widening skills gap.	**Invest in Skills Development** In the Future of Work in Autonomous Finance, both technical and soft skills will be in high demand. Technical skills will be needed to interact with and manage AI systems. In contrast, soft skills like emotional intelligence, creativity, and critical thinking will become even more valuable as they are difficult to automate. Organizations should invest in comprehensive reskilling and upskilling initiatives to prepare for this future. This could involve partnering with educational institutions to provide relevant training programs, offering online courses, or setting up internal training departments. Investing in skill development helps employees stay relevant in the changing workplace and benefits organizations by ensuring they have the skills needed to thrive in the era of Autonomous Finance. (World Economic Forum, 2020a)
Shifting Organizational Structures As roles and skills evolve, so too will the structures within which they operate. Traditional hierarchical structures may give way to more flexible and collaborative team-based structures. Research by Deloitte suggests that *"instead of a series of linear career paths, they [organizations] are shifting toward a more dynamic approach to work"* (Deloitte Insights, 2019). This shift toward fluidity and flexibility presents challenges, including maintaining clear communication, ensuring accountability, and fostering a cohesive organizational culture in an increasingly virtual and dispersed work environment.	**Rethink Organizational Structures** The integration of AI in finance will necessitate rethinking organizational structures. Traditional hierarchical models may give way to more flexible and adaptive structures that facilitate collaboration between humans and machines. For instance, instead of having separate teams for AI and human workers, organizations might consider integrating AI systems across different teams and functions. This enhances the capabilities of diverse groups and facilitates the seamless coordination of workflows.
Dancing with Regulatory Frameworks As Autonomous Finance stands on the cutting edge of technology, it presents unique regulatory challenges. The need is for compliance with existing laws and regulations and advocacy for new regulations supporting innovation in Autonomous Finance. The rapid pace of technology change often means regulatory frameworks are constantly in flux, creating uncertainty. A study by PwC found that 78% of CEOs consider over-regulation a significant threat to their organization's growth (PwC, 2021). The challenge lies in maintaining regulatory agility and shaping a legal environment conducive to the growth of Autonomous Finance.	**Navigating the Regulatory Ballet with Agility** Regulatory compliance is a significant challenge in the face of rapidly evolving AI technologies. To stay agile in a changing regulatory landscape, organizations should stay informed about new developments in the field and advocate for supportive legislation. This might involve hiring legal experts specializing in AI law, participating in industry forums, or even engaging with policymakers to promote regulations supporting innovation in Autonomous Finance.

THE CHALLENGES	STRATEGY TO ADDRESS THE CHALLENGE
Cybersecurity: The Hidden Dance Partner The increased data flow and interconnectivity in Autonomous Finance bring cybersecurity to the forefront. The World Economic Forum's Global Risks Report 2021 ranks cyber threats among the most significant risks (World Economic Forum, 2021). Protecting sensitive financial data, ensuring the integrity of systems and networks, and fostering a culture of cybersecurity awareness pose considerable challenges in this new dance.	**Strengthen Cybersecurity Measures** With increased data flow and interconnectivity in Autonomous Finance, robust cybersecurity measures are essential to protect sensitive financial data. Organizations should implement advanced security protocols, educate employees about potential threats, and promote a culture of cybersecurity awareness. These measures can help fortify the organization's defenses against potential cyber threats. (World Economic Forum, 2020a)
Data Privacy: The Unseen Choreographer In an era where data is the new oil, data privacy has become increasingly important. Research by Cisco showed that 84% of consumers want more control over how their personal information is used. Balancing the need for data to fuel AI systems with the need to respect individuals' privacy rights is a significant challenge in the world of Autonomous Finance.	**Prioritize Data Security and Privacy** The protection of personal data and privacy is paramount in the age of AI. Organizations should adhere to data privacy regulations and implement secure data handling procedures to ensure financial data safety and personal and sensitive information. Furthermore, educating employees about the importance of data privacy can help cultivate a culture of respect for privacy and confidentiality. (IBM, 2020)
The Delicate Balance of AI Governance The rise of AI in finance necessitates robust governance of AI models. This involves ensuring ethical AI decisions, transparency, and accountability in AI-driven decision-making and managing the risks associated with AI, such as bias and unfairness. As the World Economic Forum's White Paper on AI Governance suggests, AI governance should be characterized by *"transparency, explainability, fairness, and accountability"* (World Economic Forum, 2020a). However, translating these principles into practice requires vigilance and continuous oversight.	**Implement Effective AI Governance** Implementing effective AI governance is crucial to ensure that AI systems are used ethically and responsibly. Organizations should establish guidelines for ethical AI use, implement transparency measures, and set up ethics review boards. These measures can help prevent the misuse of AI and ensure that AI technologies are used in a way that benefits everyone.

The dance between Autonomous Finance and the Future of Work is complex and ever evolving. It is characterized by a continuous interplay of challenges and opportunities that require a dynamic and flexible approach. By understanding these challenges and adopting the strategies outlined in this section, organizations can not only navigate the dance successfully but also lead the way in shaping the future of finance. The journey may be intricate, and the dance complex, but the rhythm can be found with the right strategies, and a harmonious performance can be achieved. This can help drive innovation, enhance productivity, and foster a more inclusive and fulfilling future of work.

15.6 A Case Study of Accounts Receivables

As the exploration of Autonomous Finance and the Future of Work continues, the abstract concepts under study begin to crystallize when applied to a specific function within finance. The A/R function is a powerful lens to visualize the potential shifts and transformations looming ahead.

This part of the discussion aims to analyze the expected changes in roles, responsibilities, and skills within the A/R function, hypothesize the potential evolution of the organizational structure, and contemplate the implications for regulatory compliance, cybersecurity, data security, and privacy. The objective is to preview how the nuanced dance between Autonomous Finance and the Future of Work might unfold in reality. Moreover, the critical role of change management will be highlighted as a common theme straddling all these areas.

Let us spotlight the A/R function (see Figure 15.5) and watch this intricate dance unfold.

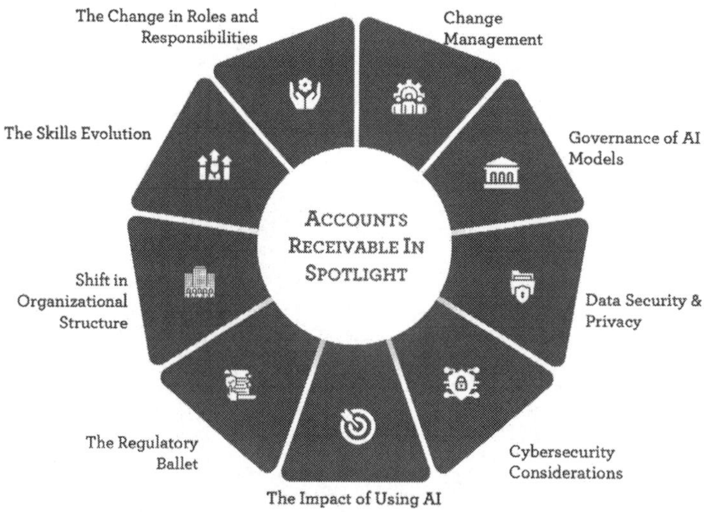

FIGURE 15.5: Accounts Receivable in Spotlight

Change in Roles and Responsibilities in A/R

In the dance of Autonomous Finance, the roles and responsibilities within the A/R team are changing, creating a symphony that brings together the best of human intellect and AI. Traditional roles, characterized by repetitive, manual tasks, are gradually being phased out. Data Entry Clerks, once the backbone of data input into A/R systems, are finding their responsibilities diminishing with the introduction of AI data extraction and RPA. The role of Billing Representatives is evolving as invoice generation and billing processes become highly automated, requiring less human input.

But as the old roles fade, new ones emerge, choreographed to the rhythm of Autonomous Finance. An AI Systems Trainer, for example, trains and optimizes AI tools for cash applications, collections, billing, and more. This role demands both technical and financial skills, signifying the growing importance of a hybrid skill set. Another rising star is the Intelligent Automation Developer, who designs and implements RPA to streamline high-volume A/R tasks.

The dance is far from over. A new ensemble of roles is taking the stage:

- Data Ethics Analyst,
- Customer Experience Designer,
- Automated Compliance Officer,
- Analytics Translator,

- Cybersecurity Specialist,
- Exception Resolver,
- AI Auditor,
- Change Manager,
- Predictive Analytics Modeler,
- NLP Engineer,
- VR Experience Developer,
- Robotics Process Automation Manager,
- Data Privacy Officer,
- Human-AI Collaboration Designer,
- Automation Ethics Auditor,
- Change Leadership Coach,
- AI Compliance Expert, and
- Automation Impact Analyst

The rhythm is clear: Roles are shifting from manual, repetitive tasks to ones that require AI oversight, strategic analysis, customer relations, and addressing exceptions. The stage is set for the human element to elevate its performance, working in harmony with AI to orchestrate the A/R function of the future.

The Skill Evolution in A/R

As the roles and responsibilities within A/R change, so will the skills required for those roles. In Autonomous Finance, the ability to work alongside AI and automation tools will be as important as understanding traditional financial operations.

Technical skills will be in high demand, including the ability to train and collaborate with AI systems, data literacy, cloud computing fundamentals, digital fluency, and the ability to interpret ML models. For instance, Credit Analysts must move beyond building models to interpreting AI model outputs, analyzing key trends, and communicating insights.

Finance-specific skills will also need to evolve. Compliance with rapidly changing regulations relating to AI systems will be critical, as will understanding and mitigating the emerging risks of these technologies. Non-routine accounting and advanced financial analysis will become more important as AI takes over more routine tasks.

Soft skills, too, will be crucial in this new environment. The ability to manage change, think creatively, and demonstrate emotional intelligence will differentiate human workers in a field increasingly dominated by AI. Leadership will rally teams through transformation and ensure that AI is used ethically and responsibly.

The rise of Autonomous Finance will also necessitate a commitment to lifelong learning and a flexible, growth mindset. Hybrid roles that require finance and technical skills will become more prevalent, and investment in continuous skills development will be key to smooth AI adoption. The next section examines how the organizational structure might need to adapt to these shifts.

A Shifting Organizational Structure in A/R

The introduction of Autonomous Finance capabilities could significantly alter the organizational structure of A/R teams. Traditionally hierarchical and rigidly structured, these teams may evolve into something much more fluid and flexible.

One potential change is a flatter structure, with fewer layers of management as AI systems help coordinate workflows and staff focus shifts to value-added activities. This could result in more direct communication and quicker decision-making processes. There might also be increased agility by forming cross-functional, agile teams rather than rigid, siloed departments. This would allow for more flexibility as tasks evolve and new challenges arise. The introduction of Autonomous Finance could also lead to the formation of new, centralized AI units within organizations. These dedicated teams would develop, implement, and monitor AI tools integrated across A/R operations.

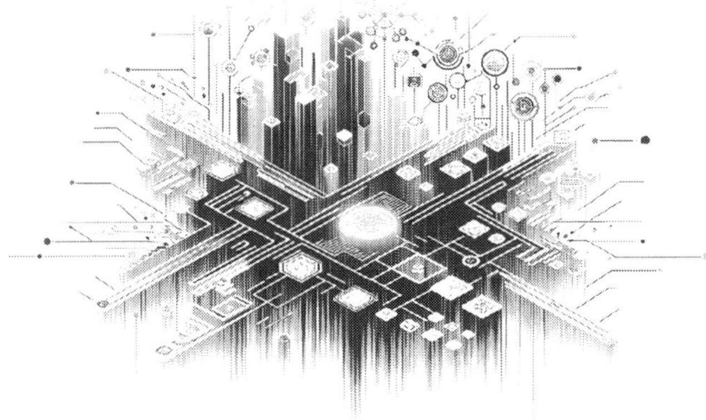

Another potential shift is the rise of hybrid roles requiring Finance and technical skills. These roles could become more prevalent as human and AI capabilities are integrated seamlessly. Centers of excellence, dedicated units with specialized expertise in ethics, analytics, automation, and other areas, may also emerge to guide A/R transformation efforts. These centers could help organizations stay at the forefront of technological developments and ensure the responsible use of AI.

Lastly, the focus on data security and privacy could create new governance structures within organizations. These could include committees and protocols to oversee ethical AI deployment and retraining programs as roles change.

The Regulatory Ballet in A/R

The complex choreography of Autonomous Finance and the Future of Work is underpinned by regulations. As the narrative leaps the regulatory ballet, the distinct challenges that Autonomous Finance presents to regulatory compliance in A/R will be probed, emphasizing how nimble adaptation is vital to staying in sync with the rhythm.

As Autonomous Finance takes the lead in the A/R function, the regulatory landscape undergoes a parallel transformation. Compliance with regulations becomes a dance in itself, with the increasing use of AI systems in finance ushering in new laws and standards. The role of Compliance Associates is evolving, shifting from comprehensive manual reviews to overseeing AI-driven monitoring for adherence to regulations and investigating higher-risk cases.

The dance, however, is not without its complexities. Navigating the regulatory ballet requires agility and precision. But with the right strategies and systems in place, the dance can be choreographed to ensure compliance while reaping the benefits of Autonomous Finance.

The Impact of Using AI in A/R

Autonomous Finance is expected to prompt significant changes in the regulatory landscape of A/R. As AI systems become more integral to A/R operations, navigating the associated regulatory challenges will become a central concern for organizations.

The use of AI systems in A/R will necessitate a robust understanding of the latest regulations relating to AI usage. With the help of AI systems, Compliance officers will need to monitor regulations and flag issues for human review continuously. The reliance on AI also necessitates a strong focus on data privacy and security. This means ensuring compliance with regulations such as GDPR and CCPA, which govern the collection and use of personal data.

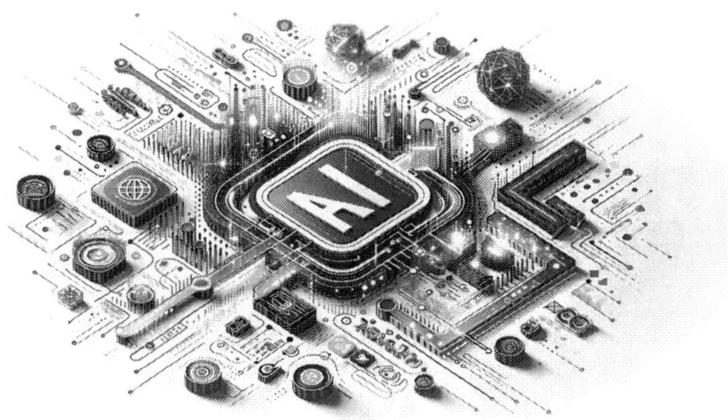

There will also be a need to ensure accountability, transparency, and ethics in A/R algorithms and data practices. This will require organizations to implement effective AI governance measures and possibly create new roles focused on AI ethics and compliance. The changing regulatory landscape also allows organizations to shape future regulations proactively. By working closely with regulators and sharing insights about the practical implications of AI in A/R, organizations can help to inform the development of balanced and effective regulations.

Cybersecurity Considerations in A/R

As the A/R function embraces Autonomous Finance, cybersecurity considerations become paramount. A/R operations involve sensitive customer data and financial transactions, making them a potential target for cyber threats. The increasing use of AI and automation in A/R operations, while bringing efficiencies, also opens up new avenues for cyber attacks. The remote nature of work, increased data sharing, and integration of AI tools can all potentially increase cybersecurity risks.

In response to these challenges, A/R teams must bolster their cybersecurity measures. This may involve adopting advanced cybersecurity tools to monitor and detect threats, implementing robust access controls, and strengthening data protections.

In addition to technological measures, there will be a need for A/R professionals to understand the importance of cybersecurity. Cybersecurity training can help employees identify threats and follow best practices to secure data.

The role of cybersecurity specialists within A/R teams may become more prominent. These professionals will oversee the security measures, ensure compliance with relevant regulations, and coordinate responses to potential cyber threats.

Incorporating cybersecurity considerations into the A/R function is about risk mitigation and fosters trust with customers and stakeholders. By demonstrating robust cybersecurity practices, organizations can assure their customers that their financial data is secure.

Data Security and Privacy in A/R

The onset of Autonomous Finance in A/R also necessitates a robust focus on data security and privacy. With customer data being used to train AI models and the increased integration of systems, protecting this data becomes crucial.

A/R teams will need to adopt advanced measures for data protection. This could include using encryption for data at rest and in transit, implementing robust access controls, and conducting regular audits for data security.

Further, data privacy becomes a key consideration. Regulations such as GDPR and CCPA impose stringent requirements for data privacy. A/R teams must ensure that using customer data for training AI models and in A/R operations complies with these regulations.

A/R teams may see the emergence of new roles, such as Data Privacy Officers, who will oversee data governance protocols and consent procedures and ensure the responsible usage of data by A/R systems. Professionals in these roles must also stay updated on the latest data privacy regulations and ensure the organization's compliance.

Additionally, A/R professionals must understand the importance of data privacy and follow best practices to safeguard customer data. Training and awareness initiatives can help inculcate these practices.

Governance of AI Models in A/R

As Autonomous Finance takes root in A/R, the governance of AI models becomes paramount. While highly efficient, AI models can make mistakes or develop biases based on the data they are trained on. Therefore, the oversight and governance of these models are critical. In the realm of A/R, this could involve setting up committees or protocols to oversee the ethical deployment of AI. It could also involve conducting regular audits of the AI models to ensure they operate as intended and do not develop biases.

Moreover, professionals in A/R teams may need to adopt new roles, such as AI Auditors, who will evaluate AI systems for accuracy, transparency, and alignment with organizational values. These professionals must have a blend of technical and financial skills to understand the workings of AI models and the implications of their decisions. Training programs will be crucial to equip professionals with the skills needed for these roles. In addition, fostering a culture of ethical AI use within the organization will be essential.

Change Management in A/R

The evolution of A/R to incorporate Autonomous Finance is not just a technological shift. It is a transformation involving people, processes, and culture, so change management is essential. The transition to Autonomous A/R will require a significant shift in mindset and work practices. This can be challenging for employees who are used to traditional operational procedures. Here, change management is crucial in helping employees understand the new system's benefits, alleviating their fears, and empowering them to use the new tools effectively.

Effective change management strategies could include regular communication about the changes, training programs to equip employees with the new skills they need, and support systems to help them adapt to new business models, policies, and procedures. It is about fostering an environment where employees feel supported and empowered to adapt to the new A/R landscape.

Moreover, leaders will play a pivotal role in this change management process. They must champion the shift toward Autonomous Finance and guide their teams. This will require leaders to develop new skills and adapt their leadership style to this new era of work.

15.7 Conclusion

As this chapter explores the symbiotic dance between Autonomous Finance and the Future of Work draws to a close, a sense of invigorating optimism and heightened anticipation prevails. This intricate interplay of transformational forces has sculpted a proscenium for a future replete with efficiency, inclusivity, and fulfillment in professional endeavors.

Recalling the words of media theorist Marshall McLuhan, *"We shape our tools, and thereafter our tools shape us"* (TEDx Talks, 2022), the axiomatic relationship between Autonomous Finance and evolving workplace dynamics comes into sharp focus. Shaping tools and technologies in Autonomous Finance is more than an act of innovation; it defines future professional landscapes. Remarkably, this process is not unilateral. The dynamism of contemporary work environments reciprocally influences the trajectories of Autonomous Finance, engendering a cyclical cadence—a rhythm of mutual adaptation that acts as a crucible for ceaseless transformation.

As eyes set upon the horizon, a future tableau emerges, wherein the choreography between Autonomous Finance and the Future of Work attains greater fluidity. In this envisioned choreography, human acumen and machine prowess form an exquisite pas de deux. The former contributes oversight, creative ingenuity, and ethical discernment, while the latter excels in data analysis, computational complexity, and task automation.

The elucidations and recommendations in this chapter serve as theoretical guidelines and luminous beacons for this unfolding journey. They highlight the imperatives of cultivating adaptive organizational cultures, fortifying skill development paradigms, instituting stalwart security frameworks, navigating fluid regulatory terrains with agility, and fostering collaborative endeavors across diverse sectors.

As this narrative segues into the final chapter, *"Unfolding the Future: The Triad of Finance, Technology, and Human Ingenuity,"* the scope broadens to encapsulate a panoramic tableau. The subsequent chapter will synthesize the myriad influences and convergences shaping the future of Autonomous Finance, including the emergence of financial ecosystems, the paradigm shifts induced by blockchain and other disruptive technologies, and the ethical and regulatory considerations that will guide this future.

Thus, an invitation is extended to you to join this ultimate stage of exploration, peering collaboratively into a future teeming with myriad possibilities, challenges, and innovations that define the inexorable expansion of Autonomous Finance.

15.8 Key Recommendations

a) **Harness the Winds of Change:** The transformative power of Autonomous Finance is not only reshaping the financial landscape but also redefining roles and responsibilities within it. It invites professionals to leave their comfort zones and embrace the transformative tide. This entails acquiring new skills, mastering new tools, and adapting to innovative working methods. As the sun sets on some traditional roles, it dawns on new ones, creating fresh opportunities. It is essential to be proactive, not reactive, in this evolving paradigm, preparing for the dawn of these changes.

b) **Invest in Skill Development: Bridge the Gap:** The shift toward Autonomous Finance will create new skill requirements. Companies should invest in continuous learning programs that equip employees with technical skills (like working with AI tools and data analytics) and soft skills (like change management and leadership).

c) **Strengthen Cybersecurity:** Protect Your Digital Assets - With the increased use of AI and digital tools, cybersecurity becomes more critical. Companies should bolster their cybersecurity defenses and regularly update them to protect against evolving threats.

d) **Prioritize Data Security and Privacy: Respect User Rights:** Companies must ensure they are not compromising data security and privacy as more customer data is collected and analyzed. Adhere to regulations, inform customers about data usage, and implement robust data governance practices.

e) **Implement Effective AI Governance: Ensure Ethical and Fair Practices:** AI systems should be transparent, fair, and accountable. Establish a robust AI governance framework that includes regular auditing, maintains algorithmic fairness, and ensures decisions made by AI systems are explainable.

15.9 References

Capgemini. (2023). *Reinventing cybersecurity with artificial intelligence: The new frontier in digital security | Research & insight | Capgemini.* https://www.capgemini.com/research/reinventing-cybersecurity-with-artificial-intelligence/

Carmichael, J. (2016). Elon Musk Says Humanity's Survival Depends on "A.I.-Human Symbiote." *Inverse.* https://www.inverse.com/article/21157-elon-musk-ai-human-symbiote-neural-lace

Cook, T. (2018). Tim Cook: You Deserve Privacy Online. Here's How You Could Actually Get It. Time. https://time.com/collection/davos-2019/5502591/tim-cook-data-privacy/

Deloitte Insights. (2019). *Leading the social enterprise: Reinvent with a human focus. Deloitte Insights.* https://www2.deloitte.com/us/en/insights/focus/human-capital-trends/2019/leading-social-enterprise.html

Gabor, D. (1963). *Inventing the future.* London: Secker & Warburg. http://quoteinvestigator.com/2012/09/27/invent-the-future/

Gartner (2020). Gartner Survey Reveals 82% of Company Leaders Plan to Allow Employees to Work Remotely Some of the Time. Retrieved from https://www.gartner.com/en/newsroom/press-releases/2020-07-14-gartner-survey-reveals-82-percent-of-company-leaders-plan-to-allow-employees-to-work-remotely-some-of-the-time

Heffernan, M. (2012). Dare to disagree [Video]. TED Conferences. https://www.ted.com/talks/margaret_heffernan_dare_to_disagree

IBM. (2020). Cost of a Data Breach Report 2020. https://www.ibm.com/security/digital-assets/cost-data-breach-report/#/

Jobs, S. (2005, June 12). Stanford Commencement Speech. Stanford University. https://news.stanford.edu/2005/06/14/jobs-061505/

LinkedIn. (2020). Workplace Learning Report 2020. https://learning.linkedin.com/content/dam/me/learning/resources/pdfs/LinkedIn-Learning-2020-Workplace-Learning-Report.pdf

Nadella, S. (2018a). Microsoft CEO Satya Nadella: The whole world is now a computer. *Quartz.* https://qz.com/1183406/microsoft-ceo-satya-nadella-the-whole-world-is-now-a-computer/

Nadella, S. (2018b). The Future Computed: Artificial Intelligence and its role in society. *Microsoft.* https://news.microsoft.com/futurecomputed/

PwC. (2021). 24th Annual Global CEO Survey. *PwC.* https://www.pwc.com/gx/en/ceo-agenda/ceosurvey/2021.html

Schwab, K. (2016). The Fourth Industrial Revolution: what it means, how to respond. *World Economic Forum.* https://www.weforum.org/agenda/2016/01/the-fourth-industrial-revolution-what-it-means-and-how-to-respond/

TEDx Talks. (2022, July 22). *First we shape our tools and then our tools shape us | Remco Lenstra | TEDxEDHECBusinessSchool* [Video]. YouTube. https://www.youtube.com/watch?v=x80ZIf9I8-Y

World Economic Forum. (2018). *The Future of Jobs Report 2018 Centre for the New Economy and Society.* http://www3.weforum.org/docs/WEF_Future_of_Jobs_2018.pdf

World Economic Forum. (2020a). *Empowering AI Leadership: An Oversight Toolkit for Boards of Directors.* http://www3.weforum.org/docs/WEF_AI_Board_Toolkit.pdf

World Economic Forum. (2020b). *The Future of Jobs Report 2020.* https://www.weforum.org/reports/the-future-of-jobs-report-2020

World Economic Forum. (2021). *The Global Risks Report 2021.* https://about.fb.com/news/2019/03/vision-for-social-networking/

Zuckerberg, M. (2019). A Privacy-Focused Vision for Social Networking. https://about.fb.com/news/2019/03/vision-for-social-networking/

SECTION IV

The Horizons Ahead

The Future of Autonomous Finance

Section 4 concludes our exploration by casting a visionary gaze at the future of Autonomous Finance. This final section is designed to provoke thought, spark imagination, and encourage strategic planning to anticipate the upcoming shifts in the global financial landscape.

Chapter 16, *"Unfolding the Future: The Triad of Finance, Technology, and Human Ingenuity,"* serves as the apex of this intellectual journey. This climactic chapter transcends mere forecasts to delve into the confluence of emerging financial ecosystems, paradigm-shifting technologies like blockchain, and the ethical and regulatory frameworks that will govern this new financial reality. It seeks to articulate how these multitudinous elements coalesce into a transformative force capable of reshaping economies, societies, and individual lives.

By the culmination of Section 4—and indeed, the book—readers will have navigated from the historical underpinnings of finance to the frontier realms of Autonomous Finance. Armed with a nuanced, forward-looking perspective, readers will have a heightened appreciation for the immense possibilities ahead and a fortified readiness for the transformative waves that Autonomous Finance is poised to unleash upon the global landscape.

CHAPTER 16

Unfolding the Future

The Triad of Finance, Technology, and Human Ingenuity

"We always overestimate the change that will occur in the next two years and underestimate the change that will occur in the next ten."

- BILL GATES (GATES, 1995)

16.0 Key Insights

- **A Future Defined by Intelligence and Empathy:** Intelligence in Autonomous Finance transcends the algorithmic and computational. It incorporates an understanding of human behavior, emotions, and needs. According to a study by the Capgemini Research Institute, 63% of consumers prefer AI-based services that are more emotionally intelligent (Girard et al., 2019). In this light, the Autonomous Finance systems of tomorrow should aspire to create personalized financial journeys imbued with empathy, understanding, and emotional intelligence.
- **The Rise of Ethical Machines:** In an environment defined by algorithmic decision-making, ethical considerations cannot remain on the periphery. The systems must evolve to incorporate ethical reasoning, fairness, and transparency. "Ethics by design" becomes the guiding principle. Dr. Nick Bostrom, philosopher and futurist, warns, *"We need to be super careful with AI, Potentially more dangerous than nukes"* (Clifford, 2018). Such caution calls for machines that calculate and contemplate, ensuring financial decisions resonate with societal values and ethical norms.
- **Democratizing Finance:** The advent of Autonomous Finance heralds an age where high-quality financial advice and products become accessible to all, irrespective of economic status. *"Financial inclusion is not an end in itself, but a means to an end,"* opines Queen Máxima of the Netherlands, the UN Secretary-General's Special Advocate for Inclusive Finance for Development (Maxima, 2019). Autonomous Finance systems can leverage AI and big data to democratize finance, reducing social inequality and fostering economic empowerment.
- **Resilience and Adaptation:** Financial systems are not static; they are living, evolving entities. A Deloitte report highlights that 90% of financial institutions believe cloud computing will be critical for operational resilience (*Deloitte Study: 90% of Companies See Cloud Technology as Essential for Growth, Digital Transformation and Competitiveness in the Marketplace*, 2019). The inherent resilience of Autonomous Finance systems lies in their capacity to self-diagnose, rectify disruptions, and continually enhance performance. It is a future where financial systems are robust and adaptive, learning from experiences and dynamically updating to meet new challenges.
- **Catalyzing Innovation and Growth:** By automating routine tasks, Autonomous Finance paves the way for a flourishing of human creativity and strategic thinking. A report by McKinsey Global Institute suggests that automation could raise productivity growth globally by 0.8% to 1.4% annually (McKinsey, 2017). Autonomous Finance is a catalyst in a world driven by innovation, fostering an environment that thrives on agility and growth.
- **A Harmonious Symphony of Man and Machine:** The narrative of the future is not one of machines usurping human roles but of a symbiotic relationship where machines augment human intelligence, creativity, and empathy. The convergence of machine capabilities and human virtues creates a harmonious symphony, setting the stage for a future where technology serves humanity, not competes with it.

16.1 Introduction

Embarking on a journey through the labyrinthine corridors of Autonomous Finance thus far, the narrative has been a tapestry of strategic blueprints, ethical quandaries, and technological marvels. Section 3, *"Blueprint for Success: Key Considerations in Autonomous Finance,"* unfolded as a comprehensive guidebook, charting the course from a visionary outlook in Chapter 7 to the intricate topography of regulations, ethics, and cybersecurity by Chapter 15. Each chapter in this segment constructed an architectural framework, equipping the reader with the essential tools for navigating this transformative realm with finesse and foresight.

As the aperture to Section 4 of the book opens, a new vista of possibilities beckons. This chapter, the inaugural and the only entry in this exploratory section, aims to stretch the intellectual and imaginative boundaries. Its thesis occupies a unique juncture—between articulating the philosophical underpinnings that should guide Autonomous Finance and the practical challenges that lie ahead. Autonomous Finance is at a crucial inflection point no longer confined to theoretical musings. Marc Andreessen poignantly describes such a realm where *"Software is eating the world"* (Andreessen, 2011). This chapter serves as a timely discourse.

The narrative herein is organized to elucidate the philosophical and aspirational contours that should shape Autonomous Finance. Herein, the quintessence of what this domain should strive for is explored—a financial ecosystem that is not merely intelligent but empathetic, efficient, and ethically conscious. Following this illuminating foray, the discourse transitions into the landscape of challenges. Far from mere hurdles, these challenges are the crucibles in which the future will be forged. They range from technological limitations to ethical dilemmas and regulatory conundrums.

As futurist Alvin Toffler astutely observed, *"The illiterate of the 21st century will not be those who cannot read and write, but those who cannot learn, unlearn, and relearn"* (Toffler, 1999). This chapter not only sketches a future brimming with intelligent, empathetic financial systems but also raises provocative questions about this impending revolution's ethical and societal implications. It is a beacon for those navigating a world in flux—where static knowledge and fixed paradigms can harm progress.

Closing this chapter, the voyage concludes with visionary scenarios that stretch the imagination, key recommendations that serve as navigational aids, and a clarion call to action. This is not a journey for passive observers but for active participants—a collective endeavor to shape a financial future that mirrors humanity's highest virtues and aspirations.

With this kaleidoscopic lens, the reader is invited to traverse beyond the visible horizon, contemplate the realm of possibilities, and engage in a dialogue that shapes a future resonant with collective ambitions and humanistic values.

16.2 The Philosophical Underpinnings

The philosophical underpinnings of Autonomous Finance venture far beyond financial transactions; they enter the sphere of transformative, ethical, and strategic innovation. As humanity stands at the cusp of a new era, Autonomous Finance is a beacon illuminating the path toward a financial landscape infused with intelligence, empathy, and inclusivity. (See Figure 16.1 for a visual distillation of these underpinnings.)

a) **Ethical Imperatives and Fairness:** The ethical dimensions of Autonomous Finance transcend the traditional boundaries of finance and venture into the landscape of social and moral imperatives. Equitable access to financial services is no longer a pipe dream but a tangible goal within reach. With the advent of AI and ML algorithms capable of adapting to individual needs, Autonomous Finance democratizes the financial landscape. It seeks to level the playing field, making high-quality financial advice and services universally accessible, thus directly contributing to social equity.

b) **Transparency and Trust:** The currency of the future in Autonomous Finance is trust, nurtured by a commitment to transparency. As financial decisions grow increasingly automated, the algorithms driving these choices must be transparent and interpretable. The need for explainable AI becomes paramount, ensuring stakeholders can comprehend and trust the automated decisions being made on their behalf (Bostrom, 2014).

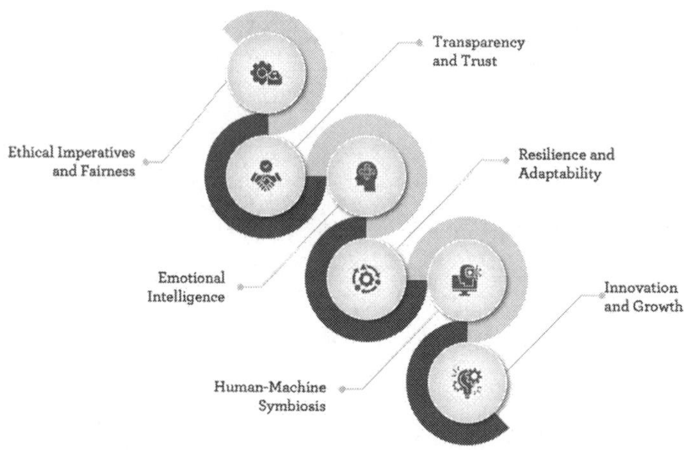

FIGURE 16.1: Impact of Adaptability on the Future of Finance

c) **Emotional Intelligence:** Contrary to the common perception of AI as devoid of emotion, the future of Autonomous Finance incorporates emotional intelligence into its algorithms. The aim is to craft personalized financial experiences that are both efficient and emotionally attuned to individual needs and aspirations (Capgemini Research Institute, 2019).

d) **Resilience and Adaptability:** Autonomous Finance will evolve into a living system capable of self-diagnosis and self-correction. It heralds a paradigm shift from static to dynamic financial models. The ability of these systems to anticipate threats, adapt to market fluctuations, and enhance performance autonomously sets new standards for resilience and adaptability (McKinsey Global Institute, 2017).

e) **Human-Machine Symbiosis:** In this visionary landscape, the conversation shifts from machine versus human to machine plus human. The future envisions a harmonious relationship where machines augment human decision-making, adding layers of efficiency, intelligence, and precision. It is a vision that transcends technological capability and taps into the essence of human values and ethics.

f) **Innovation and Growth:** The philosophical underpinnings of Autonomous Finance are incomplete without mentioning its catalytic role in fostering innovation and growth. Automating routine tasks frees human intellect for creative and strategic endeavors, laying the foundation for a culture of innovation and agility (*Introducing the Next-generation Operating Model*, 2017).

In sum, the philosophical essence of Autonomous Finance is intrinsically tied to a broader social, ethical, and human context. It is about automating transactions and aligning them with collective human values and aspirations. And in doing so, it raises the bar for what finance can achieve, redefining it as a force for good, inclusivity, and the betterment of society.

16.3 Strategic Foresight: Preparing for an Uncertain Future

Strategic foresight is not merely a predictive tool but a vital strategic asset for organizations. It is the telescope through which future landscapes are visualized, facilitating the formulation of resilient strategies. In the volatile and rapidly evolving domain of Autonomous Finance, the absence of strategic foresight is akin to navigating a labyrinth in the dark.

As shown in Figure 16.2, several vital variables demand attention in the era of Autonomous Finance:

a) **Technological Disruptions:** Advances in blockchain, quantum computing, or unforeseen innovations could radically alter the competitive arena. For instance, quantum computing has the potential to revolutionize risk assessment algorithms, setting new industry standards.

b) **Regulatory Evolution:** The ever-changing landscape of financial regulations poses a challenge and an opportunity. Organizations must remain vigilant to adapt to new compliance measures, such as GDPR or MiFID II, which have far-reaching consequences on data management and customer engagement (Zetzsche et al., 2018).

c) **Social and Ethical Norms:** The advent of ethical AI and the increasing emphasis on social responsibility are shaping consumer expectations. Companies ignoring these shifts risk obsolescence.

d) **Economic Fluctuations:** Economic variables such as inflation rates, unemployment levels, and market volatility can affect consumer behavior and, by extension, the algorithms that power Autonomous Finance systems.

e) **Geopolitical Events:** Geopolitical tensions can lead to sudden regulatory changes or market fluctuations. Being prepared for such changes is crucial for long-term success.

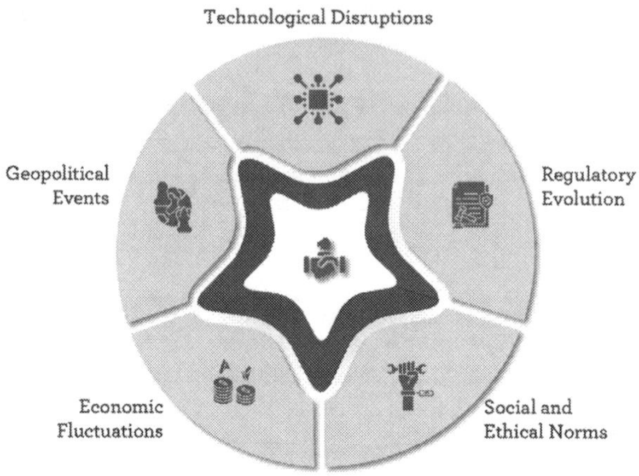

FIGURE 16.2: Preparing for an Uncertain Future using Strategic Foresight

Understanding that the future is not a predetermined outcome but a range of possibilities is crucial. Table 16.1 lists the strategies that organizations could leverage in their transformational journeys toward excellence in Autonomous Finance if presented with one or more of these potential future scenarios:

TABLE 16.1: Preparation Strategies for Potential Future Scenarios

ID	SCENARIO	PROPOSED PREPARATION STRATEGY
1	**Technological Utopia** In this idyllic future, technology and regulation evolve in a harmonious dance. AI algorithms reach unprecedented levels of efficiency, accuracy, and ethical compliance. Financial inclusivity becomes a realized vision, with people from all socioeconomic backgrounds enjoying access to personalized financial advice and services.	• **Technology Investment**: Financial institutions must make significant, strategic investments in cutting-edge technologies and ensure they are at the forefront of technological advancements. • **Talent Acquisition**: Organizations must onboard data scientists, AI experts, and blockchain developers who can navigate this utopian landscape. • **Customer Education**: In a world where technology is king, understanding the customer's relationship with technology is crucial. Comprehensive educational programs can prepare customers to engage with advanced Autonomous Finance services. • **Partnerships**: Forming alliances with technology companies, research institutions, and even competitors can facilitate the sharing of best practices and technologies.
2	**Regulatory Roadblock** In this scenario, regulators impose stringent rules that stifle technological innovation. Compliance costs skyrocket, and the pace of development slows down, affecting market competitiveness globally.	• **Regulatory Engagement**: Institutions should maintain a continuous dialogue with regulatory bodies. Engaging in the policymaking process can help shape both consumer-friendly and innovation-friendly regulations. • **Compliance Infrastructure**: Financial firms must invest in robust compliance systems capable of quickly adapting to new regulations. • **Legal Expertise**: Building an in-house team of legal experts specialized in financial regulation can offer proactive strategies to navigate the regulatory environment. • **Business Model Flexibility**: Organizations should prepare for a scenario where some services might have to be rolled back or modified to meet regulatory requirements.

ID	SCENARIO	PROPOSED PREPARATION STRATEGY
3	**Economic Downturn** An unforeseen recession shakes the global economy. Consumer trust dwindles, and spending plummets, rendering many predictive models and algorithms ineffective.	• **Financial Resilience**: Build a robust financial cushion to sustain operations during economic slumps. • **Operational Efficiency**: Leverage AI to optimize operational costs without compromising quality or service. • **Consumer Retention**: Focus on loyalty programs and personalized financial solutions to retain customers during economic hardships.
4	**Ethical Backlash** A wave of consumer activism rises against perceived ethical lapses in AI decision-making. The lack of transparency or fairness in algorithms becomes a PR nightmare, affecting brand reputation and customer trust.	• **Transparency**: Employ transparent algorithms and provide clear explanations for financial decisions. • **Consumer Education**: Launch educational campaigns to demystify the technologies used in Autonomous Finance. • **Feedback Loops**: Establish mechanisms to collect and incorporate consumer feedback for continuous improvement.
5	**Global Unrest** Political tensions escalate into trade wars or military conflicts, disrupting global financial markets. Regulatory landscapes change overnight, affecting cross-border financial services.	• **Diversification**: Diversify both operations and investments geographically to mitigate risks. • **Real-time Monitoring**: Leverage AI-powered sentiment analysis tools to monitor geopolitical sentiments and make timely adjustments. • **Scenario Planning**: Conduct regular scenario planning exercises focused on geopolitical events and their potential impacts.
6	**Data Breach Catastrophe** In this scenario, a major data breach compromises sensitive customer information, eroding public trust in digital and autonomous financial systems, which has a cascading effect on adoption rates and,w ultimately, on the financial market.	• **Cybersecurity Framework**: Adopt a multi-layered cybersecurity framework, including firewalls, intrusion detection systems, and advanced encryption standards. • **Regular Audits**: Conduct routine cybersecurity audits to identify vulnerabilities. • **Crisis Management Plan**: Develop a crisis communication plan specifying roles and responsibilities, contact information, and a step-by-step guide to manage the crisis effectively.
7	**Hypercompetitive Market** The proliferation of fintech startups and tech giants entering finance creates a hypercompetitive market. Margins are squeezed, and customer loyalty becomes fickle.	• **Unique Value Proposition**: Clearly define what sets the organization apart regarding features, customer service, and pricing. • **Consumer Insights**: Leverage advanced analytics to understand consumer behavior and tailor services accordingly. • **Agility**: Maintain a flexible organizational structure that adapts swiftly to market trends and competitor moves swiftly.
8	**AI Maturity Plateau** After rapid advances, Regulatory bodies worldwide converge on a unified set of standards and protocols for Autonomous Finance, making it easier for services to scale globally and raising the stakes of compliance. The development of AI algorithms hits a plateau. Improvements in efficiency and accuracy slow, leading to decreased investment and interest in Autonomous Finance.	• **Technology Diversification**: Invest in emerging technologies like blockchain or quantum computing to hedge against AI stagnation. • **ROI Monitoring**: Use advanced analytics to continuously assess the ROI of AI investments, making data-driven adjustments to strategy as needed.
9	**Regulatory Convergence** Regulatory bodies worldwide converge on a unified set of standards and protocols for Autonomous Finance, making it easier for services to scale globally and raising the stakes in compliance.	• **Regulatory Intelligence**: Regularly update knowledge on international financial regulations through a dedicated legal team or external consultants. • **Scalability**: Design Autonomous Finance systems with the flexibility to adapt to different regulatory environments seamlessly.

ID	SCENARIO	PROPOSED PREPARATION STRATEGY
10	**Social & Environmental Focus** Increased public scrutiny on financial activities' social and environmental impact leads to new regulations and consumer expectations around ESG (Environmental, Social, Governance) factors.	• **ESG Integration**: Incorporate ESG metrics into decision-making algorithms for investment strategies and operational practices. • **Transparency**: Regularly publish comprehensive reports on ESG performance and initiatives, making this data easily accessible to stakeholders.

In the journey towards the future, real-time analytics and agility serve as the compass and rudder, respectively. KPIs such as customer engagement levels, regulatory compliance rates, and algorithmic efficiency should be monitored diligently.

Strategic foresight provides the blueprint for not just surviving but thriving in the future landscape of Autonomous Finance. The call to action is unmistakable: *Organizations must shift from reactive spectators to proactive architects of the future.*

16.4 Navigating the Challenges

The voyage toward a fully realized Autonomous Finance ecosystem is not without its turbulent waters. While the promise of efficiency, inclusivity, and innovation beckons, an array of challenges lies in wait, each capable of altering the course of this financial revolution. Section 16.4 aims to illuminate these challenges, dissecting their intricacies to equip finance teams, policymakers, and consumers with the navigational tools necessary for this complex journey.

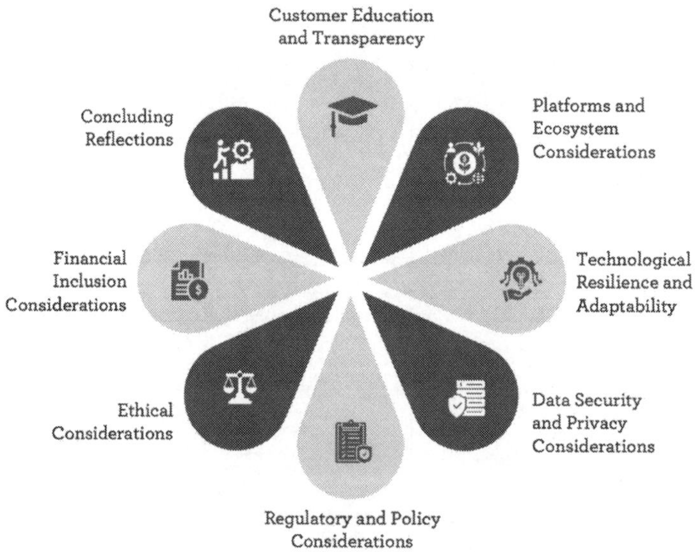

FIGURE 16.3: Navigating the Challenges

The ensuing discourse will be organized into sub-sections, each focusing on a distinct category of challenges shown in Figure 16.3. These range from technological concerns like platform compatibility and data security to softer yet equally consequential issues such as ethical considerations and customer education. These challenges are not isolated but are interrelated, often influencing one another in unforeseeable ways. Understanding these challenges in depth and breadth clears the path to mitigating them, leading to a more resilient and robust Autonomous Finance ecosystem.

Customer Education and Transparency in Autonomous Finance

Implementing Autonomous Finance is not merely a technological leap but a profound shift in the very fabric of financial interaction. As finance becomes increasingly automated and personalized, the need for customer education and transparent communication becomes important and indispensable.

In an environment where algorithms may make significant financial decisions, customers must be sufficiently educated to understand such automation's rationale, benefits, and potential drawbacks. Otherwise, there is a risk of reduced trust, especially when adverse financial events occur (Rogers et al., 2017).

Transparency goes hand-in-hand with education. Finance teams must make detailed information available about how algorithms work and ensure this information is understandable to a lay audience. Making algorithm information understandable to a lay audience is not just about disclosure but about meaningful disclosure. For instance, simplified and interactive digital dashboards could be a mechanism to achieve this (Zollmann & Wiedmaier-Pfister, 2016).

As customers become more educated and the systems more transparent, a new equilibrium between autonomy and oversight emerges. Customers may want to adjust the "level" of autonomy given to financial algorithms, thereby retaining a degree of control and personal agency (Yeung, 2017).

In the age of Autonomous Finance, customer education and transparency are pillars on which the edifice of trust is built. Organizations must adopt a culture of openness, ensuring that as their systems learn and adapt, their customer base evolves in parallel, fully aware and appreciative of the sophistication these financial technologies bring.

Platforms and Ecosystem Considerations in Autonomous Finance

The future of Autonomous Finance rests heavily upon the technological foundations laid by platforms and ecosystems. The interconnectivity of these systems has evolved into a complex dance of data, services, and user experiences, forming a seamless tapestry of financial interactions. This symbiosis extends beyond mere transactional functionalities to embrace intelligent decision-making and customer engagement.

The era of siloed technologies is over. The new age is characterized by the convergence of AI, blockchain, cloud computing, and IoT, among other technologies. This convergence amplifies the functionalities and the reach of platforms and ecosystems in Autonomous Finance (Mills, 2019).

The shift towards an integrated ecosystem introduces new complexities. Foremost among these is the issue of interoperability between heterogeneous systems. While API-first architectures have eased some of this burden, standardization across industry players remains a distant goal (Kemp, 2021).

The potential of quantum computing to disrupt existing cryptographic methodologies poses an existential threat to the security of these integrated platforms. Organizations must contemplate post-quantum cryptographic approaches to safeguard against such future vulnerabilities (Bernstein & Lange, 2017).

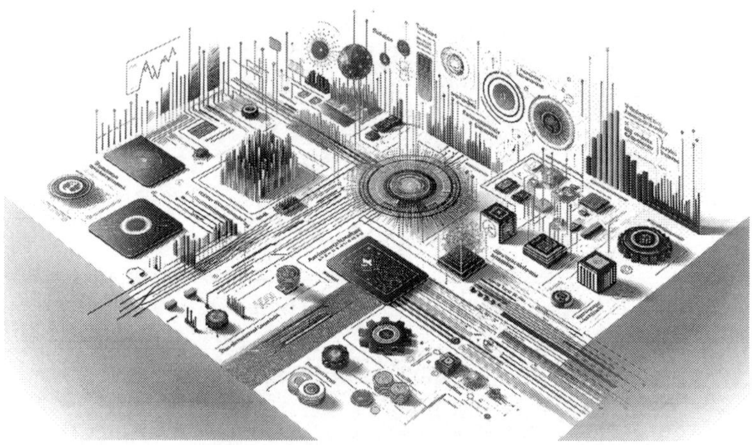

Mastery of the interdependent technological elements is not a luxury but a necessity for navigating the future of Autonomous Finance. Organizations that invest in scalable, interoperable, and secure platform architectures will survive and thrive in the impending transformation.

Technological Resilience and Adaptability in Autonomous Finance

Resilience and adaptability are more than buzzwords; they form the bedrock upon which Autonomous Finance systems must be constructed. The ever-changing landscape of technological innovations, consumer expectations, and regulatory frameworks necessitates systems that can withstand shocks and adapt and evolve in real-time.

Highly resilient systems are designed with fail-safes, redundancies, and the capacity for self-healing. Organizations must invest in fault-tolerant designs that quickly isolate problems and restore functionality. This could include multi-cloud strategies for disaster recovery and distributed ledger technologies for enhanced security and redundancy (Ahram et al., 2017).

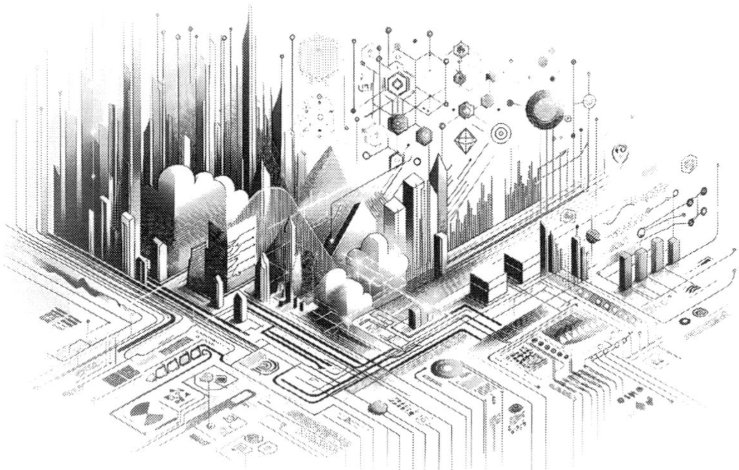

Adaptability extends beyond the technological architecture to include machine learning algorithms capable of learning from new data and scenarios. An adaptive system can recalibrate its risk models in the face of economic shocks or rapidly integrate new regulatory requirements into its compliance engines (Brynjolfsson et al., 2017).

While machines can handle many tasks autonomously, human oversight remains crucial for unexpected situations. The governance models must be designed for efficient and effective human intervention when the systems confront unanticipated challenges (Dietvorst et al., 2015).

Achieving technological resilience and adaptability in Autonomous Finance is not an endpoint but an ongoing process. It requires robust engineering, continuous learning, and dynamic governance. Organizations must recognize that resilience and adaptability are investments in long-term viability and consumer trust.

Data Security & Privacy Considerations in Autonomous Finance

In the sprawling landscape of Autonomous Finance, data security and privacy stand as immutable pillars. The global thrust towards stringent data privacy laws is not merely a regulatory compliance checkbox but a cornerstone of customer trust.

As Autonomous Finance systems become increasingly intelligent, they require more data to make more accurate predictions. This proliferation of data points expands the attack surface for potential breaches, necessitating more robust security measures (Böhme & Moore, 2018).

Compliance frameworks such as GDPR and CCPA continually evolve to counter emerging threats. Organizations must adopt a proactive approach to legal adaptability, actively participating in shaping future guidelines (Greenleaf, 2018).

Data security and privacy are not mere components but the backbone of Autonomous Finance. A proactive, rather than reactive, approach to data management and security can serve as a key differentiator for organizations in this hyper-competitive space.

Regulatory and Policy Considerations in Autonomous Finance

The specter of regulatory and policy considerations looms large over the Autonomous Finance landscape. As these systems grow in complexity and ubiquity, they inevitably attract the scrutiny of regulatory bodies, challenging traditional frameworks seldom designed with such advancements in mind.

The dynamism inherent in Autonomous Finance technologies demands an equally dynamic regulatory approach. Organizations must be prepared for a scenario where legal frameworks can change more frequently, influenced by global policy shifts and technological advancements (Zetzsche et al., 2017).

The incursion of technology firms into the finance domain raises new questions about jurisdiction and regulatory oversight. For instance, how should a social media platform offering remittance services be classified and regulated? The definitions are blurring, necessitating new paradigms in regulatory thinking (Chiu, 2019).

Regulatory agility will be paramount for organizations venturing deeper into Autonomous Finance. The confluence of technological and regulatory evolution demands an ongoing dialogue among industry players, policymakers, and regulators to forge a path that balances innovation with governance.

Ethical Considerations in Autonomous Finance

In the rapidly advancing domain of Autonomous Finance, ethics is not merely a supplementary aspect but a foundational one. As our financial systems increasingly harness machine learning and data analytics, evolving ethically becomes paramount. The intricate dance between technology and ethics necessitates the latter to be as dynamic and adaptive as the algorithms themselves.

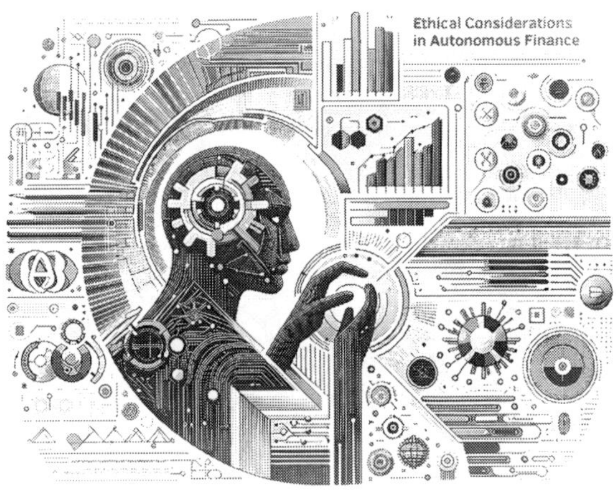

The utilization of AI algorithms in financial frameworks presents ethical challenges that transcend traditional compliance. With the automation of crucial decisions, concerns about algorithmic bias, transparency, and fairness become paramount. It's not just about the right decision but also about the 'just' decision. According to Mittelstadt et al. (2016), it's imperative for contemporary organizations to instill responsible AI practices. Such practices not only ensure regulatory adherence but, more critically, stand robust under ethical examination.

However, the horizon of Autonomous Finance doesn't depict a landscape where machines overshadow humans. Instead, it envisions a harmonized synergy between human intuition and algorithmic precision. As pointed out by Coeckelbergh in 2020, this collaboration induces vital ethical inquiries. In a realm where decision-making is a shared responsibility between humans and algorithms, who is held accountable when things go astray? And how do we delineate the boundaries of this shared accountability?

Navigating the ethical terrains of Autonomous Finance is akin to charting a dynamic and ever-evolving landscape. As we witness technological leaps, the ethical quandaries they surface are bound to grow in complexity. Organizations, therefore, must not view ethics as a one-off task. Instead, they should embrace an ongoing odyssey of ethical introspection and recalibration, ensuring that their moral compass aligns seamlessly with their technological trajectory.

Financial Inclusion Considerations in Autonomous Finance

The promise of Autonomous Finance to democratize access to financial services is both its most inspiring trait and its most daunting challenge. As these systems become increasingly prevalent, questions surrounding their role in promoting financial inclusion have gained urgency. On one hand, the digital-first nature of Autonomous Finance could vastly extend the reach of financial services. On the other, there's a risk of deepening the digital divide, leaving behind those who are digitally illiterate or lack access to technology (World Bank, 2020).

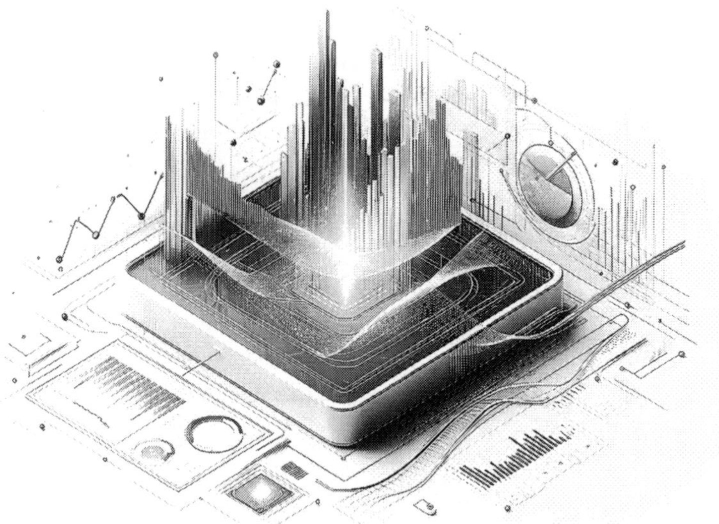

The power of AI can extend beyond Wall Street and touch the lives of the underbanked. Microloans, micro-insurance, and even personalized financial advice could become accessible, enabled by AI algorithms that assess risk and needs at unimaginable scales (Dupas & Robinson, 2013).

Although Autonomous Finance transcends many logistical barriers, it must adapt to local conditions. This includes local laws, economic conditions, and cultural factors influencing financial behavior. In certain regions, for instance, traditional banking practices may not align with cultural norms, thus requiring a more nuanced approach to financial inclusion (Morduch & Haley, 2002).

Financial inclusion in the age of Autonomous Finance is a complex but achievable aspiration. The same technology that poses risks of exclusion also offers the most effective tools for inclusion. Success will depend on navigating this paradox wisely, adopting strategies that maximize inclusivity without compromising efficiency or security.

Concluding Reflections on Navigating the Challenges

Autonomous Finance exists at the intersection of technology and humanity, a space fraught with limitless possibility and intricate complexity. As this section has elucidated, the challenges facing Autonomous Finance are multifaceted, each exerting its unique influence on the trajectory of this financial paradigm shift.

Technological challenges, like platform compatibility and data security, present immediate, tangible obstacles that demand robust technical solutions. However, beneath the veneer of technicalities lie more profound, more nebulous issues. Ethical considerations raise questions about the very soul of Autonomous Finance, urging a reevaluation of the algorithms and data practices that underpin this ecosystem. Regulatory landscapes, ever-shifting and often ambiguous, necessitate a proactive and agile approach to compliance and governance.

The human element, represented through the challenges of customer education and social inclusion, serves as a reminder that technology should be a tool for human betterment, not a force that alienates or divides. The challenge, therefore, is not merely to navigate these obstacles but to do so in a manner that aligns with society's collective ethical, social, and humanist goals.

This reflective piece serves as both a summary and a clarion call. It summarizes the inherent complexities and offers a cohesive strategy that intertwines technological prowess, ethical integrity, and regulatory understanding. It is a call to action for every stakeholder in the Autonomous Finance ecosystem, not just to be a passive spectator but an active participant in shaping a future that is as ethical as it is efficient, as humane as it is automated.

Through strategic foresight, ethical reflection, and technological innovation, the vision for Autonomous Finance can be realized, not as a utopian dream but as a tangible reality, guided by the wisdom gleaned from understanding and addressing these challenges.

16.5 The Autonomous Finance Revolution

In a world inundated with data and craving simplicity, the emergence of a Fully Autonomous Finance Ecosystem stands as a lighthouse on the horizon, signaling the advent of a new era in financial management. The previous chapters have meticulously dissected the contours of this landscape, from the philosophical underpinnings that fuel its raison d'être to the intricate challenges that beckon for innovative solutions. Yet, in envisioning the full realization of this financial paradigm, it is here that the true brilliance of Autonomous Finance shines forth.

Imagine an ecosystem where financial transactions are not just automated but intelligent, where the system learns, adapts, and fine-tunes its strategies in real-time to align with individual financial behavior and market conditions. Picture a financial universe that offers no efficiency and empathy, embedding ethical considerations into every algorithmic decision, thereby elevating financial management from a transactional exchange to a transformative experience.

- **A Symphony of Man and Machine:** The future of Autonomous Finance is not a solo performance but a symphony, a harmonious blend of technological innovation and human ingenuity. In this envisioned reality, AI algorithms and human advisors merge into a seamless entity, each augmenting the other's strengths, thereby offering financial services that are not just robust but resilient.
- **Resilience and Adaptation:** The adaptive capacity of Autonomous Finance systems will set new standards for resilience, capable of self-diagnosis and self-correction, thereby ensuring uninterrupted service even in the face of unforeseen challenges. This dynamism turns these financial platforms into evolving entities, enhancing their performance and adaptability.
- **Democratizing Financial Services:** Autonomous Finance promises to democratize access to high-quality financial advice and products by leveraging the unbounded possibilities of AI and big data. It marks a significant stride toward a more inclusive and egalitarian financial world where quality financial management is not a privilege but a right accessible to all, irrespective of economic status.
- **The Quest for Ethical Machines:** The journey toward Autonomous Finance also leads to the rise of ethical machines that calculate and contemplate. As AI evolves to incorporate ethics, fairness, and transparency, financial decisions will be efficient and honest, thereby adding a layer of conscience to computational prowess.

While the vision of a Fully Autonomous Finance Ecosystem is exhilarating, it is not without its challenges. These challenges, however, should not be seen as deterrents but as opportunities to innovate, collaborate, and sculpt a financial landscape that serves the interests of all stakeholders. This vision calls for a concerted effort from financial institutions, technology companies, regulators, and individuals to unite, think, innovate, and build a future that mirrors shared values and aspirations.

16.6 Key Recommendations

a) **Adopt and Adapt:** Organizations must proactively embrace the shift toward Autonomous Finance, invest in relevant technologies, and adapt their business models to remain competitive in this evolving landscape.

b) **Prioritize Data Privacy and Security:** Given the pivotal role of data in Autonomous Finance, robust data management practices are non-negotiable to maintain customer trust and meet regulatory requirements.

c) **Engage with Regulators and Policymakers:** Active engagement with regulatory bodies and policymakers is imperative to develop comprehensive frameworks that address the unique challenges of this evolving financial landscape.

d) **Focus on Customer Education and Transparency:** Transparent communication about these new financial models' benefits and potential risks can significantly foster customer acceptance and trust.

e) **Embrace Ethical Considerations:** As machines take on more decision-making responsibilities, integrating ethics into algorithms is crucial for ensuring that financial decisions are made with a conscience.

16.7 References

Ahram, T., Sargolzaei, A., Sargolzaei, S., Daniels, J., & Amaba, B. (2017). Blockchain technology innovations. *Procedia Computer Science, 121*, 880-889.

Andreessen, M. (2011, August 20). Why software is eating the world. *The Wall Street Journal*. https://www.wsj.com/articles/SB10001424053111903480904576512250915629460

Bernstein, D. J., & Lange, T. (2017). Post-quantum cryptography—dealing with the fallout of physics success. *IACR Cryptology ePrint Archive, 2017*, 314.

Böhme, R., & Moore, T. (2018). The "iterated weakest link" model of adaptive security investment. *Journal of Information Security, 9*(2), 73-87.

Bostrom, N. (2014). *Superintelligence: Paths, dangers, strategies*. Oxford University Press.

Brynjolfsson, E., Rock, D., & Syverson, C. (2017). Artificial intelligence and the modern productivity paradox: A clash of expectations and statistics. *National Bureau of Economic Research*.

Capgemini Research Institute. (2019). *Emotional intelligence: The essential skillset for the age of AI*. Capgemini.

Chiu, I. H. (2019). Regulatory coherence and the future of financial regulation. *Banking & Finance Law Review, 35*(1), 1-38.

Clifford, C. (2018). Elon Musk: "*Mark my words — A.I. is far more dangerous than nukes*." CNBC; CNBC. https://www.cnbc.com/2018/03/13/elon-musk-at-sxsw-a-i-is-more-dangerous-than-nuclear-weapons.html

Coeckelbergh, M. (2020). *AI Ethics*. Springer Nature.

Deloitte study: 90% of companies see cloud technology as essential for growth, digital transformation and competitiveness in the marketplace. (2019). Deloitte Romania. https://www2.deloitte.com/ro/en/pages/about-deloitte/articles/studiu-deloitte-din-companii-considera-ca-tehnologia-cloud-este-esentiala-pentru-dezvoltare-transformare-digitala-si-competitivitate-pe-piata.html

Dietvorst, B. J., Simmons, J. P., & Massey, C. (2015). Algorithm aversion: People erroneously avoid algorithms after seeing them err. *Journal of Experimental Psychology: General, 144*(1), 114.

Dupas, P., & Robinson, J. (2013). Savings constraints and microenterprise development: Evidence from a field experiment in Kenya. *American Economic Journal: Applied Economics, 5*(1), 163-192.

Gates, B. (1995). *The Road Ahead*. Viking

Girard, S., Shankavaram, D., Taylor, M., Neville, L., & Bertheau, P. (2019). Impact of AI for Customer Experience (CX) AI: re-humanizing digital customer experience. In *Capgemini*. https://www.capgemini.com/wp-content/uploads/2019/06/Point-of-view_Impact-of-AI-for-CX_Final.pdf

Greenleaf, G. (2018). Global Data Privacy Laws 2017: 120 National Data Privacy Laws, Including Indonesia and Turkey. *Privacy Laws & Business International Report*, 7-28.

Introducing the next-generation operating model. (2017). McKinsey & Company. https://www.mckinsey.com/capabilities/mckinsey-digital/our-insights/introducing-the-next-generation-operating-model

Kemp, R. (2021). API Standardization and its Impact on Financial Services. *Journal of Financial Technology, 3*(1), 48-60.

Maxima, Q. (2019). Financial Inclusion: Building on 10 years of progress. In *unsgsa.org.* https://www.unsgsa.org/sites/default/files/resources-files/2021-09/UNSGSA_Annual%20Report%202019%20digital.pdf

McKinsey Global Institute. (2017). *A future that works: Automation, employment, and productivity*. McKinsey & Company.

Mills, D. (2019). *Blockchain: Understanding Its Uses and Implications.* LinkedIn Learning.

Mittelstadt, B., Allo, P., Taddeo, M., Wachter, S., & Floridi, L. (2016). The ethics of algorithms: Mapping the debate. *Big Data & Society, 3*(2), 2053951716679679.

Morduch, J., & Haley, B. (2002). Analysis of the effects of microfinance on poverty reduction. *NYU Wagner Working Paper*, (1014).

Rogers, T., Milkman, K. L., John, L. K., & Norton, M. I. (2017). Beyond good intentions: Prompting people to make plans improves follow-through on important tasks. *Behavioral Science & Policy, 3*(2), 33-41.

Toffler, A. (1999). *Rethinking the future: The insights and analyses of the world's leading thinkers.* Nicholas Brealey Publishing.

World Bank. (2020). *The Global Findex Database 2017: Measuring Financial Inclusion and the Fintech Revolution.* World Bank Publications.

Yeung, K. (2017). 'Hypernudge': Big Data as a mode of regulation by design. *Information, Communication & Society, 20*(1), 118-136.

Zetzsche, D. A., Buckley, R. P., Arner, D. W., & Barberis, J. N. (2017). From FinTech to TechFin: The Regulatory Challenges of Data-Driven Finance. *New York University Journal of Law and Business, 14*(2), 393-431.

Zollmann, J., & Wiedmaier-Pfister, M. (2016). *Making Insurance Markets Work for the Poor: Microinsurance Policy, Regulation and Supervision*. GIZ and A2ii.

List of Abbreviations

ID	ABBREVIATION	DESCRIPTION
1	AFP	Association for Financial Professionals
2	AI	Artificial Intelligence
3	AIOps	Artificial Intelligence Operations
4	AML	Anti-Money Laundering
5	A/P	Accounts Payable
6	A/R	Accounts Receivables
7	AR	Augmented Reality
8	API	Application Programming Interface
9	B2B	Business-to-Business
10	B2C	Business-to-Consumer
11	BBVA	Banco Bilbao Vizcaya Argentaria
12	BSA	Bank Secrecy Act
13	BVS	Business Value Score
14	CAGR	Compound Annual Growth Rate
15	CAS	Cost Accounting Standards
16	CCPA	California Consumer Privacy Act
17	CEH	Certified Ethical Hacker
18	CEO	Chief Executive Officer
19	CFO	Chief Financial Officer
20	CFPB	Consumer Financial Protection Bureau
21	CFRS	Corporate Financial Reporting Standards
22	CISSP	Certified Information Systems Security Professional
23	CPU	Central Processing Unit
24	CTA	Cyber Threat Alliance
25	DeFi	Decentralized Finance
26	DSS	Data Security Standard
27	ECOA	Equal Credit Opportunity Act
28	ERC	Ethereum Request for Comment
29	EMIR	European Market Infrastructure Regulation
30	EPS	Earnings Per Share
31	ERP	Enterprise Resource Planning

ID	ABBREVIATION	DESCRIPTION
32	ESG	Environmental, Social, and Governance
33	EU	European Union
34	EY	Ernst & Young
35	FA	Fixed Assets
36	FP&A	Financial Planning and Analysis
37	FAA	Federal Aviation Administration
38	FATF	Financial Action Task Force
39	FCA	Financial Conduct Authority
40	FCRA	Fair Credit Reporting Act
41	FDA	Food and Drug Administration
42	FEMA	Foreign Exchange Management Act
43	FinTech	Financial Technology
44	FISMA	Federal Information Security Management Act
45	FIX	Financial Information Exchange
46	FRM	Financial Risk Management
47	FSB	Financial Stability Board
48	FSAP	Financial Sector Assessment Program
49	FSMA	Financial Services and Markets Act
50	FTC	Federal Trade Commission
51	GAAP	Generally Accepted Accounting Principles
52	GDPR	General Data Protection Regulation
53	GDUFE	Guangdong University of Finance and Economics
54	GE	General Electric
55	GFIN	Global Financial Innovation Network
56	GL	General Ledger
57	GLBA	Gramm-Leach-Billey Act
58	GPT	Generative Pre-trained Transformer
59	HIPAA	Health Insurance Portability and Accountability Act
60	HR	Human Resources
61	HSBC	Hongkong and Shanghai Banking Corporation
62	IBM	International Business Machines Corporation
63	IEC	International Electrical Commission
64	IFRS	International Financial Reporting Standards
65	IoT	Internet of Things
66	IOSCO	International Organization of Securities Commissions
67	IPPF	International Professional Practices Framework
68	FS-ISAC	Financial Services Information Sharing and Analysis Center
69	ISACA	Information Systems Audit and Control Association
70	ISO	International Organization for Standardization
71	JMAS	Japanese Management Accounting Standards
72	KPIs	Key Performance Indicator

ID	ABBREVIATION	DESCRIPTION
73	KPMG	Klynveld Peat Marwick Goerdeler
74	KYC	Know Your Customer
75	LIME	Local Interpretable Model-agnostic Explanations
76	LCR	Liquidity Coverage Ratio
77	M&A	Mergers And Acquisitions
78	MAR	Market Abuse Regulation
79	MAS	Monetary Authority of Singapore
80	MiFID	Markets in Financial Instruments Directive
81	MIT	Massachusetts Institute of Technology
82	ML	Machine Learning
83	NIS	Network and Information Systems Regulations
84	NIST	National Institute of Standards and Technology
85	NLP	Natural Language Processing
86	NYCRR	New York's Cybersecurity Regulation
87	OCR	Optical Character Recognition
88	OECD	Organization for Economic Cooperation and Development
89	OMB	Office of Management and Budget
90	PCI	Payment Card Industry
91	PDPA	Personal Data Protection Act
92	PIPEDA	Personal Information Protection and Electronic Documents Act
93	POCs	Proof of Concepts
94	PSD2	Payment Services Directive Two
95	RegTech	Regulation Technology
96	ROE	Return on Equity
97	ROI	Return on Investment
98	RPA	Robotic Process Automation
99	RTP	Real-Time Payments
100	SaaS	Software-as-a-service
101	SEC	Securities and Exchange Commission
102	SHAP	Shapley Additive Explanations
103	SIFMA	Securities Industry and Financial Markets Association
104	SOX	Sarbanes-Oxley Act
105	SSC	Security Standards Council
106	TCH	The Clearing House
107	TFCD	Task Force on Climate-related Financial Disclosures
108	TKP	Team Kanban Practitioner
109	UI	User Interface
110	UK	United Kingdom
111	UN	United Nations
112	US	United States
113	VR	Virtual Reality

Index

A

A/R, 340-347
  Change Management in, 347
  Cybersecurity Considerations in, 345
  Data Security and Privacy in, 345
  Governance of AI Models in, 346
  Impact of Using AI in, 344
  Regulatory Ballet in, 343
  Shifting Organizational Structure in, 342
Accessibility, The Power of, 259
Accounts Payable (A/P), 93, 110
Accounts Receivable (A/R), 93, 111, 339
Adaptability, 47-50, 81, 356, 363
Adaptive Strategies in Finance, 3.3
  Case Studies Illustrating Adaptability, 49
Adaptive Automation, 337
Advanced Analytics, 77, 146, 230
Advanced Technologies, 317
Advanced Technology Integration, 38, 39
Agile Finance, 29
  Core Elements of, 30-34
    Continuous Learning, 32
    Cross-functional Collaboration, 31
    Real-time Data and Analytics, 30
    Technology Integration, 30
  Future Trends & Opportunities in, 38-40
    Advanced Technology Integration, 39
    Compliance with Evolving Regulations, 40
    Continuous Learning and Development, 40
    Embracing Growth & Innovation, 40
    Expansion of Cross-Functional Roles, 39
    Emphasis on Data Literacy, 39
    Growing Focus on ESG Factors, 39
  implementation steps of, 34
    Assess Organizational Readiness, 35
    Building Agile Teams, 35
    Driving Cultural Change, 37
    Implementing Agile Methodologies and Tools, 36
  Measuring Agility: KPIs and Metrics, 33
    Cycle Time, 33
    Employee Engagement, 34
    Financial Forecast Accuracy, 33
    Percentage of Automated Processes, 33
Agile finance, 26, 31
Agile methodologies, 36, 41, 51, 65
AI & ML, 142, 298
  Role of, 142, 143
  Rules, 298, 299
AI Chatbots, 263, 264
AI Ethical Dilemma, 273
  Algorithmic Accountability Act, 273
  Anti-Discrimination Act, 273
  Automated individual decision-making and Profiling, 273
  Equality Act, 273
  FCRA, 273
AI Ethics, 300
AI Governance, 339, 348
AI-Driven Customer Support, 267
Algorithmic Accountability Act, 273, 299
Amazon, 51, 60, 264, 301
API economy, 229
APIs, 224
Apple, 265, 301
Audit Support, 104, 136
Augmented Reality, 158
Automation, Complexity of, 79
Autonomous Compliance, 135
Autonomous Finance,
  adopting, 78, 80, 176, 284, 311
  Advantages of, 80, 81
  AI support in, 264
  and the future of work, 331, 333

APIs in, 149, 225
automation in, 76
autonomic finance and, 160, 165, 166
blockchain's transformative impact on, 144
building blocks of, 75
cloud computing in, 147
Concept and Origin of, 4.2
core technologies in, 142, 150
cultural values and, 283
cybersecurity threats in, 307, 313
cybersecurity's role in, 146
data & analytics in, 233
DeFi and, 164
Dimensions of, 181
  Data Intelligence, 183, 197
  Human Capital Development, 183, 201
  Operational Architecture, 183, 202
  Scalable Adoption, 184, 204
  Strategic Alignment, 182, 195
  Technological Enablement, 183, 199
ecosystem considerations in, 362
embedded finance and, 163
ethical compass in, 282
ethical considerations in, 365
guiding principles for, 282
human-machine dance in, 335
impact of, 333
implementation of, 189
integration and interoperability in, 222
levels of integration in, 223
multifaceted dynamics of, 191
need for, 72, 73, 335
open finance and, 161
philosophical underpinnings of, 356
principles of, 73
regulatory and policy considerations in, 364
Revolution, 367
roadmap to compliance in, 284
solutions to barriers of, 82
the age of, 175
the barriers to, 82
the evolutionary leap in, 143
the fundamentals of, 73
the future of, 84, 351
the role of SaaS in, 148
Threat Landscape, 308-312
  Externally Sourced Threats, 309
  Internal Threat, 309
  Social Engineering Attacks, 309
  Technology-Based Threats, 309
Top 10 Cybersecurity Threats, 312
Transformation, 177
Transparency in, 361
Unseen Tech in, 154
Use Cases, The Future of, 106

B

B2B financial setup, 257
Balancing Innovation, 256
Bank Secrecy Act (BSA), 272, 277, 279
Banking and Financial Services, 277, 278
Banking data, 161
Battling Bias, 297, 298
Beauty of Feedback Loops, 336
Bias Mitigation, 294
  See also Mitigate Bias
Bias, 83
  AI Algorithm, 170
  AI Model, 152
  AI Systems, 273, 282
    Decision Making, 83
    Historical Data, 83
    Training Data, 143, 295
    Opportunities, 10.10
Bitcoin, 227
Blockchain, 77, 144, 318, 357
Blockchain Technology, 31, 144, 145, 153, 228
Budgeting and Forecasting, 95, 114, 118
building financial resilience, 18
Business Resilience, 12
Business Value Score (BVS), 186, 371

C

Capital Budgeting, 101, 128
Capital Structure Management, 101, 129
Cash Management (CM), 97,121
ceaseless transformation, 7, 347
Certified Ethical Hacker (CEH), 323
Certified Information Systems Security Professional (CISSP), 323
Champion continuous learning, 42
changing regulatory landscape, 338, 344
Cloud Computing, 60, 140-148
Cohesive Matrix, 167
collaborative Measures, 320, 324, 325
Compliance, 9.5
  and Security, 191, 229
  frameworks, 230, 364
  Measures, 276-291, 357
  Monitoring, 78, 100-104, 136, 245, 292
  Risk Management, 100, 127

compliance strategies, 100, 104, 333
Continuous improvement, 42, 170-178, 191-194, 253
Continuous Learning, 32, 42, 170
Cost Accounting, 95, 113
COVID-19 pandemic, 7-14, 28
Credit Risk Management, 99, 125
critical regulations, 271
  CCPA - US, 271
  Data Protection Act, 271
  GDPR - EU, 271
  Personal Data Protection Act (PDPA), 271
  Personal Information Protection and Electronic Documents Act (PIPEDA), 271
Cross-Functional
  collaboration, 26-31, 37, 41, 42, 51, 53
  roles, 39
Customer Education, 358-369
Customer Experience, 253-257, 265
Cyber Threat Alliance (CTA), 325
Cybersecurity, 314-323
  Awareness, 306, 320-322, 339
  Conscious Culture, 320-321
  Education, 323
  Framework, 318-322
  Strategies for, 314, 315
cybersecurity regulations, 274, 277, 287
  Federal Information Security Management Act (FISMA), 274
  Health Insurance Portability and Accountability Act (HIPAA), 274
  Network and Information Systems Regulations (NIS), 274
  New York's Cybersecurity Regulation (23 NYCRR 500), 274
  Payment Card Industry Data Security Standard (PCI DSS), 274
cybersecurity threats, sources of, 310
  Hacktivist groups, 310
  Individual Hackers, 310
  Insider threats, 310
  Organized crime groups, 310
  State-sponsored attackers, 310

D

Data
  -Driven Approach, 230
  Auditing, 238
  breaches, 77-83, 148-157, 313, 316
  Cleansing, 237
  Disposal, 247, 248
  Enrichment, 237
  Governance, 239-241
  Integrity Management, 237
  literacy, 39, 341
  Management, 235
    Acquisition, 235
    Analytics and Metadata Management, 236
    Best Practices, 231
    Governance, 236
    Integration, 235
    Master Data Management, 236
    Quality Management, 235
    Security and Privacy, 236
    Storage and Management, 236
  Minimization, 247
  Quality Management, aspects of, 236
    Monitoring, 237
    Profiling, 237
    Standardization, 237
    Validation, 237
  Visualization, The Periscope of, 242
data privacy regulations, 243
  CCPA, 226-244, 265-281, 319-321
  GDPR, 6-11, 226-244, 277-287
  HIPAA, 244, 274-283, 372
Debt,
  Financing, 101, 130
  Management, 98, 123
Decentralization, 77, 226, 227
Decentralized Finance (DeFi), 77, 153-155, 229
Decision Support, 152
Digital
  acceleration, 50, 51, 52, 54
  Savviness, 53, 65
  transformation, 53, 65, 86, 141, 193
Disruptions, 5–9
  Case Studies of, 8, 9
  Domino Effects of, 7
  economic, 5
  environmental, 7
  ethical, 6
  Examples of Past, 8, 9
  legal, 6
  political, 6
  social, 5
  technological, 6
  Types of, 6
Dividend Policy, 101, 130
Drucker, Peter, 33, 37, 152, 177

E

Economic Calculus, 167

Edison's era, 4
Embracing growth, 40, 42
Enterprise Resource Planning (ERP), 30
Enterprise Risk Management, 100, 127
Equally compelling, 149
Equity, 5, 115, 123, 129, 280, 303, 356
Equity & Debt Financing, 101, 130
ESG factors, 39, 42, 288, 360
Ethical,
  Compliance, 358
  Conduct, 9, 16, 51, 270, 276, 282
  Machines, 86, 354, 368
  Mindset, 53
EU's GDPR, 270, 299, 315
evaluating potential partners, 214, 216
Experimentation, 176-178, 187-194
explainability, 153, 174, 270, 298-303, 339
Explainability, 82, 282, 294,

F

Fair Credit Reporting Act (FCRA), 270
Fairness amid Bias, 273
Feedback Loops, The Beauty of, 336
Finance Team's Dilemma, 10
Finance-specific skills, 342
Financial
  accounting, 92, 93, 94, 110
  Action Task Force (FATF), 281
  adaptability, 14, 15
  Compliance, 103, 135, 156
  Compliance, The Sine Qua Non of, 156
  Democratization Movement, 153
  Evolution, 286, 287
    innovation, 14
    Leadership, 180
    management solutions, 55
    Modeling, 96, 117
    Planning & Analysis (FP&A), 96, 116
    Policy Enforcement, 104, 105, 137
    Potential, 180
    Process Automation, 102, 132
  regulations, 272, 287, 357
    Bank Secrecy Act (BSA), 272
    Basel III international regulatory framework, 272
    Dodd-Frank Wall Street Reform and Consumer Protection Act, 272
    FCRA, 272
    Payment Services Directive (PSD2), 272
  Reporting, 11, 94, 112
  Resilience, 13-20
    Barriers to, 17, 18
    Strategies for, 15, 16, 17
  Security, 155-156
  Services Information Sharing and Analysis Center (FS-ISAC), 372
  Stability, 14
  Stability Board (FSB), 280
  systems, pathways of, 102
    data management, 102
    financial process automation, 102
    implementation, 102
    IT governance, 102
    reporting, 102
    system integration, 102
Fiscal Fortitude, 15
Fixed Assets (FA) accounting, 94
FOCUS Framework, 181, 182
  C: Converging, 181
  F: Formalizing, 181
  O: Organizing, 181
  S: Synchronizing, 182
  U: Unifying, 182
FOCUS framework, Dimensions of the, 195-206
  Data Intelligence, 183, 197, 198, 199
  Human Capital Development, 183, 201, 202
  Operational Architecture, 183, 202, 203, 204
  Scalable Adoption, 184, 204, 205, 206
  Strategic Alignment, 182, 195, 196, 197
  Technology Enablement, 183, 199, 200
Forecasting, 251
  Cash Flow Forecasting, 251
  Customer Behavior Prediction, 251
  Market Movement Forecasting, 251
Foreign Exchange Management, 98, 123
Friedrich Nietzsche, 21
Function-specific challenges, 27
Future of finance, 50-52
Future-Proofing, 170
Future-Ready finance, characterizing, 56-61
  Adaptability, 59
  Agility, 58
  Autonomy, 59, 60
  Diversity, 61
  Modularity, 60
  Prudence, 60
  Resilience, 58
  Systemic Integration, 3.6

G

Gap Dilemma, 338

Garbage In, Garbage Out, 238
General Electric (GE), 4
General Ledger (GL) accounting, 92
Generative AI, 151-153
GL accounting, 92, 110
Global Financial Terrain, 277
Globalization, 46, 63, 286, 287
Good Governance, 298
Governance,
  AI, 332, 339, 348
  Data, 239, 240, 346
  Ecosystems, 280
Gramm-Leach-Bliley Act (GLBA), 315

H

Hackers, 305, 309, 310, 317
Human Capital Development, 183, 201, 202
Human-AI Symbiosis, 169
Human-Machine Dance, 335–338
Hyper-Personalization, 228-230, 253, 256, 266, 267

I

IBM's Cyber Range, 323
Immersive Financial Experience, 158
  Augmented Reality, 158
  Virtual Reality, 158
Industry Vertical, 278, 281
Infrastructure, 309, 311, 317, 358
Insurance Considerations in Cybersecurity, 320, 323
Integration, 223-224
Application-Level, 223
Data-Level, 223
  Future Tech in, 223
  Power of Intelligent, 224
  Process-Level, 223
  UI-Level, 223
Integration and Interoperability, 241
  Unification of, 241
Interconnected Ecosystems, 160-166
  Autonomic, 165
  DeFi, 164
  Embedded Finance, 163
  Open Banking, 161
  Open Data, 162
  Open Finance, 161
  Internal Controls, 9, 104, 136
International Organization of Securities Commissions (IOSCO), 280
Internet of Things (IoT), 155
Interoperability, 78, 83
Interoperability, guidelines for, 275
  Financial Information eXchange (FIX) Protocol, 275
  ISO 20022 Standard, 226, 275
  Open Banking Standard, 275, 280
  PSD2, 211, 226
  The Clearing House (TCH) Real-Time Payments (RTP) System, 275
Investment Management, 124
Investor Relations, 101, 115, 131

J

Jobs, Steve, 230, 231, 234, 334, 335

K

Kennedy, John F., 45, 65
Key Performance Indicators (KPIs), 33, 185-188
Keynes, John Maynard, 27
KPI Monitoring, 97, 119

L

Liquidity Management, 97, 121
Liquidity Risk Management, 100, 126
Localization, 287, 291

M

Management Accounting, 94, 95, 113, 279
Management Reporting, 96, 119
Market Risk Management, 99, 125, 250
Measuring Agility, 33
  KPIs in, 33
  Metrics in, 33
Measuring success, 188
Mergers and Acquisitions (M&A), 101, 129
Mitigate Bias, 303
modern financial systems, 103, 148, 249

N

National Institute of Standards and Technology (NIST), 316
Natural Language Processing (NLP), 142
Niche Navigators, 279

O

Ongoing Support, 190-191
  Compliance and Security, 191
  MLOps and AIOps, 191
  Technical Support, 190
  User Support and Training, 191
Operational Efficiency:
  Cost reduction strategies, 116
  Enhancing productivity, 80, 85, 339
  See also Strategic Implications
Operational Risk Management, 13, 100,126

Organization for Economic Co-operation and Development (OECD), 280
organizational decision-making, 94

P

PACE Framework, 50-54
  Advocate of Ethics and Sustainability, 51
  Catalyst of Digital Acceleration, 51
  Enabler of Change, 51
  Pioneer of Value, 51
  See also Finance Transformation in Autonomous Finance
Panic of 1893, 4
Partnerships in Cybersecurity, 324
Pas de Deux of Shared Control, 336
PCI Security Standards Council (SSC), 281
Performance Measurement, 95, 115
Personalization, The Power of, 81, 258
Pilot Projects, 178, 187-194
POCs, 178, 187, 188, 373
Potential Future Scenarios, Strategies for, 358
predictive analytics, 250, 251
  Algorithmic Trading, 250
  Credit Scoring, 250
  Market Risk Management, 250
  Portfolio Optimization, 250
Privacy Threats, 316
Proactive Financial Management, 261, 262
Process Automation, 252
  Financial, 102, 132
  Robotic, 171
Profitability Analysis, 95, 115
Proof of Concepts, 177, 178
  See POCs
Public Sector, Use Cases in the, 105

Q

Quantum Computing, 157

R

Real-Time data, 30
RegTech, 32, 77-78, 156-157
Regulatory Compliance, 270, 286
regulatory compliance, aspects of, 275
  AML directives, 275
  Dodd-Frank Wall Street Reform and Consumer Protection Act, 275
  Financial Conduct Authority (FCA) regulations, 275
  Monetary Authority of Singapore (MAS), 275
  The Sarbanes-Oxley Act (SOX), 275
Regulatory Developments, 286
Regulatory Landscape, 276-281
  by Functional Areas within Finance, 279, 281
  by Geography, 277, 281
  by Governing Ecosystems, 280, 281
  by Industry Vertical, 277, 281
  See Regulatory Terrain
Regulatory Navigation, 33
Regulatory Reporting, 105, 135
Regulatory Technology, 156
Regulatory Terrain, 279, 281
  See also Regulatory Landscape
Regulatory and Ethical Compliance, Future of, 286, 289
Resilience in Business, Aspects of, 12-13
  Financial, 13
  Operational, 12
  Organizational, 13
  Social, 13
  Strategic, 12
Resource Allocation, 170, 180, 187
Risk management (FRM), 99-100
  compliance, 100
  credit, 99
  enterprise, 100
  financial liquidity, 100
  market, 99
  operational, 100
Robust Cybersecurity Framework, Elements of a, 320
  Advanced Technical Controls, 320
  Authentication, 321
  Collaborative Cybersecurity Measures, 321
  Cybersecurity Conscious Culture, 321
  Data Protection, 321
  Insurance Considerations in, 322
  Proactive Risk Management, 320
  Regulatory Compliance, 322
  Secure Infrastructure Design, 320
  Secure Transactions, 321
Robust Cybersecurity Risk Management, 306
ROI Assessment, 170
Rollouts, 178, 188, 189, 193

S

Satya Nadella, 21, 293, 334, 335
Scalability, 82, 226, 227, 359
  Decentralization and, 226, 227
Scalability vs Complexity, 167
Scaling, 178, 188, 189
Scenario Planning, 17, 97, 120, 359

Secure Transactions, 317, 319, 321
Securities Industry and Financial Markets Association (SIFMA), 325
Security, 82, 103, 220, 225, 243, 264
Soft skills, 334,338, 342, 348
Standardization, 225, 226, 237
Strategic Partnerships, 209-217
  Building & Managing, 216, 217
  Dimensions of, 209
  Academic Partnerships, 211
  Co-creation, 212
  Fintech Partners, 209
  Incubators, 213
  Industry Networks, 210
  Open Banking, 211
  Regulatory, 210
  Risk Management in, 215
  Social Impact Partnerships, 213
  Talent Acquisition, 212
  Technology Alliances, 209
  Value Proposition of, 213, 214
    Economic, 213
    Operational, 214
    Strategic, 213
Strategic Planning, 8, 96, 116, 195
Strategic Prioritization and Planning, 177, 186, 187
Strategic Technology Integration, 170
sustainability, 51
Sustainable Development Goals (SDGs), 280

T

Tax Accounting, 94, 113
Tech Giants on the Horizon
  APIs, 224
  Blockchain, 225
  Exploring Beyond, 225
Technological,
  challenges, 82, 226, 367
  limitations, 55, 355
  Resilience, 363
Technology,
  Responsible Use of, 168, 169
  Selection & Integration, 167
  Third-party risk management, 243, 249, 253
The Future of Work, 331-347
  Autonomous Finance and, 331-347
Thomas Edison, 4
  Edison's contributions, 4
  General Electric case study, 4
  Seizing opportunities, 4
traditional finance models, 31
traditional finance skills, 52
Transactions,
  Journey Beyond, 257
  Secure, 314, 317, 319, 321
Transparency, 74, 82, 169, 228, 248, 282
  Customer Education and, 361, 369
  security and, 31, 144
Treasury management, 97-99, 121, 279

U

Uber's ML system, 20, 301, 304
Underpinnings,
  historical, 352
  Philosophical, 355-357, 367
  Technological, 142, 224
United States economy, 4
Use Cases:
  Private Sector, 90, 91
  Public Sector, 105
  Regulatory Reporting, 135

V

Variance Analysis, 95, 114
Vertical Spectrum, 277
Virtual Reality, 158
vulnerability, 307, 208, 311-316
  AI and ML Models, 312
  APIs, 312
  Areas of, 311
  Data, 311
  Hardware and Infrastructure, 311
  People and Processes, 312
  security audits and, 316
  Software Applications, 311
  Third-Party Vendors, 312

W

Walmart, 19
Winston Churchill, 29
Working Capital Management, 15, 97, 109, 122, 268

Z

Zero-based budgeting, 16
Zuckerberg, Mark, 335

Printed by Amazon Italia Logistica S.r.l.
Torrazza Piemonte (TO), Italy

57970337R00232